Applied Mathematical Sciences
Volume 144

T0192197

Springer
New York
Berlin
Heidelberg
Barcelona
Hong Kong
London
Milan
Paris
Singapore
Tokyo

Applied Mathematical Sciences

1. *John:* Partial Differential Equations, 4th ed.
2. *Sirovich:* Techniques of Asymptotic Analysis.
3. *Hale:* Theory of Functional Differential Equations, 2nd ed.
4. *Percus:* Combinatorial Methods.
5. *von Mises/Friedrichs:* Fluid Dynamics.
6. *Freiberger/Grenander:* A Short Course in Computational Probability and Statistics.
7. *Pipkin:* Lectures on Viscoelasticity Theory.
8. *Giacaglia:* Perturbation Methods in Non-linear Systems.
9. *Friedrichs:* Spectral Theory of Operators in Hilbert Space.
10. *Stroud:* Numerical Quadrature and Solution of Ordinary Differential Equations.
11. *Wolovich:* Linear Multivariable Systems.
12. *Berkovitz:* Optimal Control Theory.
13. *Bluman/Cole:* Similarity Methods for Differential Equations.
14. *Yoshizawa:* Stability Theory and the Existence of Periodic Solution and Almost Periodic Solutions.
15. *Braun:* Differential Equations and Their Applications, 3rd ed.
16. *Lefschetz:* Applications of Algebraic Topology.
17. *Collatz/Wetterling:* Optimization Problems.
18. *Grenander:* Pattern Synthesis: Lectures in Pattern Theory, Vol. I.
19. *Marsden/McCracken:* Hopf Bifurcation and Its Applications.
20. *Driver:* Ordinary and Delay Differential Equations.
21. *Courant/Friedrichs:* Supersonic Flow and Shock Waves.
22. *Rouche/Habets/Laloy:* Stability Theory by Liapunov's Direct Method.
23. *Lamperti:* Stochastic Processes: A Survey of the Mathematical Theory.
24. *Grenander:* Pattern Analysis: Lectures in Pattern Theory, Vol. II.
25. *Davies:* Integral Transforms and Their Applications, 2nd ed.
26. *Kushner/Clark:* Stochastic Approximation Methods for Constrained and Unconstrained Systems.
27. *de Boor:* A Practical Guide to Splines.
28. *Keilson:* Markov Chain Models—Rarity and Exponentiality.
29. *de Veubeke:* A Course in Elasticity.
30. *niatycki:* Geometric Quantization and Quantum Mechanics.
31. *Reid:* Sturmian Theory for Ordinary Differential Equations.
32. *Meis/Markowitz:* Numerical Solution of Partial Differential Equations.
33. *Grenander:* Regular Structures: Lectures in Pattern Theory, Vol. III.
34. *Kevorkian/Cole:* Perturbation Methods in Applied Mathematics.
35. *Carr:* Applications of Centre Manifold Theory.
36. *Bengtsson/Ghil/Källén:* Dynamic Meteorology: Data Assimilation Methods.
37. *Saperstone:* Semidynamical Systems in Infinite Dimensional Spaces.
38. *Lichtenberg/Lieberman:* Regular and Chaotic Dynamics, 2nd ed.
39. *Piccini/Stampacchia/Vidossich:* Ordinary Differential Equations in $\mathbf{R}^n$.
40. *Naylor/Sell:* Linear Operator Theory in Engineering and Science.
41. *Sparrow:* The Lorenz Equations: Bifurcations, Chaos, and Strange Attractors.
42. *Guckenheimer/Holmes:* Nonlinear Oscillations, Dynamical Systems, and Bifurcations of Vector Fields.
43. *Ockendon/Taylor:* Inviscid Fluid Flows.
44. *Pazy:* Semigroups of Linear Operators and Applications to Partial Differential Equations.
45. *Glashoff/Gustafson:* Linear Operations and Approximation: An Introduction to the Theoretical Analysis and Numerical Treatment of Semi-Infinite Programs.
46. *Wilcox:* Scattering Theory for Diffraction Gratings.
47. *Hale et al:* An Introduction to Infinite Dimensional Dynamical Systems—Geometric Theory.
48. *Murray:* Asymptotic Analysis.
49. *Ladyzhenskaya:* The Boundary-Value Problems of Mathematical Physics.
50. *Wilcox:* Sound Propagation in Stratified Fluids.
51. *Golubitsky/Schaeffer:* Bifurcation and Groups in Bifurcation Theory, Vol. I.
52. *Chipot:* Variational Inequalities and Flow in Porous Media.
53. *Majda:* Compressible Fluid Flow and System of Conservation Laws in Several Space Variables.
54. *Wasow:* Linear Turning Point Theory.
55. *Yosida:* Operational Calculus: A Theory of Hyperfunctions.
56. *Chang/Howes:* Nonlinear Singular Perturbation Phenomena: Theory and Applications.
57. *Reinhardt:* Analysis of Approximation Methods for Differential and Integral Equations.
58. *Dwoyer/Hussaini/Voigt (eds):* Theoretical Approaches to Turbulence.
59. *Sanders/Verhulst:* Averaging Methods in Nonlinear Dynamical Systems.
60. *Ghil/Childress:* Topics in Geophysical Dynamics: Atmospheric Dynamics, Dynamo Theory and Climate Dynamics.

(continued following index)

Jean-Claude Nédélec

Acoustic and Electromagnetic Equations

Integral Representations
for Harmonic Problems

Springer

Jean-Claude Nédélec
Centre de Mathématiques Appliquées
Ecole Polytechnique
91228 Palaiseau Cedex
France
nedelec@cmapx.polytechnique.fr

Editors

J.E. Marsden
Control and Dynamical Systems, 107-81
California Institute of Technology
Pasadena, CA 91125
USA

L. Sirovich
Division of Applied Mathematics
Brown University
Providence, RI 02912
USA

With 2 figures.

Mathematics Subject Classification (2000): 35C15, 35L05, 35S05, 45A05, 45B05, 45E05, 45F15, 45P05

Library of Congress Cataloging-in-Publication Data
Nédélec, Jean-Claude.
    Acoustic and electromagnetic equations: integral representations for harmonic
problems/Jean-Claude Nédélec.
        p. cm. — (Applied mathematical sciences; 144)
    Includes bibliographical references and index.
    ISBN 978-1-4419-2889-4
        1. Wave equations—Numerical solutions.   2. Maxwell equations—Numerical solutions.
I. Title.   II. Applied mathematical sciences (Springer-Verlag New York, Inc.); v. 144.
QA927 .N38 2000
530.12′4—dc21   00-045038

Printed on acid-free paper.

Printed in the United States of America.

9 8 7 6 5 4 3 2 1

Springer-Verlag   New York Berlin Heidelberg
A member of BertelsmannSpringer Science+Business Media GmbH

Preface

This book is devoted to the study of the acoustic wave equation and of the Maxwell system, the two most common wave equations encountered in physics or in engineering. The main goal is to present a detailed analysis of their mathematical and physical properties.

Wave equations are time dependent. However, use of the Fourier transform reduces their study to that of harmonic systems: the harmonic Helmholtz equation, in the case of the acoustic equation, or the harmonic Maxwell system. This book concentrates on the study of these harmonic problems, which are a first step toward the study of more general time-dependent problems.

In each case, we give a mathematical setting that allows us to prove existence and uniqueness theorems. We have systematically chosen the use of variational formulations related to considerations of physical energy.

We study the integral representations of the solutions. These representations yield several integral equations. We analyze their essential properties. We introduce variational formulations for these integral equations, which are the basis of most numerical approximations.

Different parts of this book were taught for at least ten years by the author at the post-graduate level at Ecole Polytechnique and the University of Paris 6, to students in applied mathematics. The actual presentation has been tested on them. I wish to thank them for their active and constructive participation, which has been extremely useful, and I apologize for forcing them to learn some geometry of surfaces.

A large part of the material contained in this book would not have been in the present state without the work of my many students. Among the ones who have been working on these subjects, I especially want to thank in historical order M.N. Leroux, J. Giroire, M. Djaoua, A. Bendali, A. Da Costa Sequeira, T. Ha Duong, F. Rogier, A. Mazari, Z. Benjelloum Touimi El-Dabaghi, E. Bécache, V. Levillain, F. Starling, T. Abboud, A. Morelot, A. De La Bourdonnaye, I. Terrasse, M. Filipe, B. Zhou, V. Mathis, H. Ammari, D. Barbier, N. Vialle-Béreux, P. Ferreira, and C. Latiri-Grouz.

The author wishes to thank E. Bonnetier and especially S. Christiansen, who greatly helped him in the hard task of translating this book into English starting from a previous version in French.

I also want to thank the persons who were always my main support, my wife, Henriette, especially for her patience, and my children, Laurence and François.

Palaiseau, France   Jean-Claude Nédélec
                                                           February 2000

Contents

Preface   **v**

1 Some Wave Equations   **1**
  1.1   Introduction   1
  1.2   Physical Background   3
        1.2.1   The acoustic equation   3
        1.2.2   The Maxwell equations   4
        1.2.3   Elastic waves   6

2 The Helmholtz Equation   **9**
  2.1   Introduction   9
  2.2   Harmonic Solutions   10
  2.3   Fundamental Solutions   12
  2.4   The Case of the Sphere in $\mathbb{R}^3$ 13
        2.4.1   Spherical harmonics   15
        2.4.2   Legendre polynomials   17
        2.4.3   Associated Legendre functions   22
        2.4.4   Vectorial spherical harmonics   35
  2.5   The Laplace Equation in $\mathbb{R}^3$ 40
        2.5.1   The sphere   40
        2.5.2   Surfaces and Sobolev spaces   47
        2.5.3   Interior problems: Variational formulations   55
        2.5.4   Exterior problems   59
        2.5.5   Regularity properties of solutions in $\mathbb{R}^n$ 65

	2.5.6	Elementary differential geometry	67
	2.5.7	Regularity properties	79
2.6	The Helmholtz Equation in $\mathbb{R}^3$	84
	2.6.1	The spherical Bessel functions	85
	2.6.2	Dirichlet and Neumann problems for a sphere . .	90
	2.6.3	The capacity operator T	97
	2.6.4	The case of a plane wave	97
	2.6.5	The exterior problem for the Helmholtz equation	102

3 Integral Representations and Integral Equations   **110**
3.1	Integral Representations	110
3.2	Integral Equations for Helmholtz Problems	117
	3.2.1	Equations for the single layer potential	117
	3.2.2	Equations for the double layer potential	119
	3.2.3	The spherical case	120
	3.2.4	The far field	122
	3.2.5	The physical optics approximation for the sphere	130
3.3	Integral Equations for the Laplace Problem	135
3.4	Variational Formulations for the Helmholtz Problems . .	141
	3.4.1	The operator S	141
	3.4.2	Fredholm operators	142
	3.4.3	The operator N	143
	3.4.4	Formulation with the far field	146

4 Singular Integral Operators   **150**
4.1	The Hilbert Transform	150
4.2	Singular Integral Operators in $\mathbb{R}^n$	157
	4.2.1	Odd kernels	158
	4.2.2	The M. Riesz transforms	160
	4.2.3	Adjoint operators	166
4.3	Application to Integral Equations	167
	4.3.1	Introduction	167
	4.3.2	Homogeneous kernels	168
	4.3.3	Pseudo-homogeneous kernels	173
4.4	Application to Integral Equations	175

5 Maxwell Equations and Electromagnetic Waves   **177**
5.1	Introduction	177
5.2	Fundamental Solution and Radiation Conditions	179
5.3	Multipole Solutions	185
	5.3.1	Multipoles	185
	5.3.2	The capacity operator	200

5.4   Exterior Problems   204

    5.4.1   Trace and lifting associated
              with the space $H(\mathrm{curl})$ 205

    5.4.2   Variational formulations for the
              perfect conductor problem   214

    5.4.3   Coupled variational formulations
              for impedance conditions   226

5.5   Integral Representations   234

5.6   Integral Equations   243

    5.6.1   The perfect conductor   243

    5.6.2   The zero frequency limit   250

    5.6.3   The dielectric case   253

    5.6.4   The infinite conductivity limit:
              The perfect conductor   266

5.7   The Far Field   288

    5.7.1   Far field and scattering amplitude   288

    5.7.2   Integral equations and far field   298

References   **301**

Index   **313**

1
Some Wave Equations

1.1 Introduction

This book is devoted to the study of the acoustic wave equation and of the Maxwell system, the two most common wave equations that are encountered in physics or in engineering. Our objective is to present a detailed analysis of their mathematical and physical properties.

Waves usually propagate in unbounded domains. They describe such phenomena as the sound emitted by a loud speaker or the electromagnetic field generated by a radar antenna. Other examples are the diffraction of an acoustic wave by a building or of an electromagnetic wave by an airplane, and there are many more. Due to this unboundedness, we will be concerned mostly with the study of these equations in exterior domains.

Wave equations are time dependent. However, by Fourier transform their study is reduced to the study of harmonic systems; the use of the harmonic Helmholtz equation in the case of the acoustic equation, or the harmonic Maxwell system. This book concentrates on the study of these harmonic problems, which are a first step toward the study of more general time-dependent problems.

In each case, we give a mathematical setting that allows us to prove existence and uniqueness theorems. We have systematically chosen the use of variational formulations. They are related to considerations of physical energy. They are also the basis of most numerical approximations.

The natural spaces that appear in this setting are Sobolev spaces, and we describe them in some detail. We try to be elementary and complete in order to help the reader and avoid the need for specialized books on this subject.

The notion of outgoing radiation condition appears naturally when we study the exterior harmonic Helmholtz equation. We describe analytically the solutions of this problem in the special case of a spherical object. This leads to the introduction of spherical harmonics and spherical Bessel functions. The descriptions of these functions are quite technical, but they provide a great deal of insight into the notion of radiation condition and the behavior of the solution of the Helmholtz equation at infinity. This approach also yields a quite simple variational formulation of this equation in the general case of a problem in the exterior of a bounded object, and as a byproduct an elegant proof of existence and uniqueness.

We study the integral representations of the solutions to these Helmholtz problems. These representations yield several integral equations. We analyse their essential properties, especially in the case of the Laplace equation. We give a detailed presentation of the integral equations with a hypersingular kernel. Upon introducing special differential operators on the boundary of the domain, we derive the corresponding variational formulations. These tools are also necessary in the case of the Maxwell equation. We choose to present them in an elementary, but non-canonical, way that avoids all the setting of differential geometry. Yet it is clear that all these operators are well defined as differential forms on a Riemanian manifold. We introduce the variational formulation of integral equations. We give some coercivity results in the case of the Laplace equation. We then show how the extension to the Helmholtz equation leads to compact perturbations of the previous case.

Chapter 4 is devoted to a description of pseudo-homogeneous operators and of some of their essential properties. This is a powerful tool to prove both regularity results and compactness properties for integral operators.

The last chapters are devoted to the case of the exterior harmonic Maxwell equations. Using vectorial spherical harmonics, we can solve explicitly the problem of the perfect conductor boundary condition in the exterior of a sphere. This will serve us as a guide to introduce the different forms of the radiation condition. We give several variational formulations and some results of uniqueness, regularity and existence. These variational formulations appear to be saddle-point formulations.

Using the previous results on the Helmholtz case, we present the corresponding integral representation and the associated integral equations. We introduce some variational formulations for these equations, from which we derive some asymptotics. In particular, we describe the derivation of a perfect conductor model starting from a highly conductive object.

Although we have decided not to speak of the numerical approximations of these equations, the tools that we have introduced here constitute a good starting point to understand the extensive literature on this important subject.

1.2   Physical Background

1.2.1   The acoustic equation

The most studied wave equation is the scalar acoustic equation that describes the propagation of sound in a homogeneous media in the space $\mathbb{R}^n$. It takes the form of a hyperbolic equation

$$\frac{\partial^2 p}{\partial t^2} - c^2 \Delta p = 0, \quad x \in \mathbb{R}^n, \tag{1.2.1}$$

where c is the speed of sound, and p is the induced pressure.

In a space of dimension 1, all the regular solutions are of the form

$$p(t, x) = f(x - ct) + g(x + ct) \tag{1.2.2}$$

where f and g are arbitrary functions.

This expression shows that if the functions f and g have compact support, the solution propagates at a finite speed equal to c. Finite speed propagation is one of the essential characteristics of hyperbolic equations.

A harmonic solution of the wave equation (1.2.1) is a function of the form

$$p(t, x) = \Re\left(u(x)e^{-i\omega t}\right). \tag{1.2.3}$$

The quantity ω is called the **pulsation** of this harmonic wave. The function u is now a solution to the Helmholtz equation

$$\Delta u + k^2 u = 0, \quad x \in \mathbb{R}^n, \quad k = \frac{\omega}{c}. \tag{1.2.4}$$

The number k is called the wave number. The quantity $f = \omega/2\pi$ is called the **frequency**.

The Helmholtz equation has a very special family of solutions called **plane waves**. Up to a multiplicative factor, they are the complex-valued functions of the form

$$u(x) = e^{i(\vec{k}.\vec{x})}, \quad |\vec{k}| = k. \tag{1.2.5}$$

The vector $\vec{k}$ can be real, in which case these solutions are of modulus 1. When the vector $\vec{k}$ is complex and such that $(\vec{k} \cdot \vec{k}) = k^2$, the solutions are exponentially decreasing in a half-space determined by the imaginary part of the vector $\vec{k}$ and exponentially increasing in the other half-space.

Another remarkable group of solutions is the family of spherical waves, which depend only on the radial variable r. For example, in dimension 3, we can verify that the following function satisfies (1.2.4)

$$u(x) = \frac{\sin kr}{r}, \quad r = \sqrt{x_1^2 + x_2^2 + x_3^2}. \tag{1.2.6}$$

We notice that r times a spherical wave is asymptotically a "plane wave" for large value of r. This explains why such a plane wave can be used as a model that describes a remote punctual acoustic source. We will use it constantly in the following and give a special treatment for problems involving plane wave excitation.

1.2.2   The Maxwell equations

The other wave equations encountered very commonly are the Maxwell's equations that describe the propagation of electromagnetic fields. We consider the case of a homogeneous medium with electric permittivity ϵ and magnetic permeability μ. The unknowns are the electric field E and the magnetic field H. The equations are

$$\begin{cases} -\varepsilon \dfrac{\partial E}{\partial t} + \text{curl}H = 0, \\[2mm] \mu \dfrac{\partial H}{\partial t} + \text{curl}E = 0. \end{cases} \tag{1.2.7}$$

Taking the divergence of the equations (1.2.7) yields

$$\frac{\partial}{\partial t}\text{div}E = \frac{\partial}{\partial t}\text{div}H = 0, \tag{1.2.8}$$

which indicates that solutions satisfy

$$\text{div}E = \text{div}H = 0; \tag{1.2.9}$$

at least if this relation is true at time zero.

The vectorial identity

$$\Delta E = \nabla \text{div}E - \text{curl curl}E \tag{1.2.10}$$

easily shows that E and H are also solutions to the wave equations

$$\begin{cases} \dfrac{\partial^2 E}{\partial t^2} - c^2 \Delta E = 0, \\[2mm] \dfrac{\partial^2 H}{\partial t^2} - c^2 \Delta H = 0, \end{cases} \tag{1.2.11}$$

where the speed of light c in this medium is

$$c = \frac{1}{\sqrt{\varepsilon \mu}}. \tag{1.2.12}$$

We notice though, that the system of six independent wave equations (1.2.11) may not be equivalent to the Maxwell system (1.2.7), the solutions of which must satisfy (1.2.9).

We also observe this when looking for **plane waves**: they are the harmonic solutions to the Maxwell equations and take the form

$$\begin{cases} E(t,x) = \Re\left(E(x)e^{-i\omega t}\right), \\[2mm] H(t,x) = \Re\left(H(x)e^{-i\omega t}\right). \end{cases} \tag{1.2.13}$$

The couple E, H is now a solution to the harmonic Maxwell equations

$$\begin{cases} i\omega\varepsilon E + \text{curl}H = 0, \\ -i\omega\mu H + \text{curl}E = 0. \end{cases} \tag{1.2.14}$$

Simple computations, similar to the ones above, show that E, H satisfy (1.2.9) and also

$$\begin{cases} \Delta E + k^2 E = 0, \\ \Delta H + k^2 H = 0, \quad k = \omega\sqrt{\varepsilon\mu}. \end{cases} \tag{1.2.15}$$

The plane waves are associated with a scalar wave equation. Looking for E, H such that

$$\begin{cases} E(x) = \vec{e}\,e^{i(\vec{k}.\vec{x})}, \\ H(x) = \vec{h}\,e^{i(\vec{k}.\vec{x})}, \quad \text{with } |\vec{k}| = k \end{cases} \tag{1.2.16}$$

leads to the equations

$$\begin{cases} \omega\varepsilon\vec{e} + \vec{k} \wedge \vec{h} = 0, \\ -\omega\mu\vec{h} + \vec{k} \wedge \vec{e} = 0. \end{cases} \tag{1.2.17}$$

We notice that the three vectors $\vec{e}, \vec{h}, \vec{k}$ are mutually orthogonal, and the relations (1.2.17) now describe all the plane wave solutions. The vector $\vec{k}$ gives the direction of propagation of the wave. We see that the electric and the magnetic fields lie in the plane orthogonal to $\vec{k}$, called the **phase plane**. These three vectors can be real or complex. When $\vec{k}$ is real and $\vec{e}$ is complex of the form $\vec{e} = \vec{\alpha}_1 + i\vec{\alpha}_2$ with $(\vec{\alpha}_1 \cdot \vec{\alpha}_2) = 0$, we speak of a **circular plane wave**.

The case of an axis invariance

Although the Maxwell equation by nature lives in three-dimensional spaces, it is interesting and useful to study the case where the fields E and H has a given exponential variation in the coordinate x_3 of the form $e^{i\gamma x_3}$. This situation appears in the propagation of such fields in very long objects, such as a fiber optic guide. Notice that the components of the fields along the third axis are not zero. The curl operator reduces to

$$\text{curl}H = e^{i\gamma x_3} \begin{cases} \dfrac{\partial H_3}{\partial x_2} - i\gamma H_2, \\[2mm] i\gamma H_1 - \dfrac{\partial H_3}{\partial x_1}, \\[2mm] \dfrac{\partial H_2}{\partial x_1} - \dfrac{\partial H_1}{\partial x_2}. \end{cases} \tag{1.2.18}$$

The harmonic Maxwell system is now decoupled in two independent subsystems. The solutions to the first one:

$$\begin{cases} i\omega\varepsilon \left(1 - \dfrac{\gamma^2}{\omega^2\varepsilon\mu}\right) E_1 + \dfrac{\partial H_3}{\partial x_2} = 0, \\[2mm] i\omega\varepsilon \left(1 - \dfrac{\gamma^2}{\omega^2\varepsilon\mu}\right) E_2 - \dfrac{\partial H_3}{\partial x_1} = 0, \\[2mm] -i\omega\mu H_3 + \dfrac{\partial E_2}{\partial x_1} - \dfrac{\partial E_1}{\partial x_2} = 0 \end{cases} \tag{1.2.19}$$

are called **transverse electric** ($E_3 = 0$). The solutions to the second one:

$$\begin{cases} i\omega\varepsilon E_3 + \dfrac{\partial H_2}{\partial x_1} - \dfrac{\partial H_1}{\partial x_2} = 0, \\[2mm] -i\omega\mu \left(1 - \dfrac{\gamma^2}{\omega^2\varepsilon\mu}\right) H_1 + \dfrac{\partial E_3}{\partial x_2} = 0, \\[2mm] -i\omega\mu \left(1 - \dfrac{\gamma^2}{\omega^2\varepsilon\mu}\right) H_2 - \dfrac{\partial E_3}{\partial x_1} = 0 \end{cases} \tag{1.2.20}$$

are called **transverse magnetic** ($H_3 = 0$). In both cases, solutions can be computed from a unique scalar function (resp. H_3 or E_3) which satisfies the two-dimensional Helmholtz equation

$$\begin{cases} (\omega^2\varepsilon\mu - \gamma^2) H_3 + \left(\dfrac{\partial^2 H_3}{\partial x_1^2} + \dfrac{\partial^2 H_3}{\partial x_2^2}\right) = 0, \\[2mm] (\omega^2\varepsilon\mu - \gamma^2) E_3 + \left(\dfrac{\partial^2 E_3}{\partial x_1^2} + \dfrac{\partial^2 E_3}{\partial x_2^2}\right) = 0. \end{cases} \tag{1.2.21}$$

1.2.3   Elastic waves

The equations of elastic waves describe the displacement of an elastic object in the linear approximation of small displacements. The unknowns are the three displacements (u_1, u_2, u_3) of the object. The strain tensor ε is defined by

$$\varepsilon_{ij} = \frac{1}{2}\left(\frac{\partial u_i}{\partial x_j} + \frac{\partial u_j}{\partial x_i}\right), \tag{1.2.22}$$

and is related to the stress tensor σ via the constitutive equation

$$\sigma_{ij} = 2\mu\,\varepsilon_{ij} + \lambda\,\text{trace}(\varepsilon)\,\delta_i^j. \tag{1.2.23}$$

The Lamé coefficients μ and λ are positive constants characteristic of the medium and so is the density ρ.

The elastic equation is

$$\rho\frac{\partial^2 u_i}{\partial t^2} - \sum_{j=1}^{3}\frac{\partial}{\partial x_j}\sigma_{ij}(u) = 0. \tag{1.2.24}$$

The harmonic solutions of the form

$$u_i(t, x) = \Re\left(u_i(x)e^{-i\omega t}\right) \qquad (1.2.25)$$

are solutions of the harmonic elastic equations

$$\rho\omega^2 u_i + \sum_{j=1}^{3} \frac{\partial}{\partial x_j}\sigma_{ij}(u) = 0 \qquad (1.2.26)$$

which are also, in vectorial notation,

$$\rho\omega^2 u + \mu\Delta u + (\lambda + \mu)\nabla\mathrm{div}u = 0. \qquad (1.2.27)$$

Taking the divergence of equation (1.2.27), we have

$$(\lambda + 2\mu)\Delta\mathrm{div}u + \rho\omega^2\mathrm{div}u = 0. \qquad (1.2.28)$$

Taking the curl of equation (1.2.27) gives

$$\rho\omega^2\mathrm{curl}u + \mu\Delta\mathrm{curl}u = 0. \qquad (1.2.29)$$

From these two equations, we see that in the isotropic elastic system, waves can propagate with two different velocities. This is also the case when looking for **plane waves**.

These waves are of two types: those associated with the divergence of u, and those associated with the curl of u.

The P waves or pressure waves

We obtain this type as plane wave solutions with vanishing curl. The formula (1.2.10) allows us to rewrite (1.2.27) as

$$\rho\omega^2 u + (\lambda + 2\mu)\mathrm{grad}\ \mathrm{div}u - \mu\mathrm{curl}\ \mathrm{curl}u = 0. \qquad (1.2.30)$$

If u has the form

$$u = \nabla\varphi, \qquad (1.2.31)$$

then φ necessarily satisfies

$$\rho\omega^2\varphi + (\lambda + 2\mu)\Delta\varphi = 0, \qquad (1.2.32)$$

i.e., φ is a plane wave for the scalar equation (1.2.32), with velocity

$$C_p = \frac{\sqrt{\lambda + 2\mu}}{\sqrt{\rho}\ \omega}, \quad k_p = \frac{1}{C_p}. \qquad (1.2.33)$$

The pressure waves P are of the form

$$u_p = \vec{k}e^{i(\vec{k}.\vec{x})}, \quad |\vec{k}| = k_p. \qquad (1.2.34)$$

They oscillate along the direction of propagation $\vec{k}$.

The S waves or shearing waves

We obtain this type as plane wave solutions with zero divergence, with speed

$$C_s = \sqrt{\frac{\mu}{\rho\omega^2}}, \quad k_s = \frac{1}{C_s}. \tag{1.2.35}$$

They satisfy

$$\rho\omega^2 u + \mu\Delta u = 0. \tag{1.2.36}$$

The shearing waves S are of the form

$$u_s = \vec{\alpha}e^{i(\vec{k}.\vec{x})}, \qquad |\vec{k}| = k_s. \tag{1.2.37}$$

The divergence of such a vector is

$$\text{div}u_s = i(\vec{\alpha}.\vec{k})e^{i(\vec{k}.\vec{x})} \tag{1.2.38}$$

which is zero, if $\vec{\alpha}$ is orthogonal to the propagation vector $\vec{k}$. These waves oscillate in a direction orthogonal to the direction of propagation. The vector $\vec{\alpha}$ can have complex values, in which case we speak of polarized waves.

The above equations model most practical industrial problems where wave propagation is an important phenomenon. For example, in the case of the sound emitted by a loud speaker, as in the case of the electromagnetic field generated by a radar antenna, waves are created by a local source. In diffraction problems, e.g., of an acoustic wave by a building or of an electromagnetic wave by an airplane, waves are the result of an interaction with an incoming plane wave. These are all exterior problems.

Piezo-electric crystal devices, which are commonly used to make electronic filters, give rise to a coupled problem involving both elastic waves and an electrostatic field.

2
The Helmholtz Equation

2.1 Introduction

The acoustic wave equation described the propagation of the sound in a medium like the air. It results, e.g., from the equation of the compressible gas dynamic, also called the compressible Navier-Stokes equations. In the case of small displacements of the gas, a linearization of these equations leads to an equation for the displacement and the small variation of the pressure in the gas. In the case of a homogeneous medium of mean density ρ_0, this is the well-known above system of equations

$$\begin{cases} \rho_0 \dfrac{\partial \vec{u}}{\partial t} + \nabla p = 0, \\[2mm] \dfrac{1}{c^2}\dfrac{\partial p}{\partial t} + \rho_0 \operatorname{div} \vec{u} = 0 \end{cases} \tag{2.1.1}$$

where c is the speed of sound in the medium, $\vec{u}$ the speed of displacement in the medium and p the pressure, which is supposed to be isotropic. Eliminating the displacement $\vec{u}$, we obtain a scalar wave equation for the pressure p:

$$\frac{1}{c^2}\frac{\partial^2 p}{\partial t^2} - \Delta p = 0. \tag{2.1.2}$$

In order to set up the problem, we must specify the domain Ω in which we want to solve our equation, and give boundary conditions.

The domain will be either an **interior domain** denoted by Ω or Ω_i or its complement Ω_e, which we will call an **exterior domain**. The two domains are contained in $\mathbb{R}^2$ or $\mathbb{R}^3$, and their common boundary Γ is a bounded regular curve or surface. The **unit normal** n to Γ is defined to be exterior to Ω_i.

The two classical boundary conditions are the **Dirichlet problem**, for which the boundary data is

$$p \mid_\Gamma = 0 \tag{2.1.3}$$

and the **Neumann problem** for which the boundary data is

$$\frac{\partial p}{\partial n} \mid_\Gamma = 0. \tag{2.1.4}$$

A complete set of data is obtained upon prescribing initial values for p and its time derivative $\partial p / \partial t$. Up to a multiplicative constant corresponding to a **choice of physical units**, we have to solve

$$\begin{cases} \dfrac{\partial^2 p}{\partial t^2} - \Delta p = 0, & x \in \Omega, \quad t > 0; \\[2mm] p(0, x) = p_0(x), & x \in \Omega; \\[2mm] \dfrac{\partial p}{\partial t}(0, x) = p_1(x), & x \in \Omega; \\[2mm] p(t, x) \mid_\Gamma = 0, & \left(\text{or } \dfrac{\partial p}{\partial n}(t, x) \mid_\Gamma = 0 \right). \end{cases} \tag{2.1.5}$$

2.2 Harmonic Solutions

It is usual to look for time harmonic solutions of the equation (2.1.2). We look for them in the form

$$p(t, x) = \Re\big(u(x)\, e^{-i\omega t}\big) \tag{2.2.1}$$

where ω is the pulsation of the wave. Here, we have selected a sign (we could have chosen the other sign convention $\epsilon^{i\omega t}$).

The function u now has complex values. Denoting by u_d the Dirichlet data and by u_n the Neumann data, equation (2.1.2) becomes

$$\begin{cases} \Delta u + k^2 u = 0, & x \in \Omega, \quad k = \dfrac{\omega}{c}; \\[2mm] u_{|\Gamma} = u_d & \left(\text{or } \dfrac{\partial u}{\partial n} \mid_\Gamma = u_n \right) \end{cases} \tag{2.2.2}$$

where k is called the wave number.

When we are dealing with an **interior problem**, we speak of **stationary wave solutions**. In that case, we know that the operator $-\Delta$, with its boundary conditions, is self-adjoint and has a compact resolvent in the space $L^2(\Omega)$, admitting a **spectral decomposition** with positive eigenvalues of finite order.

The **Fredholm alternative** tells us then that

$$\begin{cases} \text{- either } k^2 \text{ is not an eigenvalue and there is a unique} \\ \text{solution to the equation (2.2.2),} \\ \text{- or } k^2 \text{ is an eigenvalue and the corresponding eigenfunctions} \\ \text{are solutions of equation (2.2.2) with a zero right-hand side.} \end{cases} \quad (2.2.3)$$

When we are dealing with an **exterior problem**, we speak of **progressive wave solutions**. This situation is extremely different from the previous case. The operator $-\Delta$, with its boundary conditions on Γ, is not self-adjoint, nor does it have a compact inverse in $L^2(\Omega_e)$. Considering the fact that plane waves are solutions to the homogeneous system in the whole space, it is natural to impose additional boundary conditions at infinity to equation (2.2.2) in order to guarantee uniqueness. To eliminate the plane wave, it is sufficient to look for solutions u that decrease at infinity as $1/r$, where $r^2 = x_1^2 + x_2^2 + x_3^2$ is the distance from the origin. Yet, this is not enough to guarantee uniqueness. For example

$$u = \frac{\sin(kr)}{r} \quad (2.2.4)$$

is a non-zero solution of (2.2.2) in free space.

Thus, we add the extra **Sommerfeld radiation condition** or **outgoing wave condition**

$$\left| \frac{\partial u}{\partial r} - iku \right| \leq \frac{c}{r^2} \qquad \text{at infinity.} \quad (2.2.5)$$

We will see later that such a solution behaves at infinity like

$$\begin{cases} \dfrac{e^{ikr}}{r} \alpha \left(\dfrac{\vec{r}}{r} \right) & \text{if } n = 3, \\[3mm] \dfrac{e^{ikr}}{\sqrt{r}} \alpha \left(\dfrac{\vec{r}}{r} \right) & \text{if } n = 2, \end{cases} \quad (2.2.6)$$

where α is a function of the vector $\vec{r}/r$ (i.e, a function defined on the unit sphere).

Returning to the original wave equation and using (2.2.1), we find that, when n=3,

$$p(t, x) \sim \alpha \left(\frac{\vec{r}}{r} \right) \frac{1}{r} \Re \left(e^{ik(r - t)} \right) \quad (2.2.7)$$

which clearly describes a wave that is moving away from the origin. It is clear that a change of convention in equation (2.2.5) yields a radiation condition with the opposite sign

$$\left| \frac{\partial u}{\partial r} + iku \right| \leq \frac{c}{r^2}. \quad (2.2.8)$$

We expect that this **ingoing Sommerfeld condition** will, as well, lead to a well-posed problem. A well-posed problem (we are going to prove this!)

is: find u such that

$$\begin{cases} \Delta u + k^2 u = 0, & \text{in} \quad \Omega_e; \\[2mm] u_{|\Gamma} = u_d, \quad (\text{or } \dfrac{\partial u}{\partial n}|_\Gamma = u_n); \\[2mm] \left| \dfrac{\partial u}{\partial r} - iku \right| \le \dfrac{c}{r^2}, \text{ at infinity.} \end{cases} \qquad (2.2.9)$$

2.3   Fundamental Solutions

The fundamental solutions are the solutions in the whole space and in the sense of distributions of the equation

$$\Delta E + k^2 E = -\delta_0, \qquad (2.3.1)$$

where δ_0 is the Dirac mass at the origin. Solutions are not unique, since we can add to a solution any plane wave or any combination of such plane waves. We need to specify the behavior of the solutions at infinity. It is natural to look for **radial solutions** of the form $E(r)$.

  If n $= 3$, equation (2.3.1) becomes

$$\frac{1}{r^2}\frac{d}{dr}r^2\frac{dE}{dr} + k^2 E = 0, \quad r > 0, \qquad (2.3.2)$$

whose solution is

$$E(r) = c_1 \frac{1}{r}e^{ikr} + c_2 \frac{1}{r}e^{-ikr}. \qquad (2.3.3)$$

It is easy to check that the radiation condition (2.2.5) leads to $c_2 = 0$ (whereas (2.2.8) leads to $c_1 = 0$). **The unique radial outgoing fundamental solution of equation (2.3.1) in $\mathbb{R}^3$ is**

$$E(r) = \frac{1}{4\pi r}e^{ikr}. \qquad (2.3.4)$$

  If n $= 2$, equation (2.3.4) becomes

$$\frac{1}{r}\frac{d}{dr}r\frac{dE}{dr} + k^2 E = 0, \quad r > 0. \qquad (2.3.5)$$

This is a **Bessel equation** whose solutions are not elementary functions. It is known that the linear space of solutions has dimension 2.

  The **Hankel function** of order zero, denoted $H_0^{(1)}(r)$, is a solution of (2.3.5). At infinity, it behaves like

$$H_0^{(1)}(r) \sim \sqrt{\frac{2}{\pi}}\frac{e^{i(r - \pi/4)}}{\sqrt{r}}. \qquad (2.3.6)$$

The Hankel function behaves at the origin like

$$H_0^{(1)}(r) \sim \frac{2i}{\pi}\text{Log}r. \qquad (2.3.7)$$

The unique radial outgoing fundamental solution of equation (2.3.1), if $n = 2$, is

$$E(r) = -\frac{1}{4i} H_0^{(1)}(kr). \qquad (2.3.8)$$

The ingoing solution is the complex conjugate $\overline{E}$ where

$$\overline{E}(r) = \frac{1}{4i} H_0^{(2)}(kr). \qquad (2.3.9)$$

2.4 The Case of the Sphere in $\mathbb{R}^3$

We consider the unit sphere **S** in $\mathbb{R}^3$ (Fig. 1). The case of a sphere of arbitrary radius follows by a change of scale. It is natural, in this geometry, to use spherical coordinates (r, θ, φ), where r is the radius and θ, φ the two Euler angles

$$\begin{cases} x_1 = r \sin\theta \cos\varphi, \\ x_2 = r \sin\theta \sin\varphi, \\ x_3 = r \cos\theta. \end{cases} \qquad (2.4.1)$$

The vectors $\vec{e}_\theta$ and $\vec{e}_\varphi$ are unitary.

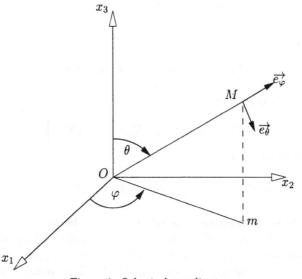

Figure 1: *Spherical coordinates.*

In these coordinates, the Laplace operator has the expression

$$\Delta u = \frac{1}{r^2}\frac{\partial}{\partial r}\left(r^2\frac{\partial u}{\partial r}\right) + \frac{1}{r^2}\left(\frac{1}{\sin^2\theta}\frac{\partial^2 u}{\partial\varphi^2} + \frac{1}{\sin\theta}\frac{\partial}{\partial\theta}(\sin\theta\frac{\partial u}{\partial\theta})\right). \quad (2.4.2)$$

We will denote by Δ_S the **Laplace-Beltrami operator** on the unit sphere **S**, defined by

$$\Delta_S u = \frac{1}{\sin^2\theta}\frac{\partial^2 u}{\partial\varphi^2} + \frac{1}{\sin\theta}\frac{\partial}{\partial\theta}\left(\sin\theta\frac{\partial u}{\partial\theta}\right). \quad (2.4.3)$$

The area element on the sphere is : $d\sigma = \sin\theta d\theta d\varphi$. The operator Δ_S is self-adjoint for the hermitian product in $L^2(S)$ given by

$$\int_S u\bar{v}d\sigma = \int_0^{2\pi}\int_0^\pi u(\theta,\varphi)\bar{v}(\theta,\varphi)\sin\theta d\theta d\varphi. \quad (2.4.4)$$

This can be seen using an integration by parts

$$\begin{cases} \int_S \Delta_S u\bar{v}d\sigma = -\int_0^{2\pi}\int_0^\pi\left(\frac{1}{\sin\theta}\frac{\partial u}{\partial\varphi}\frac{\partial\bar{v}}{\partial\varphi} + \sin\theta\frac{\partial u}{\partial\theta}\frac{\partial\bar{v}}{\partial\theta}\right)d\theta d\varphi \\[2mm] \qquad = -\int_S\left[\left(\frac{1}{\sin\theta}\frac{\partial u}{\partial\varphi}\frac{1}{\sin\theta}\frac{\partial\bar{v}}{\partial\varphi}\right) + \left(\frac{\partial u}{\partial\theta}\frac{\partial\bar{v}}{\partial\theta}\right)\right]\sin\theta d\theta d\varphi \quad (2.4.5) \\[2mm] \qquad = \int_S u\Delta_S\bar{v}d\sigma. \end{cases}$$

The **surfacic gradient** of the function u, denoted $\nabla_S u$, is defined by

$$\nabla_S u = \frac{1}{\sin\theta}\frac{\partial u}{\partial\varphi}\vec{e}_\varphi + \frac{\partial u}{\partial\theta}\vec{e}_\theta. \quad (2.4.6)$$

We have

$$\int_S(\nabla_S u.\nabla_S\bar{v})d\sigma = -\int_S\Delta_S u\bar{v}d\sigma. \quad (2.4.7)$$

Let $H^1(S)$ be **the Hilbert space**

$$H^1(S) = \left\{u\in L^2(S),\nabla_S u\in\left(L^2(S)\right)^2\right\}$$

with its hermitian product

$$(u,v)_{H^1(S)} = \frac{1}{4}\int_S u\bar{v}d\sigma + \int_S(\nabla_S u.\nabla_S\bar{v})d\sigma. \quad (2.4.8)$$

The Laplace-Beltrami operator is self-adjoint in the space $L^2(S)$ and it is coercive on the space $H^1(S)/\mathbb{R}$. It admits a family of eigenfunctions which constitutes an orthogonal basis for the space $L^2(S)$. This basis is also orthogonal for the scalar product in $H^1(S)$. These eigenfunctions are called **spherical harmonics**. They are described in detail in the next section.

2.4.1 Spherical harmonics

Let $\mathbb{P}_l$ be the space of complex-valued polynomials of total degree less than l in the three variables. We have

$$\dim(\mathbb{P}_l) = \frac{(l+3)(l+2)(l+1)}{6}. \qquad (2.4.9)$$

Let $\widetilde{\mathbb{P}}_l$ be the space of homogeneous polynomials of degree l in the three variables. Let $\mathcal{H}_l$ be the space of homogeneous polynomials of degree l that are moreover harmonic, i.e., that satisfy

$$\Delta p = 0. \qquad (2.4.10)$$

Finally, let $\mathcal{Y}_l$ denote the space of the restrictions to the unit sphere of polynomials in $\mathcal{H}_l$.

We will use the following **notation**

$$\begin{cases} \alpha = (\alpha_1, \alpha_2, \alpha_3), \\[2mm] \alpha! = \alpha_1! \alpha_2! \alpha_3! \end{cases} \qquad (2.4.11)$$

and for $p \in \mathbb{P}_l$

$$p(x) = \sum_{\alpha_1+\alpha_2+\alpha_3=l} a_\alpha (x_1)^{\alpha_1}(x_2)^{\alpha_2}(x_3)^{\alpha_3}. \qquad (2.4.12)$$

We associate with this polynomial the derivation operator

$$p\left(\frac{\partial}{\partial x}\right) = \sum_{\alpha_1+\alpha_2+\alpha_3=l} a_\alpha \left(\frac{\partial}{\partial x_1}\right)^{\alpha_1}\left(\frac{\partial}{\partial x_2}\right)^{\alpha_2}\left(\frac{\partial}{\partial x_3}\right)^{\alpha_3}. \qquad (2.4.13)$$

The hermitian product on $\widetilde{\mathbb{P}}_l$ is defined by

$$\langle p, q \rangle_l = p\left(\frac{\partial}{\partial x}\right) \bar{q}. \qquad (2.4.14)$$

If q is given by

$$q(x) = \sum_{\alpha_1+\alpha_2+\alpha_3=l} b_\alpha (x_1)^{\alpha_1}(x_2)^{\alpha_2}(x_3)^{\alpha_3}, \qquad (2.4.15)$$

then we have

$$\langle p, q \rangle_l = \sum_{\alpha_1+\alpha_2+\alpha_3=l} a_\alpha \bar{b}_\alpha \alpha!. \qquad (2.4.16)$$

This formula clearly defines an hermitian product that satisfies

$$\langle p, q \rangle_l = \overline{q\,(\partial/\partial x)\,\bar{p}}. \qquad (2.4.17)$$

Lemma 2.4.1 *The space $\widetilde{\mathbb{P}}_l$ is the direct sum of the space $\mathcal{H}_l$ and of the space $|x|^2\,\widetilde{\mathbb{P}}_{l-2}$, i.e., each p in the space $\widetilde{\mathbb{P}}_l$ can be written in a unique way* as

$$\begin{cases} p = p_1 + |x|^2 p_2, \quad p_1 \in \mathcal{H}_l, \quad p_2 \in \widetilde{\mathbb{P}}_{l-2}, \\ |x|^2 = x_1^2 + x_2^2 + x_3^2. \end{cases} \tag{2.4.18}$$

Proof

To characterize the linear space orthogonal to the subspace $|x|^2 \widetilde{\mathbb{P}}_{l-2}$, for the hermitian product given by (2.4.14), in the space $\widetilde{\mathbb{P}}_l$, we compute

$$\langle p, |x|^2 p_2 \rangle_l = \bar{p}_2 (\partial/\partial x) \Delta p = \langle \Delta p, p_2 \rangle_{l-2}. \tag{2.4.19}$$

Thus,

$$\langle p, |x|^2 p_2 \rangle_l = 0, \quad \forall p_2 \in \widetilde{\mathbb{P}}_{l-2},$$

is equivalent to

$$\Delta p = 0, \tag{2.4.20}$$

which proves the lemma.

Lemma 2.4.2 *The restriction to the unit sphere of the space $\widetilde{\mathbb{P}}_l$ is*
  if l is even

$$\widetilde{\mathbb{P}}_l = \mathcal{Y}_0 \oplus \mathcal{Y}_2 \oplus \cdots \oplus \mathcal{Y}_l, \tag{2.4.21}$$

  if l is odd

$$\widetilde{\mathbb{P}}_l = \mathcal{Y}_1 \oplus \mathcal{Y}_3 \oplus \cdots \oplus \mathcal{Y}_l. \tag{2.4.22}$$

Proof

This results from the previous lemma using an induction argument, starting from $\mathcal{H}_0$ which is equal to $\mathbb{P}_0$, or $\mathcal{H}_1$ which is equal to $\widetilde{\mathbb{P}}_1$.

Theorem 2.4.1 *Let Y_l^m, $-l \leq m \leq l$, denote an orthonormal basis of $\mathcal{Y}_l$ for the hermitian product of $L^2(S)$. The functions Y_l^m, for $l \geq 0$ and $-l \leq m \leq l$, constitute an orthogonal basis in $L^2(S)$, which is also orthogonal in $H^1(S)$. Moreover, $\mathcal{Y}_l$ coincides with the subspace spanned by the eigenfunctions of the Laplace-Beltrami operator associated with the eigenvalue $-l(l+1)$, i.e.,*

$$\Delta_S Y_l^m + l(l+1)Y_l^m = 0, \tag{2.4.23}$$

and the eigenvalue $-l(l+1)$ has multiplicity $2l + 1$.

Proof

The dimension of $\mathcal{Y}_l$ can be obtained using Lemma 2.4.2 by

$$\dim \mathcal{Y}_l = \dim \widetilde{\mathbb{P}}_l - \dim \widetilde{\mathbb{P}}_{l-2} = 2l + 1. \tag{2.4.24}$$

The orthogonality in $L^2(S)$ results from the orthogonality of the subspaces $\mathcal{Y}_{l_1}$ and $\mathcal{Y}_{l_2}$ when $l_1 \neq l_2$. Let us take p in $\mathcal{Y}_{l_1}$ and q in $\mathcal{Y}_{l_2}$. We have

$$(l_1 - l_2) \int_S p\bar{q} d\sigma = \int_S (l_1 p\bar{q} - l_2 p\bar{q}) d\sigma. \tag{2.4.25}$$

From the Euler identity for homogeneous functions, we have

$$\begin{cases} l_1 p = r\dfrac{\partial p}{\partial r}, \\[2mm] l_2 \bar{q} = r\dfrac{\partial \bar{q}}{\partial r}, \end{cases} \tag{2.4.26}$$

and so

$$\begin{cases} (l_1 - l_2) \displaystyle\int_S p\bar{q}\,d\sigma = \displaystyle\int_S \left(\dfrac{\partial p}{\partial r}\bar{q} - p\dfrac{\partial \bar{q}}{\partial r}\right) d\sigma \\[4mm] \qquad\qquad = \displaystyle\int_B (\Delta p\bar{q} - \Delta\bar{q}p)\,dx = 0. \end{cases} \tag{2.4.27}$$

This proves the orthogonality. It remains to show (2.4.23).

Let p be the harmonic polynomial in $\mathbb{R}^3$ associated with Y_l^m. Using (2.4.2), we have

$$\Delta_S Y_l^m + \frac{\partial}{\partial r} r^2 \frac{\partial p}{\partial r} = 0. \tag{2.4.28}$$

Since p is homogeneous of degree l, the Euler identity yields

$$\frac{\partial p}{\partial r} = l\frac{p}{r}, \quad\text{and so} \tag{2.4.29}$$

$$\frac{\partial}{\partial r}\left(r^2 \frac{\partial p}{\partial r}\right) = l\frac{\partial}{\partial r}(rp) = l(l+1)p. \tag{2.4.30}$$

The identity (2.4.23) follows since $p = Y_l^m$ on the unit sphere.

We remark that the orthogonality is also a direct consequence of the general properties of eigenfunctions for self-adjoint operators, and it was not necessary to prove (2.4.25). The completeness of the spherical harmonics family results from Lemma 2.4.2 and from the fact that the traces on the sphere of the polynomials $\mathbb{P}_l$ are dense in the space $L^2(S)$. ∎

We have defined the spherical harmonics and described some of their properties. Yet, we still have to choose a basis in each finite eigenspace $\mathcal{Y}_l$. Using the spherical coordinates θ, φ, we will describe the particular choice of spherical harmonics which is most commonly used.

2.4.2 Legendre polynomials

We consider the segment $]-1, +1[$ and the space $L^2(]-1, +1[)$. The **Legendre polynomials** $\mathbb{P}_l$ are the orthogonal polynomials defined on this segment for the scalar product in $L^2(]-1, 1[)$, constructed with the Gram-Schmidt orthogonalization process, when starting from the usual basis $1, x, x^2, \ldots$. The usual normalization consists in fixing $\mathbb{P}_l(1) = 1$.

The **Rodrigues formula** gives the expression of the Legendre polynomial $\mathbb{P}_l$:

$$\mathbb{P}_l(x) = \frac{(-1)^l}{2^l l!} \left(\frac{d}{dx}\right)^l (1-x^2)^l. \qquad (2.4.31)$$

We have

$$\begin{cases} \mathbb{P}_0(x) = 1, \\ \\ \mathbb{P}_1(x) = x. \end{cases} \qquad (2.4.32)$$

The coefficient of the highest degree monomial is

$$a_l = \frac{(l+1)(l+2)\cdots 2l}{2^l \ l!} = \frac{(2l)!}{2^l (l!)^2}. \qquad (2.4.33)$$

Orthogonality is easily deduced from the Rodrigues formula. It is true for degree zero

$$\int_{-1}^{+1} \mathbb{P}_l(x)\mathbb{P}_0(x)dx = 0, \quad l > 0. \qquad (2.4.34)$$

Let l_1 and l_2 be two integers with $l_1 > l_2$. We want to prove that the quantity

$$\int_{-1}^{+1} \left(\left(\frac{d}{dx}\right)^{l_1} (1-x^2)^{l_1}\right) \left(\frac{d}{dx}\right)^{l_2} (1-x^2)^{l_2} dx$$

is equal to zero. Integrating by parts (the boundary terms are zero), we obtain

$$(-1)^{l_1} \int_{-1}^{+1} \left(\left(\frac{d}{dx}\right)^{l_1+l_2} (1-x^2)^{l_2}\right) \ (1-x^2)^{l_1} dx,$$

which is clearly zero if $l_1 > l_2$.

Lemma 2.4.3 *The Legendre polynomials $\mathbb{P}_l$ satisfy the recursion formula*

$$(l+1)\mathbb{P}_{l+1}(x) - (2l+1)x\mathbb{P}_l(x) + l\mathbb{P}_{l-1}(x) = 0, \qquad (2.4.35)$$

and we have

$$\int_{-1}^{+1} (\mathbb{P}_l(x))^2 \, dx = \frac{1}{l+1/2}. \qquad (2.4.36)$$

Proof
We proceed by induction on the index l. Suppose that (2.4.36) is true up to l. We have

$$(l+1)a_{l+1} = (2l+1)a_l. \qquad (2.4.37)$$

The polynomial $\mathbb{P}_l(x)$ has the same parity as l. Its term of degree $l-1$ is equal to zero. Consider the polynomial

$$Q_l(x) = (l+1)\mathbb{P}_{l+1}(x) - (2l+1)x\mathbb{P}_l(x). \qquad (2.4.38)$$

Its term of degree $l+1$ is equal to zero by construction. Its term of degree l is equal to zero by parity. So it is of degree $l-1$ or less. It follows that

$$Q_l(x) = \sum_{i=0}^{l-1} \beta_i \mathbb{P}_i(x). \tag{2.4.39}$$

The orthogonality relation shows that

$$\frac{1}{i+1/2}\beta_i = -(2l+1)\int_{-1}^{+1} \mathbb{P}_l(x)x\mathbb{P}_i(x)dx. \tag{2.4.40}$$

The polynomial $x\mathbb{P}_i$ is of degree $i+1$ and so $\beta_i = 0$ except for $i = l-1$.
Finally, we have

$$\beta_{l-1} = -(2l+1)(l-1/2)\int_{-1}^{+1} \mathbb{P}_l(x)x\mathbb{P}_{l-1}(x)dx. \tag{2.4.41}$$

But the polynomial $x\mathbb{P}_{l-1}$ is of the form

$$x\mathbb{P}_{l-1} = \frac{a_{l-1}}{a_l}\mathbb{P}_l + \sum_{i=0}^{l-1} \delta_i \mathbb{P}_i. \tag{2.4.42}$$

The orthogonality relations and the relation (2.4.41) then give

$$\beta_{l-1} = -l\frac{(2l+1)(l-1/2)}{(2l-1)(l+1/2)} = -l, \tag{2.4.43}$$

which leads to (2.4.35).

Multiplying this relation by $\mathbb{P}_{l+1}$ then integrating, we obtain

$$\int_{-1}^{1} x\mathbb{P}_l(x)\mathbb{P}_{l+1}(x)dx = \frac{l+1}{2l+1}\int_{-1}^{+1}(\mathbb{P}_{l+1}(x))^2\,dx. \tag{2.4.44}$$

Replacing l by $l+1$ in (2.4.35), multiplying by $\mathbb{P}_l$ and integrating, we obtain

$$\int_{-1}^{1} x\mathbb{P}_l(x)\mathbb{P}_{l+1}(x) = \frac{l+1}{2l+3}\int_{-1}^{+1}(\mathbb{P}_l(x))^2\,dx$$

so that

$$\int_{-1}^{+1}(\mathbb{P}_{l+1}(x))^2\,dx = \frac{2l+1}{2l+3}\int_{-1}^{+1}(\mathbb{P}_l(x))^2\,dx, \tag{2.4.45}$$

which proves (2.4.36) for the index $l+1$.

The equality $\mathbb{P}_l(1) = 1$ is a consequence of (2.4.35).

Lemma 2.4.4 *The Legendre polynomials $\mathbb{P}_l$ satisfy the recursion relations*

$$\left\{ \begin{aligned} (1-x^2)\frac{d}{dx}\mathbb{P}_l(x) &= l\mathbb{P}_{l-1}(x) - lx\mathbb{P}_l(x) \\ &= (l+1)x\mathbb{P}_l(x) - (l+1)\mathbb{P}_{l+1}(x) \\ &= \frac{l(l+1)}{2l+1}\left(\mathbb{P}_{l-1}(x) - \mathbb{P}_{l+1}(x)\right). \end{aligned} \right. \tag{2.4.46}$$

Proof

The polynomial $(1 - x^2)(d/dx)\mathbb{P}_l$ has degree $l + 1$. It follows from the identity

$$\int_{-1}^{+1} \left((1 - x^2)\frac{d}{dx}\mathbb{P}_l \right) q(x)dx = - \int_{-1}^{+1} \mathbb{P}_l(x)\frac{d}{dx}((1 - x^2)q(x))dx \quad (2.4.47)$$

that it is orthogonal to any polynomial q of degree less than or equal to $l - 2$, and thus it is a linear combination of $\mathbb{P}_{l-1}$, $\mathbb{P}_l$ and $\mathbb{P}_{l+1}$:

$$(1 - x^2)\frac{d}{dx}\mathbb{P}_l(x) = C_{l+1}\mathbb{P}_{l+1} + C_l\mathbb{P}_l + C_{l-1}\mathbb{P}_{l-1}. \quad (2.4.48)$$

The coefficient C_l is equal to zero by parity. We find the coefficient C_{l+1} by looking at the term of highest degree, and so

$$C_{l+1} = -\frac{l(l + 1)}{2l + 1}. \quad (2.4.49)$$

The value of C_{l-1} is deduced from (2.4.47), by choosing $q = \mathbb{P}_{l-1}$, or from taking $x = 1$ in (2.4.48). Using (2.4.36), we obtain

$$\begin{cases} C_{l-1} = -(l - 1/2)\int_{-1}^{+1} \mathbb{P}_l(x)\frac{d}{dx}((1 - x^2)\mathbb{P}_{l-1}(x))dx \\ \\ = \dfrac{(l - 1/2)(l + 1)}{l + 1/2}\dfrac{a_{l-1}}{a_l} = \dfrac{l(l + 1)}{2l + 1}. \end{cases} \quad (2.4.50)$$

This proves the last relation in (2.4.46), which implies the two others, using (2.4.35).

Theorem 2.4.2 *The Legendre operator is defined by*

$$Ap = -\frac{d}{dx}(1 - x^2)\frac{d}{dx}p. \quad (2.4.51)$$

It is a self-adjoint operator in $L^2(]-1, +1[)$. The Legendre polynomials form a basis of eigenfunctions for this operator and

$$A\mathbb{P}_l = l(l + 1)\mathbb{P}_l. \quad (2.4.52)$$

They are orthogonal for the scalar product $L^2(]-1, +1[)$, and satisfy

$$\begin{cases} \int_{-1}^{+1}(1 - x^2)\frac{d}{dx}\mathbb{P}_{l_1}\frac{d}{dx}\mathbb{P}_{l_2}dx = \dfrac{l_1(l_1 + 1)}{l_1 + 1/2}\delta_{l_1}^{l_2} \\ \\ \text{(where } \delta_{l_1}^{l_2} \text{ is the Kronecker index).} \end{cases} \quad (2.4.53)$$

We have moreover

$$(1 - 2r\cos\theta + r^2)^{-1/2} = \sum_{l=0}^{\infty} r^l\mathbb{P}_l(\cos\theta), \quad r < 1. \quad (2.4.54)$$

Further, the following property holds in the sense of distributions on the segment $] - 1, +1[$

$$\sum_{l=0}^{\infty} (l + 1/2) \mathbb{P}_l(x) \mathbb{P}_l(y) = \delta(y - x), \tag{2.4.55}$$

where δ *denotes the Dirac mass at the origin.*

Proof

The Legendre polynomials form a basis for the polynomials. So they constitute a basis in $L^2(] - 1, +1[)$. Clearly the Legendre operator is self-adjoint.

Using (2.4.47) we see that the quantity

$$\begin{cases} \int_{-1}^{+1} (1 - x^2) \frac{d}{dx} \mathbb{P}_l(x) \frac{d}{dx} q(x) dx \\ = - \int_{-1}^{+1} \mathbb{P}_l(x) \frac{d}{dx} ((1 - x^2) \frac{d}{dx} q(x)) dx \end{cases} \tag{2.4.56}$$

vanishes as soon as q is of degree less than or equal to $l - 1$. This proves the equality (2.4.53) if $l_1 \neq l_2$.

When $l_1 = l_2 = l$, looking at the term of highest degree, it results from (2.4.36) that

$$\int_{-1}^{+1} \mathbb{P}_l(x) \frac{d}{dx} \left((1 - x^2) \frac{d}{dx} \mathbb{P}_l(x) \right) dx = -\frac{l(l + 1)}{l + 1/2}, \tag{2.4.57}$$

which proves (2.4.53). Integrating by parts yields (2.4.52).

To prove (2.4.54), we need to show that

$$(1 - 2r\cos\theta + r^2)^{1/2} \left(\sum_{l=0}^{\infty} r^l \mathbb{P}_l(\cos\theta) \right) = 1, \tag{2.4.58}$$

or, differentiating with respect to r, that

$$(r - \cos\theta) \sum_{l=0}^{\infty} r^l \mathbb{P}_l(\cos\theta) + (1 - 2r\cos\theta + r^2) \sum_{l=0}^{\infty} l r^{l-1} \mathbb{P}_l(\cos\theta) = 0, \tag{2.4.59}$$

an identity which is a consequence of the recursion formula (2.4.35).

Finally, any regular function φ with compact support in $] - 1, 1[$ can be represented in the form

$$\varphi(y) = \sum_{l=0}^{\infty} (l + 1/2) \left(\int_{-1}^{+1} \mathbb{P}_l(x) \varphi(x) dx \right) \mathbb{P}_l(y) \tag{2.4.60}$$

and this identity is nothing but (2.4.55).

Theorem 2.4.3 *The Legendre polynomials* $\mathbb{P}_l(\cos\theta)$ *are the only spherical harmonics which are invariant by rotations around the axis* $(0, x_3)$.

Proof

Denote $x_3 = \cos\theta$, so that $\partial/\partial x_3 = -(1/\sin\theta)\,\partial/\partial\theta$ and let u_l be one such spherical harmonic. It is independent of the variable φ and satisfies

$$\frac{1}{\sin\theta}\frac{\partial}{\partial\theta}\left(\sin\theta\frac{\partial u_l}{\partial\theta}\right) + l(l+1)u_l = 0 \qquad (2.4.61)$$

or, in the variable x_3,

$$\frac{d}{dx_3}(1-(x_3)^2)\frac{d}{dx_3}u_l + l(l+1)u_l = 0. \qquad (2.4.62)$$

The result is now a consequence of Theorem 2.4.2.

2.4.3 Associated Legendre functions

We have just seen that the Legendre polynomials are exactly the spherical harmonics invariant by rotation around the axis $(0, x_3)$. It is convenient to use the variables φ and x_3 to study all the spherical harmonics. They solve the equation

$$\frac{1}{1-x^2}\frac{\partial^2 u}{\partial\varphi^2} + \frac{\partial}{\partial x}\left((1-x^2)\frac{\partial}{\partial x}u\right) + l(l+1)u = 0. \qquad (2.4.63)$$

We show below that this equation admits a family of solutions with separate variables

$$Y_l^m(x,\varphi) = \gamma_l^m e^{im\varphi}\mathbb{P}_l^m(x) \qquad (2.4.64)$$

where the functions $\mathbb{P}_l^m(x)$ satisfy the differential equation

$$\frac{d}{dx}\left((1-x^2)\frac{d}{dx}\mathbb{P}_l^m\right) + l(l+1)\mathbb{P}_l^m - \frac{m^2}{1-x^2}\mathbb{P}_l^m = 0. \qquad (2.4.65)$$

We call **associated Legendre functions $\mathbb{P}_l^m(x)$, the solutions to equation** (2.4.64), normalized as indicated. For $m = 0$, Y_l^0 is the Legendre polynomial $\mathbb{P}_l$.

In order to describe the functions Y_l^m, we introduce new differential operators. We express them in the angles (θ, φ)

$$L_3 u = \frac{1}{i}\frac{\partial}{\partial\varphi}u. \qquad (2.4.66)$$

$$L_+ u = e^{i\varphi}\left(\frac{\partial}{\partial\theta}u + i\frac{\cos\theta}{\sin\theta}\frac{\partial}{\partial\varphi}u\right). \qquad (2.4.67)$$

$$L_- u = e^{-i\varphi}\left(-\frac{\partial}{\partial\theta}u + i\frac{\cos\theta}{\sin\theta}\frac{\partial}{\partial\varphi}u\right). \qquad (2.4.68)$$

Lemma 2.4.5 *The kinetic moments L_+, L_-, L_3, satisfy the relations of commutation:*

$$[L_+, L_-] = 2L_3, \qquad (2.4.69)$$

$$[L_3, L_+] = L_+, \qquad (2.4.70)$$

$$[L_3, L_-] = -L_-, \qquad (2.4.71)$$

$$where \quad [A, B] = AB - BA. \qquad (2.4.72)$$

Proof

Computing the expression of the two products

$$L_3 L_+ u = e^{i\varphi} \left(\frac{\partial u}{\partial \theta} + i \frac{\cos\theta}{\sin\theta} \frac{\partial u}{\partial \varphi} + \frac{1}{i} \frac{\partial^2 u}{\partial \theta \partial \varphi} + \frac{\cos\theta}{\sin\theta} \frac{\partial^2 u}{\partial \varphi^2} \right),$$

$$L_+ L_3 u = e^{i\varphi} \left(\frac{1}{i} \frac{\partial^2 u}{\partial \theta \partial \varphi} + \frac{\cos\theta}{\sin\theta} \frac{\partial^2 u}{\partial \varphi^2} \right) \qquad (2.4.73)$$

yields formula (2.4.70) by subtraction. Formula (2.4.71) is then deduced by conjugation. Computing the expression of the product $L_+ L_-$,

$$L_+ L_- u = -\frac{\partial^2 u}{\partial \theta^2} - \frac{\cos^2\theta}{\sin^2\theta} \frac{\partial^2 u}{\partial \varphi^2} + \frac{1}{i} \frac{\partial u}{\partial \varphi} - \frac{\cos\theta}{\sin\theta} \frac{\partial u}{\partial \theta}, \qquad (2.4.74)$$

we obtain by conjugation $L_- L_+$ and then subtracting, we deduce (2.4.69).

Lemma 2.4.6 *The Laplace-Beltrami operator Δ_S takes the different forms*

$$\begin{aligned}
\Delta_S &= -\tfrac{1}{2}(L_+ L_- + L_- L_+) - (L_3)^2 \\
&= -L_+ L_- - (L_3)^2 + L_3 \qquad (2.4.75) \\
&= -L_- L_+ - (L_3)^2 - L_3
\end{aligned}$$

and the following relations of commutation hold:

$$[\Delta_S, L_+] = [\Delta_S, L_-] = [\Delta_S, L_3] = 0. \qquad (2.4.76)$$

Proof

A careful comparison between the expression (2.4.3) of the operator Δ_S and the value of the product $L_+ L_-$ given by (2.4.74), leads to the first equality of (2.4.75). The second and the third equalities in (2.4.75) are then deduced from the first one and from the relation of commutation (2.4.69).

The commutation of Δ_S with L_3 results from definition (2.4.3) and from the expression of L_3. Using the expression (2.4.75) of the operator Δ_S and the relation of commutation for the kinetic moments, it follows that

$$
\begin{cases}
[\Delta_S, L_+] = -L_+L_-L_+ + L_+L_+L_- - L_3L_3L_+ + L_3L_+L_3 \\
\qquad -L_3L_+L_3 + L_+L_3L_3 + L_3L_+ - L_+L_3 \\
\qquad = 2L_+L_3 - L_3L_+ - L_+L_3 + L_3L_+ - L_+L_3 \\
\qquad = 0.
\end{cases}
\tag{2.4.77}
$$

(This is the Jacobi relation.)   ∎

The relations of commutation (2.4.76) show that **each eigenspace of the operator Δ_S is invariant by the action of the operators** L_+, L_- and L_3. We are going to use this property to express the spherical harmonics in the form (2.4.64). This is the object of the following theorem:

Theorem 2.4.4 *The spherical harmonics of order l are the $2l+1$ functions of the form*

$$
Y_l^m(\theta, \varphi) = (-1)^m \left[\frac{(l+1/2)}{2\pi} \frac{(l-m)!}{(l+m)!} \right]^{1/2} e^{im\varphi} \mathbb{P}_l^m(\cos\theta).
\tag{2.4.78}
$$

The associated Legendre functions $\mathbb{P}_l^m(\cos\theta)$ *can be computed in terms of the Legendre polynomials, with the recursion formula*

$$
\begin{cases}
if \quad 0 \le m \le l, \\
\mathbb{P}_l^m(\cos\theta) = (\sin\theta)^m \left(\dfrac{d}{dx} \right)^m \mathbb{P}_l(\cos\theta);
\end{cases}
\tag{2.4.79}
$$

$$
\begin{cases}
if \ -l \le m \le l, \quad \mathbb{P}_l^{-m}(x) = (-1)^m \dfrac{(l-m)!}{(l+m)!} \mathbb{P}_l^m(x), \\
\mathbb{P}_l^m(\cos\theta) = \dfrac{(-1)^{l+m}}{2^l l!} \dfrac{(l+m)!}{(l-m)!} (\sin\theta)^{-m} \left(\dfrac{d}{dx} \right)^{l-m} (1-x^2)^l.
\end{cases}
\tag{2.4.80}
$$

Their parity is $l+m$. For m fixed, they are mutually orthogonal, i.e.,

$$
\int_{-1}^{+1} \mathbb{P}_{l_1}^m(x) \mathbb{P}_{l_2}^m(x) dx = 0, \quad if \ l_1 \ne l_2,
\tag{2.4.81}
$$

and also satisfy the following orthogonality relations

$$
\int_{-1}^{+1} \frac{\mathbb{P}_l^{m_1}(x) \mathbb{P}_l^{m_2}(x)}{1-x^2} dx = 0, \quad if \ m_1 \ne m_2 \ and \ m_1 \ne -m_2.
\tag{2.4.82}
$$

*Further, they solve the differential equation (2.4.62), called the **Legendre equation** and satisfy*

$$
L_3 Y_l^m = m Y_l^m,
\tag{2.4.83}
$$

$$
L_+ Y_l^m = \sqrt{(l-m)(l+m+1)} Y_l^{m+1},
\tag{2.4.84}
$$

$$
L_- Y_l^m = \sqrt{(l+m)(l-m+1)} Y_l^{m-1}.
\tag{2.4.85}
$$

These spherical harmonics constitute an orthogonal basis of the space $L^2(S)$, also orthogonal in the space $H^1(S)$.

Proof

We seek spherical harmonics in the form (2.4.64). The expression of L_3 clearly shows that (2.4.83) holds. Using the relations (2.4.75), we obtain

$$L_+L_-Y_l^m = (l(l+1) - m^2 + m)Y_l^m$$
$$= (l+m)(l-m+1)Y_l^m,$$
$$L_-L_+Y_l^m = (l(l+1) - m^2 - m)Y_l^m$$
$$= (l-m)(l+m+1)Y_l^m. \tag{2.4.86}$$

It is easily checked that **the operator L_3 is hermitian and that the two operators L_+ and L_- are mutual adjoints** and thus

$$\int_S (L_+L_-Y_l^m)\,\overline{Y}_l^m\,d\sigma = \int_S |L_-Y_l^m|^2\,d\sigma,$$
$$\int_S (L_-L_+Y_l^m)\,\overline{Y}_l^m\,d\sigma = \int_S |L_+Y_l^m|^2\,d\sigma. \tag{2.4.87}$$

It follows from (2.4.86) and (2.4.87) that

$$L_-Y_l^{-l} = 0, \tag{2.4.88}$$

$$L_+Y_l^l = 0. \tag{2.4.89}$$

If we express these relations using the Legendre functions, we find that

$$\frac{d}{dx}\mathbb{P}_l^{-l} + \frac{lx}{1-x^2}\mathbb{P}_l^{-l} = 0, \tag{2.4.90}$$

$$\frac{d}{dx}\mathbb{P}_l^l + \frac{lx}{1-x^2}\mathbb{P}_l^l = 0. \tag{2.4.91}$$

These ODE's determine $\mathbb{P}_l^l$ and $\mathbb{P}_l^{-l}$ up to a multiplicative factor, and thus

$$\mathbb{P}_l^l(\cos\theta) = C_l(\sin\theta)^l,$$
$$\mathbb{P}_l^{-l}(\cos\theta) = C_{-l}(\sin\theta)^l. \tag{2.4.92}$$

Next, the commutation relations (2.4.70) and (2.4.83) give

$$L_3L_+Y_l^m = (m+1)L_+Y_l^m \tag{2.4.93}$$

which implies that $L_+Y_l^m$ is an eigenvector of the operator L_3 for the eigenvalue $m+1$. Indeed, the dimension of each of the $2l+1$ associated eigenspaces of the operator L_3 is necessarily 1, since they span the space of spherical harmonics, the dimension of which is $2l+1$. Thus $L_+Y_l^m$ is proportional to Y_l^m. The proportionality coefficient can be computed using (2.4.86) and (2.4.87). This implies (the spherical harmonics have a unit norm and so are defined up to an arbitrary multiplicative factor of modulus 1. We choose this factor to be 1)

$$L_+ Y_l^m = \sqrt{(l-m)(l+m+1)} Y_l^{m+1}, \qquad (2.4.94)$$

and so

$$L_- Y_l^m = \sqrt{(l+m)(l-m+1)} Y_l^{m-1}. \qquad (2.4.95)$$

By induction, we see that $(L_+)^n$ increases the index m by n while $(L_-)^n$ decreases the index m by n.

When expressed in the variable x_3, formulas (2.4.79) and (2.4.80) are consequences of the recursion formulas (2.4.94) and (2.4.95), starting from the value $m = 0$, as we know the expression of $Y_l^0 = \mathbb{P}_l$, the l-th Legendre polynomial. Let

$$Y_l^m = \gamma_l^m e^{im\varphi} \mathbb{P}_l^m(\cos\theta). \qquad (2.4.96)$$

We have, for $m > 0$

$$\begin{cases} L_+ Y_l^m \\ = -\gamma_l^m e^{i(m+1)\varphi} \left(\sin\theta \dfrac{d}{dx} \mathbb{P}_l^m(\cos\theta) + m\dfrac{\cos\theta}{\sin\theta} \mathbb{P}_l^m(\cos\theta) \right). \end{cases} \qquad (2.4.97)$$

A comparison between (2.4.94) and (2.4.97) yields

$$\begin{cases} \gamma_l^{m+1} \mathbb{P}_l^{m+1} \\ = -\dfrac{\gamma_l^m}{\sqrt{(l-m)(l+m+1)}} \left(\sin\theta \dfrac{d}{dx} \mathbb{P}_l^m + m\dfrac{\cos\theta}{\sin\theta} \mathbb{P}_l^m \right). \end{cases} \qquad (2.4.98)$$

Using the expression of γ_l^m given by (2.4.78), this equality can be written in the form

$$\mathbb{P}_l^{m+1} = \left(\sin\theta \dfrac{d}{dx} \mathbb{P}_l^m + m\dfrac{\cos\theta}{\sin\theta} \mathbb{P}_l^m \right). \qquad (2.4.99)$$

A simple induction then proves relation (2.4.79).

For a negative m, we use

$$\begin{cases} L_- Y_l^m \\ = \gamma_l^m e^{i(m-1)\varphi} \left(\sin\theta \dfrac{d}{dx} \mathbb{P}_l^m(\cos\theta) - m\dfrac{\cos\theta}{\sin\theta} \mathbb{P}_l^m(\cos\theta) \right), \end{cases} \qquad (2.4.100)$$

which, taking into account (2.4.95), leads to

$$\mathbb{P}_l^{m-1} = -\dfrac{1}{(l+m)(l-m+1)} \left(\sin\theta \dfrac{d}{dx} \mathbb{P}_l^m - m\dfrac{\cos\theta}{\sin\theta} \mathbb{P}_l^m \right). \qquad (2.4.101)$$

If we start from $m = 0$, this recursion formula gives the first expression in (2.4.80) by comparison with the recursion formula (2.4.79) (or (2.4.99)) which is exactly the same, except for a multiplicative factor.

Using formula (2.4.79) for $m = l$ and the expression of the coefficient of highest degree of the Legendre polynomial, we find the value of the coefficient C_l in the expression (2.4.92) and we obtain the expression

$$\mathbb{P}_l^l(\cos\theta) = \frac{(2l)!}{2^l l!}(\sin\theta)^l. \qquad (2.4.102)$$

Using formula (2.4.80), we obtain

$$\mathbb{P}_l^{-l}(\cos\theta) = (-1)^l \frac{1}{2^l l!}(\sin\theta)^l. \qquad (2.4.103)$$

The second expression in (2.4.80) is obtained using the recursion formula (2.4.101) starting from the value of $\mathbb{P}_l^l$ and this shows that this formula is true independently of the sign of m. Starting from $\mathbb{P}_l^{-l}$, we would have obtained a different formula.

The differential equation (2.4.65) for fixed l proves that the Legendre functions are the eigenfunctions of a self-adjoint operator for the scalar product in $L^2(\frac{1}{1-x^2}, S)$, associated with the eigenvalues m^2. This proves the orthogonality relations (2.4.82). For m fixed, the Legendre functions are the eigenfunctions of a self-adjoint operator for the scalar product in $L^2(S)$. This proves the orthogonality relations (2.4.81). ∎

We end this section with an additional property of spherical harmonics known as **the addition theorem for spherical harmonics**

Theorem 2.4.5 *Let $\vec{r}$ et $\vec{v}$ be two unit vectors and t their scalar product. The following addition formula holds:*

$$\sum_{m=-l}^{l} Y_l^m(\vec{r})\overline{Y}_l^m(\vec{v}) = \frac{(l+1/2)}{2\pi}\mathbb{P}_l(t), \qquad (2.4.104)$$

where Y_l^m are the spherical harmonics and $\mathbb{P}_l$ the Legendre polynomials of order l. In particular, we have

$$\sum_{m=-l}^{l} |Y_l^m(\vec{r})|^2 = \frac{(l+1/2)}{2\pi} \qquad (2.4.105)$$

and thus

$$|Y_l^m(\theta, \varphi)| \le \sqrt{\frac{(l+1/2)}{2\pi}}, \qquad (2.4.106)$$

which implies that $|\mathbb{P}_l(\cos\theta)| \le 1$.

Proof
Let Q be a rotation matrix. The function $Y_l^m(Q\vec{r})$ is again a spherical harmonic of order l because the sphere, the homogeneous polynomials and the Laplace operator are all invariant by rotation. Thus $Y_l^m(Q\vec{r})$ is a linear

combination of the $Y_l^m(\vec{\tau})$'s:

$$Y_l^m(Q\vec{\tau}) = \sum_{j=-l}^{l} \alpha_j^m Y_l^j(\vec{\tau}). \qquad (2.4.107)$$

We have

$$\int_S Y_l^m(Q\vec{\tau})\overline{Y}_l^j(Q\vec{\tau})d\sigma = \int_S Y_l^m(\vec{\tau})\overline{Y}_l^j(\vec{\tau})d\sigma = \delta_j^m. \qquad (2.4.108)$$

This proves the formula

$$\sum_{k=-l}^{l} \alpha_k^m \overline{\alpha}_k^j = \delta_j^m. \qquad (2.4.109)$$

In other words, the matrix α is unitary.

Consider now the sum

$$F(\vec{\tau}, \vec{\nu}) = \sum_{m=-l}^{l} Y_l^m(\vec{\tau})\overline{Y}_l^m(\vec{\nu}). \qquad (2.4.110)$$

The function F is a spherical harmonic in the two variables $\vec{\tau}$ and $\vec{\nu}$ separately. Using the formulas (2.4.107) and the fact that the matrix α is unitary shows that

$$F(Q\vec{\tau}, Q\vec{\nu}) = F(\vec{\tau}, \vec{\nu}). \qquad (2.4.111)$$

As there exists a rotation that maps the vector $\vec{\tau}$ onto the vector $\vec{e}_3$ and the vector $\vec{\nu}$ into the plane $(\vec{e}_1, \vec{e}_3)$, (2.4.111) shows that F depends only on the angle of these two vectors and precisely on their scalar product t. It is also an harmonic function invariant by the rotation around the axis x_3. According to Theorem 2.4.3, it must be the Legendre polynomial of degree l in the variable $x_3 = t$. The multiplicative factor is obtained using $\vec{\tau} = \vec{\nu}$, i.e., $t = 1$, then integrating on the whole sphere. We then obtain (2.4.105), as $\mathbb{P}_l(1) = 1$. ∎

Theorem 2.4.6 *The harmonic polynomial associated with the spherical harmonic function Y_l^m is given by:*
if m > 0

$$\begin{cases} H_l^m(x_1, x_2, x_3) = C_l^m(x_1 + ix_2)^m r^{l-m}(\dfrac{d}{d\xi})^{l+m}(1 - \xi^2)^l, \\ r^2 = x_1^2 + x_2^2 + x_3^2, \quad \xi = \dfrac{x_3}{r}, \end{cases} \qquad (2.4.112)$$

where the coefficient C_l^m is

$$C_l^m = (-1)^m \left(\frac{(l+1/2)}{2\pi}\frac{(l-m)!}{(l+m)!}\right)^{1/2}\frac{(-1)^l}{2^l l!}; \qquad (2.4.113)$$

if m < 0

$$Y_l^m = (-1)^m \overline{Y}_l^{-m}, \tag{2.4.114}$$

$$H_l^m(x) = (-1)^m \overline{H}_l^{-m}(x). \tag{2.4.115}$$

For m > 0, and r = 1, the derivatives of H_l^m are given by

$$\frac{\partial H_l^m}{\partial x_3} = i \sqrt{\frac{l+1/2}{l-1/2}} \sqrt{l^2 - m^2} Y_{l-1}^m, \tag{2.4.116}$$

$$\begin{cases} \dfrac{\partial H_l^m}{\partial x_1} = \dfrac{i}{2} \sqrt{\dfrac{l+1/2}{l-1/2}} \left[\sqrt{(l-m)(l-m-1)}\, Y_{l-1}^{m+1} \right. \\ \left. \qquad\qquad - \sqrt{(l+m)(l+m-1)} Y_{l-1}^{m-1} \right], \end{cases} \tag{2.4.117}$$

$$\begin{cases} \dfrac{\partial H_l^m}{\partial x_2} = \dfrac{1}{2} \sqrt{\dfrac{l+1/2}{l-1/2}} \left[\sqrt{(l-m)(l-m-1)}\, Y_{l-1}^{m+1} \right. \\ \left. \qquad\qquad + \sqrt{(l+m)(l+m-1)} Y_{l-1}^{m-1} \right]. \end{cases} \tag{2.4.118}$$

Proof
The proof of formula (2.4.112) follows from the homogeneity of the polynomial H_l^m, which must satisfy

$$H_l^m(x) = r^l Y_l^m \left(\frac{\vec{r}}{r} \right). \tag{2.4.119}$$

We then obtain (2.4.112) from the expression (2.4.78) of Y_l^m and from the relation

$$x_1 + ix_2 = re^{i\varphi}\sin\theta. \tag{2.4.120}$$

The constant C_l^m can be computed from (2.4.78) and from the Rodrigues formula (2.4.31). For $m < 0$, we use formulas (2.4.80).

The formulas (2.4.116) to (2.4.118) are direct consequences of the following lemma:

Lemma 2.4.7 *The polynomial $(1 - x^2)^l$ satisfies the recursion formulas ($m \geq 0$)*

$$\begin{cases} (l-m) \left(\dfrac{d}{dx} \right)^{l+m} (1-x^2)^l - x \left(\dfrac{d}{dx} \right)^{l+m+1} (1-x^2)^l \\ \\ = 2l \left(\dfrac{d}{dx} \right)^{l+m} (1-x^2)^{l-1}, \end{cases} \tag{2.4.121}$$

$$\begin{cases} \left(\dfrac{d}{dx}\right)^{l+m+1}(1-x^2)^l + 2lx\left(\dfrac{d}{dx}\right)^{l+m}(1-x^2)^{l-1} \\ \\ \quad = -2l(l+m)\left(\dfrac{d}{dx}\right)^{l+m-1}(1-x^2)^{l-1}, \end{cases} \tag{2.4.122}$$

$$\begin{cases} (1-x^2)\left(\dfrac{d}{dx}\right)^{l+m+1}(1-x^2)^l \\ \\ \quad + (l-m)\,x\left(\dfrac{d}{dx}\right)^{l+m}(1-x^2)^l \\ \\ \quad = -2l(l+m)\left(\dfrac{d}{dx}\right)^{l+m-1}(1-x^2)^{l-1}, \end{cases} \tag{2.4.123}$$

$$\begin{cases} 2l(1-x^2)\left(\dfrac{d}{dx}\right)^{l+m}(1-x^2)^{l-1} \\ \\ \quad + 2m\left(\dfrac{d}{dx}\right)^{l+m}(1-x^2)^l \\ \\ \quad = -2l(l+m)(l+m-1)\left(\dfrac{d}{dx}\right)^{l+m-2}(1-x^2)^{l-1}. \end{cases} \tag{2.4.124}$$

Proof
The Leibniz formula shows that

$$\left(\frac{d}{dx}\right)^k(1-x^2)^l = \sum_{p=E(\frac{k+1}{2})}^{l}(-1)^p C_l^p \frac{(2p)!}{(2p-k)!}x^{2p-k}, \tag{2.4.125}$$

$$\begin{cases} (l-m)\left(\dfrac{d}{dx}\right)^{l+m}(1-x^2)^l \\ \\ \quad = \displaystyle\sum_{p=E(\frac{l+m+1}{2})}^{l}(-1)^p \frac{l!}{p!(l-p)!}\frac{(2p)!\,(l-m)}{(2p-l-m)!}x^{2p-l-m}, \end{cases} \tag{2.4.126}$$

$$\begin{cases} x\left(\dfrac{d}{dx}\right)^{l+m+1}(1-x^2)^l \\ \\ \quad = \displaystyle\sum_{p=E(\frac{l+m+2}{2})}^{l}(-1)^p \frac{l!}{p!(l-p)!}\frac{(2p)!}{(2p-l-m-1)!}x^{2p-l-m}. \end{cases} \tag{2.4.127}$$

Further, combining the following expressions, relation (2.4.121) is obtained by subtraction. The two relations

$$
\left\{
\begin{aligned}
&\left(\frac{d}{dx}\right)^{l+m+1}(1-x^2)^l \\
&\quad = -\sum_{p=E(\frac{l+m}{2})}^{l-1}(-1)^p C_l^{p+1}\frac{(2p+2)!}{(2p-l-m+1)!}x^{2p-l-m+1},
\end{aligned}
\right.
\tag{2.4.128}
$$

$$
\left\{
\begin{aligned}
&x\left(\frac{d}{dx}\right)^{l+m}(1-x^2)^{l-1} \\
&\quad = \sum_{p=E(\frac{l+m+1}{2})}^{l-1}(-1)^p C_{l-1}^p\frac{(2p)!}{(2p-l-m)!}x^{2p-l-m+1}
\end{aligned}
\right.
\tag{2.4.129}
$$

prove (2.4.122). The formula (2.4.123) is then obtained by adding the first relation times x to the second one. Adding x times the equation (2.4.122) to (2.4.121), we find

$$
\left\{
\begin{aligned}
&(l-m)\left(\frac{d}{dx}\right)^{l+m}(1-x^2)^l \\
&\quad - 2l(1-x^2)\left(\frac{d}{dx}\right)^{l+m}(1-x^2)^{l-1} \\
&\quad + 2l(l+m)x\left(\frac{d}{dx}\right)^{l+m-1}(1-x^2)^{l-1} = 0,
\end{aligned}
\right.
\tag{2.4.130}
$$

which leads to the identity (2.4.124) using the relation (2.4.122) for the value $m-1$.

End of the proof of Theorem 2.4.6

As $\dfrac{\partial}{\partial x_3}\xi = \dfrac{r^2-x_3^2}{r^3}$, the expression of H_l^m yields

$$
\left\{
\begin{aligned}
\frac{\partial H_l^m}{\partial x_3} &= C_l^m(x_1+ix_2)^m r^{l-m-3}\left[(l-m)rx_3\left(\frac{d}{d\xi}\right)^{l+m}(1-\xi^2)^l \right.\\
&\quad \left. + (r^2-x_3^2)\left(\frac{d}{d\xi}\right)^{l+m+1}(1-\xi^2)^l\right].
\end{aligned}
\right.
\tag{2.4.131}
$$

For $r=1$, using relation (2.4.123), we obtain

$$
\left\{
\begin{aligned}
\frac{\partial H_l^m}{\partial x_3}\Big|_S &= -C_l^m(x_1+ix_2)^m 2l(l+m)\left(\frac{d}{d\xi}\right)^{l+m-1}(1-\xi^2)^{l-1} \\
&= i\sqrt{\frac{l+1/2}{l-1/2}}\sqrt{l^2-m^2}\,Y_{l-1}^m.
\end{aligned}
\right.
\tag{2.4.132}
$$

Further, since $\dfrac{\partial}{\partial x_1}\xi = -\dfrac{x_1 x_3}{r^3}$, we have

$$
\left\{
\begin{aligned}
\frac{\partial H_l^m}{\partial x_1} &= C_l^m (x_1 + ix_2)^{m-1} r^{l-m} \\[2mm]
&\times \left[\left(m + (l-m)(x_1 + ix_2)\frac{x_1}{r^2} \right) \left(\frac{d}{d\xi}\right)^{l+m} (1-\xi^2)^l \right. \\[2mm]
&\left. - \frac{x_1 x_3}{r^3}(x_1 + ix_2)\left(\frac{d}{d\xi}\right)^{l+m+1}(1-\xi^2)^l \right].
\end{aligned}
\right.
\tag{2.4.133}
$$

Replacing r by 1, we infer from (2.4.124) that

$$
\left\{
\begin{aligned}
\frac{\partial H_l^m}{\partial x_1} &= C_l^m (x_1 + ix_2)^{m-1}\left[m\left(\frac{d}{d\xi}\right)^{l+m}(1-\xi^2)^l \right. \\[2mm]
&\left. + 2l(x_1 + ix_2)x_1 \left(\frac{d}{d\xi}\right)^{l+m}(1-\xi^2)^{l-1} \right].
\end{aligned}
\right.
\tag{2.4.134}
$$

Similarly, we obtain

$$
\left\{
\begin{aligned}
\frac{\partial H_l^m}{\partial x_2} &= C_l^m (x_1 + ix_2)^{m-1}\left[im\left(\frac{d}{d\xi}\right)^{l+m}(1-\xi^2)^l \right. \\[2mm]
&\left. + 2l(x_1 + ix_2)x_2 \left(\frac{d}{d\xi}\right)^{l+m}(1-\xi^2)^{l-1} \right].
\end{aligned}
\right.
\tag{2.4.135}
$$

Combining the last two expressions yields

$$
\frac{\partial H_l^m}{\partial x_1} + i\frac{\partial H_l^m}{\partial x_2} = 2l C_l^m (x_1 + ix_2)^{m+1}\left(\frac{d}{d\xi}\right)^{l+m}(1-\xi^2)^{l-1},
\tag{2.4.136}
$$

$$
\left\{
\begin{aligned}
\frac{\partial H_l^m}{\partial x_1} - i\frac{\partial H_l^m}{\partial x_2} &= C_l^m (x_1+ix_2)^{m-1}\left[2m\left(\frac{d}{d\xi}\right)^{l+m}(1-\xi^2)^l \right. \\[2mm]
&\left. + 2l(1-x_3^2)\left(\frac{d}{d\xi}\right)^{l+m}(1-\xi^2)^{l-1} \right],
\end{aligned}
\right.
\tag{2.4.137}
$$

and in view of (2.4.112), (2.4.113) and (2.4.124),

$$
\frac{\partial H_l^m}{\partial x_1} + i\frac{\partial H_l^m}{\partial x_2} = i\sqrt{\frac{l+1/2}{l-1/2}}\sqrt{(l-m)(l-m-1)}\,H_{l-1}^{m+1},
\tag{2.4.138}
$$

$$
\left\{
\begin{aligned}
\frac{\partial H_l^m}{\partial x_1} - i\frac{\partial H_l^m}{\partial x_2} &= -C_l^m(x_1 + ix_2)^{m-1} \\
&\quad \times \left[2l(l+m)(l+m-1)\left(\frac{d}{d\xi}\right)^{l+m-2}(1-\xi^2)^{l-1} \right] \\
&= -i\sqrt{\frac{l+1/2}{l-1/2}}\sqrt{(l+m)(l+m-1)}H_{l-1}^{m-1}.
\end{aligned}
\right.
\tag{2.4.139}
$$

We then obtain the expressions of $\partial H_l^m/\partial x_1$ and $\partial H_l^m/\partial x_2$ by elimination.

Lemma 2.4.8 *For $m > 0$, the following recursion relations hold:*

$$
x_1 H_l^m(x) =
\left\{
\begin{aligned}
&\frac{i}{4}\left[\frac{r^2}{\sqrt{l^2-\frac{1}{4}}}\left[\sqrt{(l-m)(l-m-1)}H_{l-1}^{m+1}(x) \right.\right.\\
&\qquad\qquad\qquad \left.- \sqrt{(l+m)(l+m-1)}H_{l-1}^{m-1}(x)\right] \\
&\quad + \frac{1}{\sqrt{(l+1)^2-\frac{1}{4}}}\left[\sqrt{(l+m+1)(l+m+2)}H_{l+1}^{m+1}(x) \right.\\
&\qquad\qquad\qquad \left.\left.- \sqrt{(l-m+1)(l-m+2)}H_{l+1}^{m-1}(x)\right]\right],
\end{aligned}
\right.
\tag{2.4.140}
$$

$$
x_2 H_l^m(x) =
\left\{
\begin{aligned}
&\frac{1}{4}\left[\frac{r^2}{\sqrt{l^2-\frac{1}{4}}}\left[\sqrt{(l-m)(l-m-1)}H_{l-1}^{m+1}(x) \right.\right.\\
&\qquad\qquad\qquad \left.+ \sqrt{(l+m)(l+m-1)}H_{l-1}^{m-1}(x)\right] \\
&\quad + \frac{1}{\sqrt{(l+1)^2-\frac{1}{4}}}\left[\sqrt{(l+m+1)(l+m+2)}H_{l+1}^{m+1}(x) \right.\\
&\qquad\qquad\qquad \left.\left.+ \sqrt{(l-m+1)(l-m+2)}H_{l+1}^{m-1}(x)\right]\right],
\end{aligned}
\right.
\tag{2.4.141}
$$

$$
x_3 H_l^m(x) =
\left\{
\begin{aligned}
&\frac{i}{2}\left[\sqrt{\frac{l^2-m^2}{l^2-\frac{1}{4}}}\,r^2 H_{l-1}^m(x) \right.\\
&\qquad\qquad \left.- \frac{\sqrt{(l+1)^2-m^2}}{\sqrt{(l+\frac{1}{2})(l+\frac{3}{2})}}H_{l+1}^m(x)\right].
\end{aligned}
\right.
\tag{2.4.142}
$$

Proof

We multiply (2.4.112) by $(x_1 + ix_2)$ to get

$$
(x_1 + ix_2)H_l^m(x) = C_l^m(x_1 + ix_2)^{m+1}r^{l-m}\left(\frac{d}{d\xi}\right)^{l+m}(1-\xi^2)^l. \tag{2.4.143}
$$

Using (2.4.121) and (2.4.122) for the index $l + 1$, we obtain

$$
\left\{
\begin{aligned}
&(2l + 1)\left(\frac{d}{d\xi}\right)^{l+m}(1 - \xi^2)^l \\[2mm]
&\quad = 2l\left(\frac{d}{d\xi}\right)^{l+m}(1 - \xi^2)^{l-1} \\[2mm]
&\qquad - \frac{1}{2(l+1)}\left(\frac{d}{d\xi}\right)^{l+m+2}(1 - \xi^2)^{l+1}.
\end{aligned}
\right.
\tag{2.4.144}
$$

Inserting this expression in (2.4.143) leads to

$$
\left\{
\begin{aligned}
&(x_1 + ix_2)H_l^m(x) = C_l^m r^{l-m}(x_1 + ix_2)^{m+1} \\[2mm]
&\quad \times \left[\frac{2l}{2l+1}\left(\frac{d}{d\xi}\right)^{l+m}(1 - \xi^2)^{l-1}\right. \\[2mm]
&\qquad \left. - \frac{1}{(2l+2)(2l+1)}\left(\frac{d}{d\xi}\right)^{l+m+2}(1 - \xi^2)^{l+1}\right] \\[2mm]
&\quad = \frac{i}{2}\left[\sqrt{\frac{(l-m)(l-m-1)}{(l+\frac{1}{2})(l-\frac{1}{2})}}\,H_{l-1}^{m+1}(x)r^2\right. \\[2mm]
&\qquad \left. + \sqrt{\frac{(l+m+1)(l+m+2)}{(l+\frac{1}{2})(l+\frac{3}{2})}}\,H_{l+1}^{m+1}(x)\right].
\end{aligned}
\right.
\tag{2.4.145}
$$

In the same fashion, we multiply (2.4.112) by $(x_1 - ix_2)$

$$
\left\{
\begin{aligned}
&(x_1 - ix_2)H_l^m(x) = C_l^m(x_1 + ix_2)^{m-1} \\[2mm]
&\qquad \times r^{l-m+2}(1 - \xi^2)\left(\frac{d}{d\xi}\right)^{l+m}(1 - \xi^2)^l.
\end{aligned}
\right.
\tag{2.4.146}
$$

Expressing the relation (2.4.124) for the indexes $l + 1$, $m - 1$ and relation (2.4.144) for the index $m - 2$, we get the relation

$$
\left\{
\begin{aligned}
&0 = (2l + 1)(2l + 2)(1 - x^2)\left(\frac{d}{dx}\right)^{l+m}(1 - x^2)^l \\[2mm]
&\quad - (l - m + 1)(l - m + 2)\left(\frac{d}{dx}\right)^{l+m}(1 - x^2)^{l+1} \\[2mm]
&\quad + 2l(2l + 2)(l + m - 1)(l + m)\left(\frac{d}{dx}\right)^{l+m-2}(1 - x^2)^{l-1}.
\end{aligned}
\right.
\tag{2.4.147}
$$

In combination with (2.4.146), this proves

$$
\left\{
\begin{aligned}
&(x_1 - ix_2)H_l^m(x) \\
&\quad = -\frac{i}{2}\left[\sqrt{\frac{(l-m+1)(l-m+2)}{(l+\frac{1}{2})(l+\frac{3}{2})}}H_{l+1}^{m-1}(x)\right. \\
&\qquad\qquad \left. + r^2\sqrt{\frac{(l+m-1)(l+m)}{(l+\frac{1}{2})(l-\frac{1}{2})}}H_{l-1}^{m-1}(x)\right].
\end{aligned}
\right.
\tag{2.4.148}
$$

The expressions (2.4.140) and (2.4.141) are then deduced from this last relation and from (2.4.145). From (2.4.112) again, we see that

$$
x_3 H_l^m(x) = C_l^m (x_1 + ix_2)^m r^{l-m} x_3 \left(\frac{d}{d\xi}\right)^{l+m} (1-\xi^2)^l. \tag{2.4.149}
$$

Using the identity (2.4.122) for the indexes $l+1$ and $m-1$, and the identity (2.4.144) for the indexes l and $m-1$, we find the identity

$$
\left\{
\begin{aligned}
&x\left(\frac{d}{dx}\right)^{l+m}(1-x^2)^l \\
&\quad = \left[-\frac{2l(l+m)}{2l+1}\left(\frac{d}{dx}\right)^{l+m-1}(1-x^2)^{l-1}\right. \\
&\qquad\qquad \left. -\frac{l-m+1}{(2l+2)(2l+1)}\left(\frac{d}{dx}\right)^{l+m+1}(1-x^2)^{l+1}\right],
\end{aligned}
\right.
\tag{2.4.150}
$$

which inserted in (2.4.149) shows (2.4.142). ∎

2.4.4 Vectorial spherical harmonics

In order to study the Maxwell equations, we will need the vectorial spherical harmonic functions. They are defined, exactly as in the scalar case, as **the traces on the sphere S of vector fields, the three components of which are harmonic polynomials**. We denote by $(\mathcal{H}_l)^3$ the linear space, of dimension $(6l+3)$, of vectorial spherical harmonics of degree l.

In the remainder of this chapter, we construct an **orthonormal basis of vectorial spherical harmonics**. To this effect, we define the vectorial harmonic polynomials

$$
\mathcal{I}_l^m(x) = \nabla H_{l+1}^m(x), \quad l \geq 0, \quad -(l+1) \leq m \leq l+1; \tag{2.4.151}
$$

$$
\mathcal{T}_l^m(x) = \nabla H_l^m \wedge x, \quad l \geq 1, \quad -l \leq m \leq l; \tag{2.4.152}
$$

$$\begin{cases} \mathcal{N}_l^m(x) = (2l-1)H_{l-1}^m(x)x - |x|^2 \nabla H_{l-1}^m(x), \\ \\ l \geq 1, \quad -(l-1) \leq m \leq (l-1). \end{cases} \tag{2.4.153}$$

We will respectively denote by $I_l^m(x), T_l^m(x), N_l^m(x)$, the traces on the sphere S of these polynomials.

Theorem 2.4.7 *For each l, the family (I_l^m, T_l^m, N_l^m) forms an orthogonal basis of $(\mathcal{H}_l)^3$ and of $(L^2(S))^3$. Further, they satisfy*

$$\int_S |I_l^m(x)|^2 \, d\sigma = (l+1)(2l+3), \tag{2.4.154}$$

$$\int_S |T_l^m(x)|^2 \, d\sigma = l(l+1), \tag{2.4.155}$$

$$\int_S |N_l^m(x)|^2 \, d\sigma = l(2l-1). \tag{2.4.156}$$

Proof

We first check that the polynomials $\mathcal{I}_l^m, \mathcal{T}_l^m$ and $\mathcal{N}_l^m$ are harmonic. As they are also homogeneous of degree l, this shows that they belong to $(\mathcal{H}_l)^3$. We constantly use the vectorial relations

$$\begin{cases} \Delta = -\overrightarrow{\text{curl}}\overrightarrow{\text{curl}} + \nabla \text{div}; \\ \\ \text{div}\overrightarrow{\text{curl}} = 0; \\ \\ \overrightarrow{\text{curl}}\nabla = 0. \end{cases} \tag{2.4.157}$$

Study of $\mathcal{I}_l^m$

That $\mathcal{I}_l^m$ is harmonic easily follows from

$$\text{div}\, \mathcal{I}_l^m(x) = \Delta H_{l+1}^m(x) = 0, \tag{2.4.158}$$

$$\overrightarrow{\text{curl}}\mathcal{I}_l^m(x) = \overrightarrow{\text{curl}}\nabla H_{l+1}^m(x) = 0. \tag{2.4.159}$$

Study of $\mathcal{T}_l^m$

Using the vectorial relation

$$\overrightarrow{\text{curl}}(\varphi \vec{u}) = \varphi \overrightarrow{\text{curl}}\vec{u} + \nabla\varphi \wedge \vec{u}, \tag{2.4.160}$$

we have $(\overrightarrow{\text{curl}}x = \overrightarrow{\text{curl}}\nabla \dfrac{|x|^2}{2} = 0)$

$$\mathcal{T}_l^m(x) = \overrightarrow{\text{curl}}\, (H_l^m(x)x) \tag{2.4.161}$$

and hence,

$$\text{div}\, \mathcal{T}_l^m(x) = 0. \tag{2.4.162}$$

Consequently,

$$\left\{ \begin{aligned} \overrightarrow{\text{curl}}\, T_l^m(x) &= \overrightarrow{\text{curl}\,\text{curl}}\,(H_l^m(x)x) \\ &= \nabla \operatorname{div}(H_l^m(x)x) - \Delta(H_l^m(x)x). \end{aligned} \right. \tag{2.4.163}$$

Since

$$\left\{ \begin{aligned} \Delta(H_l^m(x)x_i) &= \Delta H_l^m(x)x_i + 2\left(\nabla H_l^m(x).\nabla x_i\right) \\ &= 2\frac{\partial}{\partial x_i}H_l^m(x), \end{aligned} \right. \tag{2.4.164}$$

$$\left\{ \begin{aligned} \operatorname{div}(H_l^m(x)x) &= (\nabla H_l^m(x).x) + 3H_l^m(x) \\ &= (l+3)H_l^m(x), \end{aligned} \right. \tag{2.4.165}$$

we infer that

$$\overrightarrow{\text{curl}}\, T_l^m(x) = (l+1)\nabla H_l^m(x), \tag{2.4.166}$$

which finally shows that

$$\Delta T_l^m(x) = 0. \tag{2.4.167}$$

Study of $\mathcal{N}_l^m$

We compute $\operatorname{div}\mathcal{N}_l^m$ using (2.4.165);

$$\left\{ \begin{aligned} \operatorname{div}\mathcal{N}_l^m(x) &= (2l-1)(l+2)H_{l-1}^m(x) \\ &\quad - |x|^2 \Delta H_{l-1}^m(x) - 2\left(x.\nabla H_{l-1}^m(x)\right). \end{aligned} \right. \tag{2.4.168}$$

The Euler relation for homogeneous functions reduces this expression to

$$\operatorname{div}\mathcal{N}_l^m(x) = l(2l+1)H_{l-1}^m(x). \tag{2.4.169}$$

Besides, we see in view of (2.4.166) that

$$\left\{ \begin{aligned} \overrightarrow{\text{curl}}\,\mathcal{N}_l^m(x) &= (2l-1)T_{l-1}^m(x) - 2x \wedge \nabla H_{l-1}^m(x) \\ &= (2l+1)T_{l-1}^m(x), \end{aligned} \right. \tag{2.4.170}$$

and

$$\overrightarrow{\text{curl}\,\text{curl}}\,\mathcal{N}_l^m(x) = l(2l+1)\nabla H_{l-1}^m(x). \tag{2.4.171}$$

The harmonicity of $\mathcal{N}_l^m$ results from (2.4.169) and (2.4.171).

The restrictions $I_l^m(x), T_l^m(x), N_l^m(x)$ to the sphere are spherical harmonics. If we use the surfacic gradient operator defined by (2.4.6) and the Euler relation for the normal derivatives, we obtain

$$I_l^m(x) = \nabla_S Y_{l+1}^m(x) + (l+1)Y_{l+1}^m(x)x, \tag{2.4.172}$$

$$T_l^m(x) = \nabla_S Y_l^m(x) \wedge x, \tag{2.4.173}$$

$$N_l^m(x) = -\nabla_S Y_{l-1}^m(x) + l Y_{l-1}^m(x) x. \tag{2.4.174}$$

The orthogonality properties have to be checked in each case: they result from the mutual orthogonalities of the Y_l^m's and of the $\nabla_S Y_l^m$'s, in view of (2.4.23) and (2.4.7)

$$\begin{cases} \displaystyle\int_S (\nabla_S Y_{l_1}^{m_1} . \nabla_S Y_{l_2}^{m_2})\, d\sigma = -\int_S \Delta_S Y_{l_1}^{m_1} Y_{l_2}^{m_2}\, d\sigma \\[3mm] \displaystyle\qquad\qquad = l_1(l_1+1)\int_S Y_{l_1}^{m_1} Y_{l_2}^{m_2}\, d\sigma \qquad (2.4.175) \\[3mm] \displaystyle\qquad\qquad = l_1(l_1+1)\delta_{l_1}^{l_2}\delta_{m_1}^{m_2}. \end{cases}$$

Checking the orthogonality of the vectors $T_{l_1}^{m_1}$ to the vectors $I_{l_2}^{m_2}$ or to $N_{l_2}^{m_2}$, leads to integrals of the type

$$\int_S (\nabla_S Y_{l_1}^{m_1} \wedge x . \nabla_S Y_{l_2}^{m_2})\, d\sigma = \int_S \left(\overrightarrow{\mathrm{curl}}_S Y_{l_1}^{m_1} . \nabla_S Y_{l_2}^{m_2} \right) \tag{2.4.176}$$

which are always equal to zero, as will be proved below by showing that $\mathrm{div}_S \overrightarrow{\mathrm{curl}}_S = 0$.

Next, we compute

$$\begin{cases} \displaystyle\int_S |I_l^m(x)|^2 d\sigma \\[3mm] \displaystyle\qquad = \int_S |\nabla_S Y_{l+1}^m(x)|^2 d\sigma + (l+1)^2 \int_S |Y_{l+1}^m(x)|^2 d\sigma \qquad (2.4.177) \\[3mm] \displaystyle\qquad = (l+1)(2l+3), \end{cases}$$

$$\int_S |T_l^m(x)|^2 d\sigma = \int_S |\nabla H_l^m(x)|^2 d\sigma = l(l+1), \tag{2.4.178}$$

$$\begin{cases} \displaystyle\int_S |N_l^m(x)|^2 d\sigma = \int_S |\nabla_S Y_{l-1}^m(x)|^2 d\sigma + l^2 \int_S |Y_{l-1}^m(x)|^2 d\sigma \\[3mm] \displaystyle\qquad = l(2l-1). \end{cases} \tag{2.4.179}$$

Finally, given the orthogonality of $I_l^m(x), T_l^m(x), N_l^m(x)$ and since their components are spherical harmonics, a simple counting argument shows that, for each l, they form a basis of the linear space $(\mathcal{Y}_l)^3$.   ∎

In order to prove a few extra properties of the vectorial spherical harmonics, we need to introduce some differential operators that act on scalar functions defined on the sphere S or on tangent vector fields.

A vector field v defined on S, will be represented by its coordinates in the basis $\vec{e}_\theta, \vec{e}_\varphi$ of the tangent plane

$$v = v_\theta \vec{e}_\theta + v_\varphi \vec{e}_\varphi. \tag{2.4.180}$$

We have already defined and used **the surfacic gradient**

$$\nabla_S u = \frac{1}{\sin\theta}\frac{\partial u}{\partial\varphi}\vec{e}_\varphi + \frac{\partial u}{\partial\theta}\vec{e}_\theta, \tag{2.4.181}$$

and **the vectorial surfacic rotational**

$$\overrightarrow{\text{curl}}_S u = \nabla_S u \wedge x = -\frac{\partial u}{\partial\theta}\vec{e}_\varphi + \frac{1}{\sin\theta}\frac{\partial u}{\partial\varphi}\vec{e}_\theta. \tag{2.4.182}$$

We now define **the surfacic divergence** by

$$\text{div}_S v = \frac{1}{\sin\theta}\left(\frac{\partial}{\partial\theta}(\sin\theta v_\theta) + \frac{\partial}{\partial\varphi}v_\varphi\right). \tag{2.4.183}$$

We check (c.f. (2.4.3)) that

$$\Delta_S u = \text{div}_S \nabla_S u; \tag{2.4.184}$$

and also that

$$\int_S (\nabla_S u.v)d\sigma + \int_S u\,\text{div}_S v\,d\sigma = 0. \tag{2.4.185}$$

We define **the scalar surfacic rotational** by

$$\text{curl}_S v = \frac{1}{\sin\theta}\left(\frac{\partial}{\partial\theta}(\sin\theta v_\varphi) - \frac{\partial}{\partial\varphi}v_\theta\right). \tag{2.4.186}$$

One easily checks that

$$\Delta_S u = -\text{curl}_S \overrightarrow{\text{curl}}_S u, \tag{2.4.187}$$

and that

$$\int_S \left(\overrightarrow{\text{curl}}_S u.v\right)d\sigma - \int_S u\,\text{curl}_S v\,d\sigma = 0. \tag{2.4.188}$$

Finally, we define the **vectorial Laplace-Beltrami operator**

$$\Delta_S v = \nabla_S \text{div}_S v - \overrightarrow{\text{curl}}_S \text{curl}_S v. \tag{2.4.189}$$

We denote by $TL^2(S)$ the linear space of **tangent vector fields** with square integrable modulus. The following theorem shows how the vectorial spherical harmonics span $TL^2(S)$:

Theorem 2.4.8 *The family $(\nabla_S Y_l^m, \overrightarrow{\text{curl}}_S Y_l^m)$ forms a basis of eigenvectors for the vectorial Laplace Beltrami operator on $TL^2(S)$ and thus an orthogonal basis of $TL^2(S)$, i.e.,*

$$\Delta_S \nabla_S Y_l^m + l(l+1)\nabla_S Y_l^m = 0, \tag{2.4.190}$$

$$\Delta_S \overrightarrow{\text{curl}}_S Y_l^m + l(l+1)\overrightarrow{\text{curl}}_S Y_l^m = 0, \tag{2.4.191}$$

$$\int_S |\nabla_S Y_l^m|^2 d\sigma = \int_S |\overrightarrow{\text{curl}}_S Y_l^m|^2 d\sigma = l(l+1). \tag{2.4.192}$$

Proof

Formulas (2.4.190) and (2.4.191) follow from the expressions of the scalar laplacian (2.4.184) and (2.4.187), and from the spherical harmonics property (2.4.23). We again use the relations

$$\begin{cases} \operatorname{div}_S \overrightarrow{\operatorname{curl}}_S = 0, \\[2mm] \operatorname{curl}_S \nabla_S = 0. \end{cases} \qquad (2.4.193)$$

Properties of spherical harmonics show that the family $\left(\nabla_S Y_l^m, \overrightarrow{\operatorname{curl}}_S Y_l^m\right)$ forms a basis and thus it is exactly the basis of eigenvectors associated with this operator.

2.5 The Laplace Equation in $\mathbb{R}^3$

We study here the Laplace equation. We introduce several Hilbert spaces and their main properties. This setting leads to the variational formulation for the Laplace equation which is a good introduction to the corresponding one for the Helmholtz equation.

2.5.1 The sphere

In this section we give explicit solutions to the interior and exterior problems for the Laplace equation. They are expanded on the spherical harmonics.

Let B_i and B_e, denote respectively the interior and the exterior of the sphere. Let us define **the interior and exterior Dirichlet problems**

$$\begin{cases} \Delta u = 0, \quad \text{in } B_i \text{ or } B_e, \\[2mm] u|_S = u_d. \end{cases} \qquad (2.5.1)$$

We look for a solution that consists of a sum of products of a function of r times a function of the variables (θ, φ). There exist two such families of harmonic functions. On one hand, we can check using (2.4.2) that the functions

$$H_l^m(r, \theta, \varphi) = r^l Y_l^m(\theta, \varphi) \qquad (2.5.2)$$

are harmonic, smooth at the origin and tend to infinity at infinity. On the other hand, the functions

$$K_l^m(r, \theta, \varphi) = \frac{1}{r^{l+1}} Y_l^m(\theta, \varphi) = \frac{1}{r^{2l+1}} H_l^m(x) \qquad (2.5.3)$$

are also harmonic, but are not smooth at the origin and tend to zero at infinity.

The spherical harmonic functions form a basis of the linear space $L^2(S)$, and so we can expand any given function u_d in $L^2(S)$ as a sum of spherical harmonics

$$
\begin{cases}
u_d(\theta, \varphi) = \sum_{l=0}^{\infty} \sum_{m=-l}^{l} u_l^m Y_l^m(\theta, \varphi), \\[2mm]
u_l^m = \int_S u_d(\theta, \varphi)\overline{Y_l^m}(\theta, \varphi)d\sigma.
\end{cases}
\tag{2.5.4}
$$

This expansion provides a convenient way to define the **Hilbert spaces** $H^s(S)$.

Definition
For $s > 0$, the space $H^s(S)$ is constituted of the functions in $L^2(S)$ such that the series

$$
\sum_{l=0}^{\infty} \sum_{m=-l}^{l} (l+1)^{2s}|u_l^m|^2
$$

is convergent. Its hermitian product and its norm are associated with the bilinear form

$$
((u,v))_{H^s(S)} = \sum_{l=0}^{\infty} \sum_{m=-l}^{l} (l+1)^{2s} u_l^m \bar{v}_l^m.
\tag{2.5.5}
$$

∎

For a negative s, the elements of $H^s(S)$ are not functions any more, but distributions. They belong to the space of distributions $\mathcal{D}'(S)$, the dual space of $\mathcal{D}(S)$, the space of indefinitely differentiable functions. The spherical harmonic functions belong to the space $\mathcal{D}(S)$. Thus the duality between $\mathcal{D}'$ and $\mathcal{D}$ allows us to define for any distribution u its coefficients

$$
u_l^m = \langle u, Y_l^m \rangle
\tag{2.5.6}
$$

and to extend the above definition of $H^s(S)$, to the case of a negative s, in the following way:

Definition
The space $H^s(S)$ is the space of distributions in $\mathcal{D}'(S)$ such that the series

$$
\|u\|_{H^s(S)}^2 = \sum_{l=0}^{\infty} \sum_{m=-l}^{l} (l+1)^{2s}|u_l^m|^2
$$

is convergent. This is a Hilbert space with the hermitian product (2.5.5). ∎

We have already given a definition of $H^1(S)$. The equality

$$
\int_S |\nabla_S Y_l^m|^2 d\sigma = l(l+1)
\tag{2.5.7}
$$

shows that the two definitions are equivalent and that the norm (2.4.8) and the norm associated with (2.5.5) are equivalent.

We denote by B_i the domain interior to the sphere S and by B_e the domain exterior to S. We recall that for any positive integer k, $H^k(B_i)$ is the space of functions such that all the derivatives up to the order k are in $L^2(B_i)$. We denote by $D^k u$ the operator of derivation of order k, which associates with the function u all its k-order partial derivatives. When $k = 1$, it is also the operator ∇u.

We can now give explicit solutions to these Dirichlet problems.

Theorem 2.5.1 The interior Dirichlet problem

$$\begin{cases} \Delta u = 0, & x \in B_i, \\ u|_S = u_d \end{cases} \tag{2.5.8}$$

admits the solution

$$u(r, \theta, \varphi) = \sum_{l=0}^{\infty} \sum_{m=-l}^{l} u_l^m Y_l^m(\theta, \varphi) r^l, \tag{2.5.9}$$

and this series is absolutely convergent for $r < 1$. When $u_d \in H^{k-1/2}(S)$, this solution is in the Hilbert space $H^k(B_i)$ where $H^k(B_i)$ is the space

$$H^k(B_i) = \left\{ u; u \in L^2(B_i), Du \in L^2(B_i), \dots, D^k u \in L^2(B_i) \right\}.$$

Moreover, it satisfies

$$\|u\|_{H^k(B_i)} \leq C \|u_d\|_{H^{k-1/2}(S)}, \tag{2.5.10}$$

with a constant C depending only on k.

The exterior Dirichlet problem

$$\begin{cases} \Delta u = 0, & x \in B_e, \\ u|_S = u_d \end{cases} \tag{2.5.11}$$

admits the solution

$$u(r, \theta, \varphi) = \sum_{l=0}^{\infty} \sum_{m=-l}^{l} u_l^m Y_l^m(\theta, \varphi) \frac{1}{r^{l+1}}, \tag{2.5.12}$$

*and this series is absolutely convergent for $r > 1$. When $u_d \in H^{k-1/2}(S)$, this solution is in the **Hilbert space** $W^k(B_e)$.*

$$W^k(B_e) = \left\{ u; \frac{u}{r} \in L^2(B_e), Du \in L^2(B_e), \\ \dots, r^{k-1} D^k u \in L^2(B_e) \right\} \tag{2.5.13}$$

and u satisfies,

$$\|u\|_{W^k(B_e)} \leq C \|u_d\|_{H^{k-1/2}(S)}, \tag{2.5.14}$$

where C depends only on k.

Proof

These solutions are sums of functions of the form (2.5.2) and (2.5.3), which are harmonic in $\mathbb{R}^3$. Their traces on S are effectively u_d. Only proving the convergence of these series and showing the continuity bounds is needed. For the interior problem, i.e., $r < 1$, we use the inequality (2.4.106) to check that the coefficient of r^l is bounded by

$$|\sum_{m=-l}^{l} u_l^m Y_l^m(\theta, \varphi)| \leq \sqrt{\frac{l+1/2}{2\pi}} \sum_{m=-l}^{l} |u_l^m|, \qquad (2.5.15)$$

so it converges as soon as $u_d \in H^s(S)$, for any real s.

The same proof applies to the exterior problem except that in this new situation we have $(1/r) < 1$.

The norm in $L^2(B_i)$ of the solution of the interior problem is given by

$$\int_{B_i} |u(x)|^2 dx = \int_S \left(\int_0^1 (|u(r, \theta, \varphi)|^2) r^2 dr \right) d\sigma. \qquad (2.5.16)$$

Using the orthogonality of the spherical harmonics, this integral can be evaluated in terms of the u_l^m,

$$\begin{cases} \int_{B_i} |u(x)|^2 dx = \sum_{l=0}^{\infty} \sum_{m=-l}^{l} |u_l^m|^2 \int_0^1 r^{2l+2} dr \\ \\ = \sum_{l=0}^{\infty} \sum_{m=-l}^{l} \frac{1}{2l+3} |u_l^m|^2. \end{cases} \qquad (2.5.17)$$

It follows that

$$||u||_{L^2(B_i)} \leq \frac{1}{2} ||u||_{H^{-1/2}(S)}. \qquad (2.5.18)$$

The corresponding inequality for the exterior problem is

$$\begin{cases} \int_{B_i} \frac{1}{r^2} |u(x)|^2 dx = \int_S \left(\int_1^{\infty} |u(r, \theta, \varphi)|^2 dr \right) d\sigma \\ \\ = \sum_{l=0}^{\infty} \sum_{m=-l}^{l} |u_l^m|^2 \int_1^{\infty} \frac{1}{r^{2l+2}} dr \\ \\ = \sum_{l=0}^{\infty} \sum_{m=-l}^{l} \frac{1}{2l+1} |u_l^m|^2, \end{cases} \qquad (2.5.19)$$

which yields

$$||u||_{W^0(B_r)} \leq ||u||_{H^{-\frac{1}{2}}(S)}. \qquad (2.5.20)$$

We first establish regularity in the spaces $H^1(B_i)$ and $W^1(B_e)$. To evaluate the norms $H^1(B_i)$ or $W^1(B_e)$ of the solution, we use the following expression of the gradient

$$\nabla u = \frac{\partial u}{\partial r}\vec{e}_r + \frac{1}{r}\nabla_S u \tag{2.5.21}$$

where $\nabla_S u$ is the surfacic gradient (expression (2.4.6)).

For the interior solution, we find

$$\frac{\partial u}{\partial r} = \sum_{l=0}^{\infty}\sum_{m=-l}^{l} u_l^m Y_l^m(\theta, \varphi) l r^{l-1}. \tag{2.5.22}$$

It follows, using Plancherel's theorem, that

$$\left\|\frac{\partial u}{\partial r}\right\|_{L^2(B_i)}^2 = \sum_{l=0}^{\infty}\sum_{m=-l}^{l} |u_l^m|^2 \frac{l^2}{2l+1}. \tag{2.5.23}$$

The corresponding estimate for the exterior solution is

$$\left\|\frac{\partial u}{\partial r}\right\|_{L^2(B_e)}^2 = \sum_{l=0}^{\infty}\sum_{m=-l}^{l} |u_l^m|^2 \frac{(l+1)^2}{2l+1}. \tag{2.5.24}$$

The second part of the gradient can be estimated from

$$\nabla_S u = \sum_{l=0}^{\infty}\sum_{m=-l}^{l} u_l^m r^l \nabla_S Y_l^m, \tag{2.5.25}$$

$$\int_S |\nabla_S Y_l^m|^2 d\sigma = l(l+1). \tag{2.5.26}$$

It follows, using the orthogonality in $H^1(S)$ of the spherical harmonics, that

$$\|\nabla u\|_{L^2(B_i)}^2 = \sum_{l=0}^{\infty}\sum_{m=-l}^{l} l|u_l^m|^2, \tag{2.5.27}$$

$$\|\nabla u\|_{L^2(B_e)}^2 = \sum_{l=0}^{\infty}\sum_{m=-l}^{l} (l+1)|u_l^m|^2. \tag{2.5.28}$$

These expressions show the continuity of the operator $u_d \to u$ from the space $H^{1/2}(S)$ into the space $H^1(B_i)$ (resp. $W^1(B_e)$).

We use an induction on the integer k to establish the corresponding regularity results when u_d is in the space $H^{k+1/2}(S)$. We need to prove that $\partial u/\partial x_1, \partial u/\partial x_2, \partial u/\partial x_3$, are in $H^{k-1}(B_i)$, and that $r\partial u/\partial x_1, r\partial u/\partial x_2, r\partial u/\partial x_3$ are in $W^{k-1}(B_e)$, when u_d is in $H^{k-1/2}(S)$. For the interior problem, we use (2.5.9) to obtain

$$u(x) = \sum_{l=0}^{\infty} \sum_{m=-l}^{l} u_l^m H_l^m(x), \qquad (2.5.29)$$

$$\frac{\partial u}{\partial x_i}(x) = \sum_{l=0}^{\infty} \sum_{m=-l}^{l} u_l^m \frac{\partial H_l^m}{\partial x_i}(x). \qquad (2.5.30)$$

The expression of the trace of $\partial u/\partial x_i$ on the sphere S is given by Theorem 2.4.6 and by the expressions (2.4.116) to (2.4.118). It is clear, in view of their expansions in terms of the Y_{l-1}^m and of the formulas (2.4.116) to (2.4.118) that when u_d is in the space $H^{k-1/2}(S)$, $\partial u/\partial x_i$ belongs to $H^{k-3/2}(S)$. This yields the result for the interior problem.

For the exterior problem, we use (2.4.107) to obtain

$$u(x) = \sum_{l=0}^{\infty} \sum_{m=-l}^{l} u_l^m \frac{1}{r^{2l+1}} H_l^m(x), \qquad (2.5.31)$$

and so

$$r\frac{\partial u}{\partial x_i}(x) = \sum_{l=0}^{\infty} \sum_{m=-l}^{l} u_l^m \left[\frac{1}{r^{2l}}\frac{\partial H_l^m}{\partial x_i}(x) - (2l+1)\frac{x_i}{r^{2l+2}}H_l^m(x) \right]. \qquad (2.5.32)$$

It follows that on the unit sphere

$$\frac{\partial u}{\partial x_i} = \sum_{l=0}^{\infty} \sum_{m=-l}^{l} \left[u_l^m \frac{\partial H_l^m}{\partial x_i}(x) - (2l+1)u_l^m x_i H_l^m(x) \right]. \qquad (2.5.33)$$

Lemma 2.4.8 gives the expansion in spherical harmonics of the traces on the sphere of the functions $x_i H_l^m(x)$. The proof is a consequence of the boundedness of all the coefficients in these formulas. ∎

We now consider **the interior and exterior Neumann problems**

$$\begin{cases} \Delta u = 0, & \text{in } B_i \text{ or } B_e, \\ \dfrac{\partial u}{\partial n}\Big|_S = u_n. \end{cases} \qquad (2.5.34)$$

Expanding the data u_n in the spherical harmonics

$$u_n(\theta, \varphi) = \sum_{l=0}^{\infty} \sum_{m=-l}^{l} v_l^m Y_l^m(\theta, \varphi), \qquad (2.5.35)$$

we exhibit the solutions of the interior and exterior Neumann problems, observing that

$$\frac{\partial}{\partial n} H_l^m(x)|_S = \frac{1}{r}(\vec{x}.\nabla H_l^m(x))|_S = lY_l^m, \qquad (2.5.36)$$

$$\begin{cases} \dfrac{\partial}{\partial n} K_l^m |_S = \dfrac{1}{r^{2l+2}} \left[(\vec{x}.\nabla H_l^m(x)) - (2l+1)H_l^m(x) \right] |_S \\[2mm] = -(l+1)Y_l^m. \end{cases} \qquad (2.5.37)$$

(H_l^m and K_l^m are defined by (2.5.2) and (2.5.3)).

The following theorem holds:

Theorem 2.5.2 *The* **interior Neumann problem** *(2.5.34)* *admits a solution if*

$$v_0^0 = 0. \qquad (2.5.38)$$

The solution is equal to zero at the origin and is given by the expression

$$u(r,\theta,\varphi) = \sum_{l=1}^{\infty} \sum_{m=-l}^{l} \frac{v_l^m}{l} r^l Y_l^m(\theta,\varphi). \qquad (2.5.39)$$

If u_n lies in the space $H^{k-3/2}(S)$, the solution is in the space $H^k(B_i)$, and satisfies

$$\begin{cases} \|u\|_{H^k(B_i)} \le C \|u_n\|_{H^{k-3/2}(S)}, \\[2mm] \text{with a constant } C \text{ depending only on } k. \end{cases} \qquad (2.5.40)$$

The **exterior Neumann problem** *(2.5.34) admits a solution given by*

$$u(r,\theta,\varphi) = -\sum_{l=0}^{\infty} \sum_{m=-l}^{l} \frac{v_l^m}{l+1} \frac{1}{r^{l+1}} Y_l^m(\theta,\varphi). \qquad (2.5.41)$$

If u_n lies in the space $H^{k-3/2}(S)$, then the solution is in the space $W^k(B_e)$ and satisfies

$$\begin{cases} \|u\|_{W^k(B_e)} \le C \|u_n\|_{H^{k-3/2}(S)}, \\[2mm] \text{with a constant } C \text{ depending only on } k. \end{cases} \qquad (2.5.42)$$

Proof
Using the formulas (2.5.36) and (2.5.37), we can in each case compute the value of u on the surface S, using its coefficients v_l^m/l or $-v_l^m/(l+1)$. The result follows then from Theorem 2.5.1.

Definition
Let's $\mathbf{T_{ext}}$ denote the **Dirichlet-to-Neumann operator**, which, to a function u in the space $H^s(S)$, associates the normal derivative $\partial u/\partial n$ of the exterior Dirichlet problem with data u. When u is given by

$$u(\theta,\varphi) = \sum_{l=0}^{\infty} \sum_{m=-l}^{l} u_l^m Y_l^m(\theta,\varphi), \qquad (2.5.43)$$

$\mathbf{T_{ext}}$ admits the expression

$$T_{ext}u = -\sum_{l=0}^{\infty}\sum_{m=-l}^{l}(l+1)u_l^m Y_l^m(\theta,\varphi) \qquad (2.5.44)$$

and thus

$$-\int_S (T_{ext}u)\bar{u}\,d\sigma = \sum_{l=0}^{\infty}\sum_{m=-l}^{l}(l+1)|u_l^m|^2 = ||u||^2_{H^{1/2}(S)}. \qquad (2.5.45)$$

Theorems 2.5.1 and 2.5.2 show that, for each real s, the operator T_{ext} is an isomorphism of $H^{s+1}(S)$ onto $H^s(S)$ and moreover that for $s = -\frac{1}{2}$,

$$-\int_S (T_{ext}^{-1}u)\bar{u}\,d\sigma = \sum_{l=0}^{\infty}\sum_{m=-l}^{l}\frac{1}{l+1}|u_l^m|^2 = ||u||^2_{H^{-1/2}(S)}. \qquad (2.5.46)$$

2.5.2 Surfaces and Sobolev spaces

We want to solve the interior boundary problems in an open domain Ω_i of $\mathbb{R}^3$, which is bounded and regular. We also want to solve exterior boundary problems in the domain Ω_e. Consider the domain Ω_i and let Γ be its boundary. At this point, we need to give a precise meaning to the words "domain" and "regular." The regularity of the surface Γ (and of the domain) is expressed in the following way.

Consider a covering of Ω_i by a finite union of open sets ω_i, for $0 \le i \le p$. The surface Γ is enclosed in the set $\bigcup_{i=1}^p \omega_i$ and the set ω_0 does not intersect the surface Γ. We say that the surface Γ is regular, if for each index i, there exists a diffeomorphism ϕ_i that maps the set ω_i onto the unit ball Q, such that $\Gamma \cap \omega_i$ is mapped into the equatorial plane $z = 0$ of the unit ball and such that $\omega_i \cap \Omega_i$ is mapped into the region $z \le 0$ below the equatorial plane, while $\omega_i \cap \Omega_e$ is mapped into the region $z \ge 0$ above the equatorial plane. We say Γ is of class C^k, if each of the diffeomorphisms as well as their inverses are functions of class C^k. The integer k will always be big enough and adjusted to fit the context. Such a covering of Ω_i and the corresponding diffeomorphisms ϕ_i are called an **atlas** and the pairs (ω_i, ϕ_i) are called **charts** (see Fig. 2).

Notice that, when the domain is not simply connected, we need more than one set ω_0 to describe the interior of this domain, the number of such sets being related to the actual topology of the domain. We will avoid this aspect in the following.

Corresponding to the above covering of the domain Ω_i, we can define a partition of unity, i.e., a set of C^k positive functions $\lambda_i, i = 0$ to p, with compact support, such that

$$\sum_{i=0}^{p}\lambda_i(x) \equiv 1 \text{ on } \Omega_i \qquad \text{and support } (\lambda_i) \subset \omega_i$$

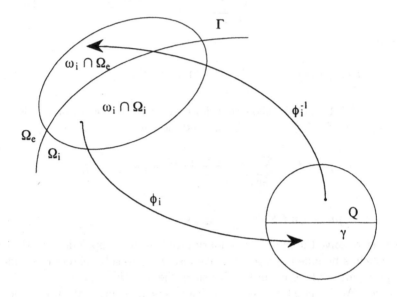

Figure 2: *The charts.*

For a chart (ω_i, ϕ_i), we denote by Γ_i the piece of surface $\Gamma \cap \omega_i$ and the equatorial plan of the ball Q will be called γ. Let $\mathbb{R}^{3+}$ be the upper half space $(z > 0)$ and $\mathbb{R}^{3-}$ be its complement $(z < 0)$.

Definition
The Sobolev space $H^m(\Omega_i)$ is the space of functions in $L^2(\Omega_i)$, whose derivatives (in the sense of distribution) up to the order m, are also in $L^2(\Omega_i)$. It is a Hilbert space with the norm

$$\|u\|^2_{H^m(\Omega_i)} = \sum_{l=0}^{m} |u|^2_{H^l(\Omega_i)} \tag{2.5.47}$$

where $|u|^2_{H^l(\Omega_i)}$ is the following semi-norm:

Notation

$$\begin{cases} \alpha = (\alpha_1, \alpha_2, \alpha_3), |\alpha| = \alpha_1 + \alpha_2 + \alpha_3, \\ \left(\dfrac{\partial}{\partial x}\right)^\alpha = \left(\dfrac{\partial}{\partial x_1}\right)^{\alpha_1} \left(\dfrac{\partial}{\partial x_2}\right)^{\alpha_2} \left(\dfrac{\partial}{\partial x_3}\right)^{\alpha_3}, \\ |u|^2_{H^l(\Omega_i)} = \sum_{|\alpha|=l} \left| \left(\dfrac{\partial}{\partial x}\right)^\alpha u \right|^2_{L^2(\Omega_i)}. \end{cases} \tag{2.5.48}$$

We admit the following result (which can be found, e.g., in the book of Adams [8] on Sobolev spaces): the space of indefinitely differentiable functions on the closure of the domain Ω_i is dense in the space $H^m(\Omega_i)$.

Alternatively $H^m(\Omega_i)$ can be defined using an atlas (ω_i, ϕ_i): for each set ω_i (including ω_0), the function $\tilde{u}_i = (\lambda_i u) \circ \phi_i^{-1}$ is in the space $H^m(\mathbb{R}^{3-})$. It follows from the Leibniz formulas for derivation, and from the regularity of the diffeomorphisms ϕ_i, that the norm

$$||u||^2 = \sum_{i=0}^{p} ||\lambda_i u \circ \phi_i^{-1}||^2_{H^m(\mathbb{R}^{3-})} \tag{2.5.49}$$

is a norm on $H^m(\Omega_i)$ equivalent to the norm (2.5.47).

In order to define Sobolev spaces in the domain $\mathbb{R}^2$, we introduce the Fourier transform

$$\hat{u}(\xi) = \frac{1}{2\pi} \int_{\mathbb{R}^2} e^{-i(x_1\xi_1 + x_2\xi_2)} u(x_1, x_2) dx_1 dx_2. \tag{2.5.50}$$

The Fourier transform is an isometry on $L^2(\mathbb{R}^2)$, i.e.,

$$\int_{\mathbb{R}^2} |\hat{u}(\xi)|^2 d\xi = \int_{\mathbb{R}^2} |u(x)|^2 dx. \tag{2.5.51}$$

The inverse operator is its conjugate denoted by *. The inversion formula is

$$u(x) = \frac{1}{2\pi} \int_{\mathbb{R}^2} e^{i(x.\xi)} \hat{u}(\xi) d\xi, \tag{2.5.52}$$

in the sense of Bochner integrals. Any regular function with compact support satisfies

$$\begin{cases} \dfrac{\widehat{du}}{dx_j} = i\xi_j \hat{u}, & j = 1 \text{ or } 2; \\[2mm] \dfrac{\partial}{\partial \xi_j} \hat{u} = -i\widehat{x_j u}, & j = 1 \text{ or } 2. \end{cases} \tag{2.5.53}$$

These properties suggest the following definition: **the space $H^s(\mathbb{R}^2)$ for s real and positive**, is the Hilbert space of functions in $L^2(\mathbb{R}^2)$ such that $(1 + |\xi|^2)^{s/2} \hat{u}(\xi)$ is in $L^2(\mathbb{R}^2)$.

Using the properties (2.5.51) and (2.5.53), we can see that this definition of $H^m(\mathbb{R}^2)$ is equivalent to the definition given previously, when m is an integer.

A function is said to be rapidly decreasing when this function and all its partial derivatives decrease at infinity more rapidly than any positive power of the variable.

Let $\mathcal{S}(\mathbb{R}^2)$ be the space of indefinitely differentiable functions rapidly decreasing at infinity. The dual space $\mathcal{S}'(\mathbb{R}^2)$ is the space of tempered distributions (also called slowly increasing distributions).

Relations (2.5.53) show that the Fourier transform maps $S(\mathbb{R}^2)$ onto $S(\mathbb{R}^2)$. In view of the Plancherel formula

$$\int_{\mathbf{R}^2} \widehat{u}(\xi) v(\xi) d\xi = \int_{\mathbf{R}^2} u(x) \widehat{v}(x) dx \tag{2.5.54}$$

we can define an extension of the Fourier transform to the space $S'(\mathbb{R}^2)$ by setting

$$\langle \widehat{u}, v \rangle = \langle u, \widehat{v} \rangle, \; \forall\, v \in S(\mathbb{R}^2), \text{ for } u \text{ in } S'(\mathbb{R}^2).$$

We have defined an extension of the Fourier transform to the space $S'(\mathbb{R}^2)$ which coincides with the usual Fourier transform in $L^2(\mathbb{R}^2)$. This extension allows us to define $H^s(\mathbb{R}^2)$ for s **real positive and negative as well** by

$$H^s(\mathbb{R}^2) = \left\{ u \in S'(\mathbb{R}^2); \int_{\mathbf{R}^2} (1 + |\xi|^2)^s |\widehat{u}(\xi)|^2 d\xi < \infty \right\}.$$

The associated norm is

$$\|u\|^2_{H^s(\mathbf{R}^2)} = \int_{\mathbf{R}^2} (1 + |\xi|^2)^s |\widehat{u}(\xi)|^2 d\xi. \tag{2.5.55}$$

It is quite clear that the dual space associated with $H^s(\mathbb{R}^2)$ is the space $H^{-s}(\mathbb{R}^2)$. We are now in a position to give the **definition of $H^s(\Gamma)$ for a real s, when Γ is an indefinitely differentiable surface.** A function u defined on the surface Γ is in $H^s(\Gamma)(s > 0)$ if for each set ω_i, the function $\widetilde{u}_i = \lambda_i u \circ \phi_i^{-1}$ is in $H^s(\mathbb{R}^2)$ (we extend it by zero outside of γ).

The above definition can be extended to negative s. Let $S(\Gamma)$ **denote the space of indefinitely differentiable functions on Γ. Its dual space is $S'(\Gamma)$** (which is also $(\mathcal{D}'(\Gamma))$). Every distribution in $S'(\Gamma)$ can be mapped by restriction and transport on γ to a distribution $\widetilde{u}_i$ in $S'(\mathbb{R}^2)$. Thus, the above definition makes sense when $s < 0$. The dual space of $H^s(\Gamma)$ is the space $H^{-s}(\Gamma)$.

We now have all the necessary tools to prove **the trace theorem**, which specifies the links between a function on Ω_i and its trace on Γ. We will use the space $\mathcal{D}(\overline{\Omega_i})$ of functions indefinitely differentiable up to the boundary of the set Ω_i. Let n be the exterior normal to the surface Γ.

Theorem 2.5.3 (The trace theorem) *The trace mapping, which associates with a function u in $\mathcal{D}(\overline{\Omega_i})$ the restriction on Γ of its derivatives,*

$$u \xrightarrow{\;\gamma\;} u_{|\Gamma}, \frac{\partial u}{\partial n}|_\Gamma, \ldots, \left(\frac{\partial}{\partial n}\right)^{m-1} u_{|\Gamma} \tag{2.5.56}$$

can be extended as a continuous linear mapping of the Hilbert space $H^m(\Omega_i)$ into respectively $H^{m-1/2}(\Gamma)$, $H^{m-3/2}(\Gamma)$, ..., $H^{1/2}(\Gamma)$. It admits a left inverse denoted by $\mathcal{R}$, which is a continuous mapping of the product space $\prod_{l=1}^m H^{m-l+1/2}(\Gamma)$ into $H^m(\Omega_i)$. The lifting $\mathcal{R}$, which associates with

$\vec{v} = (v^0, \ldots, v^{m-1})$ *the function* $\mathcal{R}\vec{v}$, *satisfies*

$$\left(\frac{\partial}{\partial n}\right)^l \mathcal{R}\vec{v} = v^l, \qquad 0 \le l \le m - 1. \tag{2.5.57}$$

Proof

In order to prove this theorem, we will use extensively the transport to the half space. We mean by localization the use of the partition functions $\lambda_i, i = 0 \cdots p$. Using the partition of unity, we only have to prove the corresponding results for the functions $\tilde{u}_i$. We will establish several preliminary lemmas.

Lemma 2.5.1 *Let* $u(x_1, x_2, z)$ *be a function defined on the complex half space* $z < 0$. *Let* $\hat{u}(\xi, z)$ *be its partial Fourier transform in the variables* (x_1, x_2). *Then* u *is in the space* $H^m(\mathbb{R}^{3-})$ *if and only if:*

$$\left(\frac{\partial}{\partial z}\right)^m \hat{u} \in L^2(\mathbb{R}^{3-}), (1 + |\xi|^2)^{1/2}\left(\frac{\partial}{\partial z}\right)^{m-1} \hat{u} \in L^2(\mathbb{R}^{3-}),$$
$$\ldots, (1 + |\xi|^2)^{m/2}\hat{u} \in L^2(\mathbb{R}^{3-}).$$

The proof is straightforward, using the properties of the Fourier transform, and ordering the partial derivatives.

Remark

It follows from Lemma 2.5.1 that if u is in $H^m(\mathbb{R}^{3-})$, then $\partial u/\partial z$ is in $H^{m-1}(\mathbb{R}^{3-})$.

Lemma 2.5.2 *For every* u *in* $H^m(\mathbb{R}^{3-})$, *we have* $(m \ge 1)$

$$\|u\|_{H^{m-1/2}(\mathbf{R}^2)} \le \|u\|_{H^m(\mathbf{R}^{3-})}. \tag{2.5.58}$$

Proof

A direct integration yields

$$\begin{cases} (1 + |\xi|^2)^{m-1/2}|\hat{u}(\xi, 0)|^2 \\ \quad = 2\Re \int_{-\infty}^0 (1 + |\xi|^2)^{m-1/2}\overline{\hat{u}}(\xi, z)\frac{\partial \hat{u}}{\partial z}(\xi, z)dz. \end{cases} \tag{2.5.59}$$

We integrate (2.5.59) in $\mathbb{R}^2$ and we bound the right-hand side using the Cauchy-Schwartz inequality. It follows that

$$\begin{cases} \|u\|^2_{H^{m-1/2}(\mathbf{R}^2)} \\ \quad \le \|(1 + |\xi|^2)^{\frac{m}{2}}\hat{u}\|^2_{L^2(\mathbf{R}^{3-})} + \|(1 + |\xi|^2)^{\frac{m-1}{2}}\frac{\partial \hat{u}}{\partial z}\|^2_{L^2(\mathbf{R}^{3-})}, \end{cases} \tag{2.5.60}$$

which is (2.5.58). ∎

The continuity for the first trace is now proved by localization and transport using the diffeomorphisms ϕ_i.

The case of the following traces is more complex. Let us look at $\partial u/\partial n$. By localization, we only have to bound $(\partial/\partial n)(\lambda_i u)$ on the piece of surface

Γ_i, or equivalently to consider the case of a function u whose support is ω_i. We denote by ϕ the associated diffeomorphism. We have

$$\frac{\partial}{\partial z}(u \circ \phi^{-1}) = \sum_{j=1}^{3} \frac{\partial}{\partial x_j} u \cdot \frac{\partial \phi^{-1}}{\partial z}. \tag{2.5.61}$$

The vector $(\partial/\partial z)\phi^{-1}$ is transverse to the surface Γ_i, i.e., its component β along the vector $\vec{n}$ does not vanish. Using (2.5.61), the normal derivative on the piece of surface Γ_i, is $(n_j = (\vec{n} \cdot \vec{e}_j))$

$$\frac{\partial u}{\partial n} = \frac{1}{\beta} \left[\left(\frac{\partial}{\partial z}(u \circ \phi^{-1}) - \sum_{j=1}^{3} \frac{\partial}{\partial x_j} u \cdot \left(\frac{\partial \phi_j^{-1}}{\partial z} - \beta n_j \right) \right) \right]. \tag{2.5.62}$$

$\partial u/\partial n$ is the trace on Γ_i of the function of the right-hand side. It follows from the above remark that $(\partial/\partial z)(u \circ \phi^{-1})$ is an element of $H^{m-1}(\mathbb{R}^3)$. It is also the case of the right-hand side. This gives the continuity of the second trace.

For the next traces, we proceed by induction. For example, computing $(\partial^2/\partial z^2)(u \circ \phi^{-1})$, we can express $(\partial^2/\partial n^2)u$ as the trace of a function of $H^{m-2}, \ldots$. Let's now build the lifting $\mathcal{R}$ in the case of the first trace. Again we only have to express it in the half space and then use the transport.

Lemma 2.5.3 *Let v^0 be in the space $H^{1/2}(\mathbb{R}^2)$. The function u whose partial Fourier transform $\hat{u}(\xi, z)$ is given by*

$$\hat{u}(\xi, z) = \hat{v}^0(\xi)e^{(1 + |\xi|^2)^{1/2}z} \tag{2.5.63}$$

has v^0 for trace and satisfies

$$\|u\|_{H^m(\mathbb{R}^{3-})} \leq \sqrt{(m+1)/2}\|v^0\|_{H^{m-1/2}(\mathbb{R}^2)}; \; m \geq 1. \tag{2.5.64}$$

Proof
A direct integration yields

$$\begin{cases} \displaystyle\int_{\mathbb{R}^{3-}} (1 + |\xi|^2)^{m-l} \left| \frac{\partial^l}{\partial z^l} \hat{u} \right|^2 d\xi dz \\ \qquad = \displaystyle\int_{\mathbb{R}^2} \frac{1}{2}(1 + |\xi|^2)^{m-1/2}|\hat{v}^0(\xi)|^2 d\xi, \end{cases} \tag{2.5.65}$$

$$\int_{\mathbb{R}^{3-}} (1 + |\xi|^2)^m |\hat{u}|^2 d\xi dz = \int_{\mathbb{R}^2} \frac{1}{2}(1 + |\xi|^2)^{m-1/2}|\hat{v}^0(\xi)|^2 d\xi. \tag{2.5.66}$$

The lemma follows using Lemma 2.5.1. ∎

Remark
The lifting $\mathcal{R}v^0$ is indeed the solution, in the lower half space, to the equation

$$\begin{cases} -\Delta u + u = 0, \\ u_{|\Gamma} = v^0. \end{cases} \tag{2.5.67}$$

The construction of the lifting for the next traces is more complex. We will consider the case of $\partial u/\partial n$. Upon using the lifting operator for the first trace constructed in Lemma 2.5.3, we can assume that $v^0 = 0$. We then localize using the partition of unity. The normal derivative of the lifting u can be computed using the transport, the vanishing of v^0 and (2.5.61). It follows that

$$\frac{\partial u}{\partial n} = \frac{1}{\beta}\frac{\partial}{\partial z}(u \circ \phi^{-1}), \qquad u \circ \phi^{-1}|_\gamma = 0. \tag{2.5.68}$$

We only need to build the lifting in the lower half space, for a known $\partial u/\partial z$ and a vanishing u.

Lemma 2.5.4 Let $v^1 \in H^{m-3/2}(\mathbb{R}^2)$. The function u, whose partial Fourier transform $\hat{u}$ is given by

$$\hat{u}(\xi, z) = z\hat{v}^1(\xi)e^{(1 + |\xi|^2)^{1/2}z}, \tag{2.5.69}$$

admits v^1 for normal derivative and satisfies

$$\|u\|_{H^m(\mathbb{R}^{3-})} \leq \sqrt{(m + 1)((m - 1/2)^2 + 1/2)/3}\,\|\hat{v}^1\|_{H^{m-3/2}(\mathbb{R}^2)}. \tag{2.5.70}$$

Proof
A direct integration and some computations yield

$$\int_{\mathbb{R}^{3-}}(1 + |\xi|^2)^m|\hat{u}|^2d\xi dz = \frac{1}{4}\int_{\mathbb{R}^2}(1 + |\xi|^2)^{m-3/2}|v^1(\xi)|^2d\xi, \tag{2.5.71}$$

$$\begin{cases} \displaystyle\int_{\mathbb{R}^{3-}}(1 + |\xi|^2)^{m-l}\left(\frac{d}{dz}\right)^l\hat{u} \\ \displaystyle\quad = \hat{v}^1(\xi)(1 + |\xi|^2)^{\frac{l-1}{2}}\left((1 + |\xi|^2)^{1/2}z + l\right)e^{(1 + |\xi|^2)^{1/2}z}, \end{cases} \tag{2.5.72}$$

$$\begin{cases} \displaystyle\int_{\mathbb{R}^{3-}}(1 + |\xi|^2)^{m-l}\left|\left(\frac{d}{dz}\right)^l\hat{u}\right|^2d\xi dz \\ \displaystyle\quad = \frac{(l^2 - l + 1/2)}{2}\int_{\mathbb{R}^2}(1 + |\xi|^2)^{m-3/2}|v^1(\xi)|^2d\xi. \end{cases} \tag{2.5.73}$$

∎

We have built a lifting operator for the first two traces. We can now proceed by induction. Using the previous liftings, we can choose v^0 and v^1 to be zero, from which we have:

$$\frac{\partial^2 u}{\partial n^2} = \frac{1}{\beta^2}\frac{\partial^2}{\partial z^2}(u \circ \phi^{-1}). \tag{2.5.74}$$

We only have to consider the case of the half space for which we introduce the operator (here $l = 2$)

$$\widehat{u}(\xi, z) = \frac{1}{l!} z^l \widehat{v}^l(\xi) e^{(1 + |\xi|^2)^{1/2} z}, \tag{2.5.75}$$

which is easily seen to be continuous in the adequate norms.   ∎

Remarks

- In view of the above proof, it is clear that the trace theorem is a local property of the Sobolev spaces. The lifting $\mathcal{R}$ can be either interior or exterior to the surface Γ. We can choose its support to be as close as we want to the surface Γ, by using a cut-off function whose value is 1 on Γ.

- Incidentally, the trace theorem shows that in the case of the sphere S, the Sobolev spaces $H^{m-1/2}(S)$ defined in the case of the sphere in Section 2.5.1 for $m \geq 1$ are the same as those we have introduced here. This is true for every real s. In that case, the explicit solution to the Dirichlet problem is a particular lifting which is called the harmonic lifting.

- The above results were established in the case of a domain in $\mathbb{R}^3$ delimited by a surface. It is quite clear that, changing slightly the definitions, they can be adapted to domains in $\mathbb{R}^2$ delimited by plane curves, or to domains in $\mathbb{R}^n$ delimited by hypersurfaces of dimension $n - 1$.

We will now give further properties of the Sobolev spaces.

Theorem 2.5.4 *There exists an extension operator* $\mathbb{P}$, *continuous from* $H^m(\Omega_i)$ *to* $H^m(\mathbb{R}^3)$. *More explicitly, there exists for each u in* $H^m(\Omega_i)$, *an element* $\mathbb{P}u$ *in* $H^m(\mathbb{R}^3)$, *which we can choose with compact support, such that* $\mathbb{P}u = u$ *on the set* Ω_i, *and*

$$\begin{cases} ||\mathbb{P}u||_{H^m(\mathbf{R}^3)} \leq C||u||_{H^m(\Omega_i)}, \\ \text{with a constant } C \text{ depending only on } m. \end{cases} \tag{2.5.76}$$

Proof
We use the trace Theorem 2.5.3. The traces of u on Γ, coming from the interior domain, can be lifted, using the exterior domain lifting operator $\mathcal{R}$. The function whose value is u in the set Ω_i and $\mathcal{R}u$ in the exterior domain has continuous traces across the surface. We can check, using again the localization and transport, that such a function belongs to $H^m(\Omega_i)$.   ∎

Let $H_0^m(\Omega_i)$ be the **closure** of the space $\mathcal{D}(\Omega_i)$ of indefinitely differentiable functions with compact support in the open set Ω_i, in the space $H^m(\Omega_i)$.

Theorem 2.5.5 *The space* $H_0^m(\Omega_i)$ *is exactly the space of functions u in* $H^m(\Omega_i)$, *the traces* $u, \partial u/\partial n, \ldots, (\partial/\partial n)^{m-1}u$ *of which vanish on the surface* Γ; *or equivalently it is the space of functions in* $H^m(\Omega_i)$, *whose extension by zero to* $\mathbb{R}^3$ *belongs to* $H^m(\mathbb{R}^3)$.

Proof

Again, we localize and transport to the half space and then use a partial Fourier transform.

Theorem 2.5.6 *The injection of $H_K^s(\mathbb{R}^n)$ into $H^{s-\varepsilon}(\mathbb{R}^n)$ is compact for every $\varepsilon > 0$, where $H_K^s(\mathbb{R}^n)$ denotes the set of functions in $H^s(\mathbb{R}^n)$ whose support is in a fixed bounded closed set K.*

Proof

Let u_n be a bounded sequence of functions in H^s with support in K. Suppose first that s is positive. Then u_n is in $L^1(K)$ and the functions $\widehat{u}_n$ form a sequence of continuous functions such that

$$|\widehat{u}_n(\xi) - \widehat{u}_n(\eta)| \leq C \sup_{x \in K} \left| e^{-i\xi x} - e^{-i\eta x} \right|. \qquad (2.5.77)$$

Using the Ascoli theorem, we can extract a subsequence which converges uniformly on each ball. On the exterior of a ball of radius R, we have

$$\int_{B_R^c} (1 + |\xi|^2)^{s-\varepsilon} |\widehat{u}(\xi)|^2 d\xi \leq \frac{1}{(1 + R^2)^\varepsilon} \|u\|_{H^s}^2 (\mathbb{R}^n). \qquad (2.5.78)$$

This shows the result for $s > 0$. The result is extended to negative s by duality.

Theorem 2.5.7 *The injection of $H^s(\Gamma)$ into $H^{s-\varepsilon}(\Gamma)$ is compact for $\varepsilon > 0$ and so is the injection of $H^m(\Omega_i)$ into $H^{m-1}(\Omega_i)$.*

Proof

The case of $H^s(\Gamma)$ follows from Theorem 2.5.6 using localization and transport. The case of $H^m(\Omega_i)$ follows from Theorem 2.5.4.   ∎

Remark

The compact injection result is wrong in the space $H^s(\mathbb{R}^2)$ and also in the space $H^m(\Omega_e)$. Indeed, as these domains are non-bounded, we can exhibit simple counter-examples, based on sequences of functions with unit norm and support going to infinity, that do not converge.

2.5.3 Interior problems: Variational formulations

Let us first consider the **Dirichlet problem**, which consists in solving

$$\begin{cases} -\Delta u(x) = f(x), & x \in \Omega_i, \\ \\ u_{|\Gamma} = u_d. \end{cases} \qquad (2.5.79)$$

We seek a solution in the space $H^1(\Omega)$. In this space, this equation does not directly make sense. The trace theorem implies that u_d must be in the space $H^{1/2}(\Gamma)$. The lifting described in the trace theorem, allows us to change our equation to:

$$\begin{cases} -\Delta v(x) = g(x), & x \in \Omega_i, \\ \\ v_{|\Gamma} = 0, \end{cases} \tag{2.5.80}$$

upon choosing

$$\begin{cases} v = u - \mathcal{R}u_d, \\ \\ g = f + \Delta\mathcal{R}u_d. \end{cases} \tag{2.5.81}$$

It is now natural to look for v in the space $H_0^1(\Omega_i)$ of functions in $H^1(\Omega_i)$ which vanish on the surface Γ. The trace theorem shows that it is a Hilbert space. Starting from the equation (2.5.80), multiplied by a test function vanishing on Γ, then using the following Green formula

$$\int_{\Omega_i} \left[\Delta v(x)u(x) + \sum_{i=1}^{3} \frac{\partial u}{\partial x_i}(x)\frac{\partial v}{\partial x_i}(x) \right] dx = \int_{\Gamma} u\frac{\partial v}{\partial n}d\gamma, \tag{2.5.82}$$

we obtain the variational formulation

$$\int_{\Omega_i} (\nabla v(x) \cdot \nabla w(x))dx = \int_{\Omega_i} g(x)w(x)dx, \quad \forall w \in H_0^1(\Omega_i). \tag{2.5.83}$$

We will show that this new formulation admits a unique solution as a consequence of the following abstract theorem

Theorem 2.5.8 (Lax-Milgram theorem) *Let V be a Hilbert space and a be a continuous bilinear form on V which satisfies*

$$|a(v,v)| \geq \alpha\|v\|_V^2, \quad \alpha > 0, \tag{2.5.84}$$

then the problem

$$a(u,v) = \langle g,v \rangle, \quad \forall v \in V, \tag{2.5.85}$$

admits a unique solution $u \in V$, if g belongs to the dual space V^*.

Proof
The uniqueness results from (2.5.84).

Let A be the operator from V into V^* defined by

$$\langle Au, v \rangle = a(u,v). \tag{2.5.86}$$

It follows from (2.5.84), that the adjoint operator A^* is such that $A^*(V)$ is dense in V^*. Let g be in this image. We define u in V by

$$\langle u, f \rangle = \langle g, v \rangle, \quad \forall v \text{ with } f = A^*v. \tag{2.5.87}$$

This linear form is continuous on V^* and well-defined on a dense subset of V^*. Thus, from the Hahn-Banach theorem, it admits a unique extension as a linear form on V^* which is an element of V. This is the meaning of the equation (2.5.85).

Lemma 2.5.5 (Poincaré's inequality) *Consider two parallel planes which enclose the open domain Ω_i and let $d(\Omega_i)$ be the distance between these two parallel planes. Then, each u in $H_0^1(\Omega_i)$ satisfies the inequality*

$$\int_{\Omega_i} |u(y)|^2 dy \leq (d(\Omega_i))^2 \int_{\Omega_i} |\nabla u(y)|^2 dy. \tag{2.5.88}$$

Proof
We extend u by zero outside the open set Ω_i. If n is the normal direction to the planes, any point $y \in \Omega_i$ can be written $y = x + sn$, where x denotes the projection of y on one of the planes.

$$u(y) = \int_0^s \frac{\partial u}{\partial n}(x + tn)dt. \tag{2.5.89}$$

Using the Cauchy-Schwarz inequality and integrating, it follows that

$$\int_{\Omega_i} |u(y)|^2 dy \leq (d(\Omega_i))^2 \int_{\Omega_i} \left| \frac{\partial u}{\partial n} \right|^2 dy, \tag{2.5.90}$$

from which we infer (2.5.88).   ∎

We are ready to prove

Theorem 2.5.9 *The Dirichlet problem (2.5.79) has a unique solution in the space $H^1(\Omega_i)$, if u_d is in $H^{1/2}(\Gamma)$ and if f has the expression*

$$f = \sum_{i=1}^{3} \frac{\partial}{\partial x_i} g_i, \quad g_i \in L^2(\Omega_i). \tag{2.5.91}$$

The corresponding mapping is continuous from $H^{1/2}(\Gamma) \otimes H^{-1}(\Omega_i)$ into $H^1(\Omega_i)$.

Proof
Lemma 2.5.5 shows that the bilinear form (2.5.83) satisfies the hypothesis of the Lax-Milgram Theorem 2.5.8. Thus, the problem (2.5.83) admits a unique solution in the space $H_0^1(\Omega_i)$, if f is in the dual space of $H_0^1(\Omega_i)$, which is clearly the case when (2.5.91) is satisfied. The continuity then follows from the properties of the lifting $\mathcal{R}u_d$.   ∎

Let us now consider the **Neumann problem**, whose equation is

$$\begin{cases} -\Delta u(x) = f(x), \quad x \in \Omega_i, \\ \\ \dfrac{\partial u}{\partial n}\big|_\Gamma = u_n. \end{cases} \tag{2.5.92}$$

A variational formulation is

$$
\left\{
\begin{aligned}
&\int_{\Omega_i} (\nabla u(x) \cdot \nabla v(x)) dx \\
&= \int_{\Omega_i} f(x)v(x) dx + \int_{\Gamma} u_n v d\gamma, \quad \forall v \in H^1(\Omega_i).
\end{aligned}
\right.
\tag{2.5.93}
$$

Theorem 2.5.10 *The Neumann problem (2.5.92) has a unique solution in the Hilbert space $H^1(\Omega_i)$, up to an additive constant (when Ω_i is connected), when u_n is in the space $H^{-1/2}(\Gamma)$, and f has the expression*

$$
f = \sum_{i=1}^{3} \frac{\partial}{\partial x_i} g_i + g_0, \qquad g_i \in L^2(\Omega_i);
\tag{2.5.94}
$$

and when u_n and f satisfy the constraint

$$
\langle f, 1 \rangle + \langle u_n, 1 \rangle = 0.
\tag{2.5.95}
$$

The corresponding mapping is continuous $H^{-1/2}(\Gamma) \otimes H^{-1}(\Omega_i)$ into $H^1(\Omega_i)$.

Proof
Choosing v to be equal to 1 yields condition (2.5.95), which is thus necessary. To prove the result, we only have to prove the coercivity inequality (2.5.84), in the space $H^1(\Omega_i)$ quotiented by the linear space of constant functions. This is the object of the following lemma:

Lemma 2.5.6 *If the open set Ω_i is connected, there exists a constant C, such that for every u in the space $H^1(\Omega_i)$, we have*

$$
\left\{
\begin{aligned}
&\int_{\Omega_i} |u - \alpha|^2 dx \leq C \int_{\Omega_i} |\nabla u(x)|^2 dx, \\
&\alpha = \frac{1}{mes(\Omega_i)} \int_{\Omega_i} u(x) dx.
\end{aligned}
\right.
\tag{2.5.96}
$$

Proof
The proof proceeds by contradiction. If the result were wrong, there would exist a sequence of functions u_n such that

$$
\int_{\Omega_i} |u_n - \alpha_n|^2 \, dx = 1,
\tag{2.5.97}
$$

$$
\int_{\Omega_i} |\nabla u_n(x)|^2 \, dx \leq \frac{1}{n}.
\tag{2.5.98}
$$

The sequence $u_n - \alpha_n$ is bounded in $H^1(\Omega_i)$. Theorem 2.5.6 shows that we can extract a subsequence such that $u_n - \alpha_n$ converges in $L^2(\Omega_i)$. From (2.5.98) and the connectedness of the domain, the limit must be a constant function. Thus, $u - \alpha$ vanishes, which is contradictory with equality (2.5.97).

Remark

It is possible to exhibit a more explicit constant C for specific geometries. For example if the domain is convex, we can choose the square of the diameter of this set. For ad hoc geometries, the square of the constant $d(\Omega_i)$ introduced in Lemma 2.5.5 is suitable.

General remark

The above problems and their variational formulations are not specific to the dimension of the space. These results can be extended with obvious modifications to the case of open sets in $\mathbb{R}^n$ (in particular in $\mathbb{R}^2$).

2.5.4 Exterior problems

We will examine here the Dirichlet and Neumann problems in the exterior domain Ω_e. In the case of the domain exterior to the sphere S, we have seen that solutions to these problems are in the weighted Sobolev spaces $W^k(B_e)$. It is natural to introduce the corresponding weighted Sobolev spaces in the case of the domain Ω_e.

Let $W^k(\Omega_e)$ denote the **Hilbert spaces**:

$$W^k(\Omega_e) = \left\{ u; \frac{u}{(1+r^2)^{1/2}} \in L^2(\Omega_e), \nabla u \in L^2(\Omega_e), \right.$$

$$\left. \ldots, (1+r^2)^{\frac{k-1}{2}} D^k u \in L^2(\Omega_e) \right\}.$$

The change of the weight r to $(1+r^2)^{1/2}$ allows us to extend the definition to the case where Ω_e is the whole domain $\mathbb{R}^3$. Notice that on each bounded part of the open set Ω_e, the space $W^k(\Omega_e)$ coincides with $H^k_{loc}(\Omega_e)$. Functions in these two spaces differ only by their behavior at infinity. In particular, the trace theorem applies to the functions in the space $W^k(\Omega_e)$. The corresponding extension can be chosen with bounded support.

We denote by $\mathcal{D}(\bar{\Omega}_e)$ the space of indefinitely differentiable functions, which are the restriction to the closed domain $\bar{\Omega}_e$ of functions in $\mathcal{D}(\mathbb{R}^3)$. We admit that $\mathcal{D}(\bar{\Omega}_e)$ is dense in the space $W^k(\Omega_e)$. For these spaces, we will prove some results akin to Poincaré's inequality. They are all based on the following inequality:

Lemma 2.5.7 (Hardy's inequality) Let f be in $\mathcal{D}(]A, \infty[)$. Then f satisfies

$$a) \quad \text{if} \quad \beta \neq -1, \quad \gamma \in \mathbb{R} \quad \text{and} \quad A \geq exp\left(\frac{2|\gamma|}{|\beta+1|}\right), \tag{2.5.99}$$

$$\begin{cases} \displaystyle\int_A^\infty |f(r)|^2 r^\beta (Logr)^\gamma dr \\ \displaystyle\qquad \leq \left(\frac{4}{|\beta+1|}\right)^2 \int_A^\infty \left|\frac{df}{dr}\right|^2 r^{\beta+2}(Logr)^\gamma dr; \end{cases} \tag{2.5.100}$$

$$b) \quad \text{if} \quad \gamma \neq -1 \quad \text{and} \quad A \geq 1, \tag{2.5.101}$$

$$\int_A^\infty |f(r)|^2 \frac{1}{r} (Logr)^\gamma dr \leq \left(\frac{2}{|\gamma + 1|} \right)^2 \int_A^\infty \left| \frac{df}{dr} \right|^2 r (Logr)^{\gamma+2} dr; \tag{2.5.102}$$

$$c) \quad \text{if} \quad \beta \neq -1, \tag{2.5.103}$$

$$\int_0^\infty |f(r)|^2 r^\beta dr \leq \left(\frac{2}{|\beta + 1|} \right)^2 \int_0^\infty \left| \frac{df}{dr} \right|^2 r^{\beta+2} dr. \tag{2.5.104}$$

Proof

a) We integrate by parts the left-hand side of (2.5.100) using the inequality

$$\frac{d}{dr} \left(r^{\beta+1} (Logr)^\gamma \right) = r^\beta (Logr)^\gamma \left(\beta + 1 + \frac{\gamma}{Logr} \right). \tag{2.5.105}$$

The hypothesis on A implies that for $r \geq A$,

$$r^\beta (Logr)^\gamma \left(|\beta + 1 + \frac{\gamma}{Logr}| \right) \geq \frac{|\beta + 1|}{2} r^\beta (Logr)^\gamma. \tag{2.5.106}$$

Thus, integrating, we obtain

$$\left\{ \begin{aligned} &\int_A^\infty |f(r)|^2 r^\beta (Logr)^\gamma dr \\ &\qquad \leq \frac{2}{|\beta + 1|} \int_A^\infty |f(r)|^2 \frac{d}{dr} \left(r^{\beta+1} (Logr)^\gamma \right) dr \\ &\qquad = -\frac{4}{|\beta + 1|} \int_A^\infty f(r) \frac{df}{dr}(r) r^{\beta+1} (Logr)^\gamma dr. \end{aligned} \right. \tag{2.5.107}$$

Applying the Cauchy-Schwarz inequality leads to

$$\left\{ \begin{aligned} &\int_A^\infty |f(r)|^2 r^\beta (Logr)^\gamma dr \\ &\qquad \leq \frac{4}{|\beta + 1|} \left(\int_A^\infty |f(r)|^2 r^\beta (Logr)^\gamma dr \right)^{1/2} \\ &\qquad\quad \times \left(\int_A^\infty \left| \frac{df}{dr}(r) \right|^2 r^{\beta+2} (Logr)^\gamma dr \right)^{1/2}. \end{aligned} \right. \tag{2.5.108}$$

When $\gamma = 0$, it is possible to choose $A = 0$ and then use directly (2.5.105).

b) If $\beta = -1$ and $\gamma \neq -1$, we start from the identity

$$\frac{1}{r} (Logr)^\gamma = \frac{1}{\gamma + 1} \frac{d}{dr} (Logr)^{\gamma+1} \tag{2.5.109}$$

and follow the same steps.   ∎

Let Ω_e be an open set of $\mathbb{R}^n$, the complement of which is bounded in $\mathbb{R}^n$. We **define the weighted Sobolev spaces** $W^{k,\alpha}(\Omega_e)$:
a) when $-(\alpha + n/2)$ is not an integer equal to $0, 1, \ldots, k-1$,

$$W^{k,\alpha}(\Omega_e) = \left\{ u, (1+r^2)^{\alpha/2} u \in L^2(\Omega_e), \ldots, (1+r^2)^{\frac{\alpha+k}{2}} D^k u \in L^2(\Omega_e) \right\};$$

b) when $-(\alpha + n/2) = l$, $l = 0, 1, \ldots,$ or $k-1$,

$$\left\{ \begin{array}{l} W^{k,\alpha}(\Omega_e) = \left\{ u; \quad \dfrac{(1+r^2)^{\alpha/2}}{\mathrm{Log}(2+r^2)} u \in L^2(\Omega_e), \right. \\[3mm] \qquad \ldots, \dfrac{(1+r^2)^{\frac{\alpha+l}{2}}}{\mathrm{Log}(2+r^2)} D^l u \in L^2(\Omega_e), (1+r^2)^{\frac{\alpha+l+1}{2}} D^{l+1} u \in L^2(\Omega_e), \\[3mm] \qquad \left. \ldots, (1+r^2)^{\frac{\alpha+k}{2}} D^k u \in L^2(\Omega_e) \right\}. \end{array} \right.$$

When $n = 3$ and $\alpha = -1$, and for the domain B_e, these spaces coincide with the spaces $W^k(B_e)$ already introduced in (2.5.13).

Lemma 2.5.8 *If $-(\alpha + n/2) > 0$, the polynomials of total degree l lie in $W^{k,\alpha}(\Omega_e)$ for the following values of l:*

$$l = E(-\alpha - n/2) \qquad \textit{in case a}),$$

$$l = -\alpha - n/2 \qquad \textit{in case b});$$

otherwise, the space $W^{k,\alpha}(\Omega_e)$ does not contain any polynomial.

It is easy to check the different cases.
We **define** $W_0^{k,\alpha}(\Omega_e)$ as the space of functions in $W^{k,\alpha}(\Omega_e)$ with vanishing traces on the surface Γ. When this surface Γ is **the unit sphere** S, the following property holds:

Theorem 2.5.11 *The semi-norm*

$$|u|_{k,\alpha,B_e}^2 = \int_{B_e} r^{\alpha+k} ||D^k u||^2 dx \qquad (2.5.110)$$

is a norm on the space $W_0^{k,\alpha}(B_e)$.

Proof
Using a density argument, we only need to prove (2.5.110) for functions in the space $\mathcal{D}(B_e)$. We introduce (r, θ), the spherical coordinates in $\mathbb{R}^n$, where r is the euclidian distance to the center of the sphere S and where θ stands for all the angles that define a point on S. We denote by $\psi(\theta) r^{n-1} dr d\theta$ the volume element in spherical coordinates. The function $\psi(\theta)$ is bounded. For any order of derivation λ, we have

$$\frac{\partial^{|\lambda|} u}{\partial x^\lambda}(r, \theta) = \int_1^r \frac{\partial}{\partial r} \left(\frac{\partial^{|\lambda|} u}{\partial x^\lambda} \right)(\rho, \theta) d\rho. \qquad (2.5.111)$$

We consider the following semi-norms:
in case a) and in case b) when $m > l$,

$$|u|^2_{m,\Omega_e} = \int \psi(\theta)d\theta \int_1^\infty \left(\sum_{|\lambda|=m} \left| \frac{\partial^{|\lambda|}u}{\partial x^\lambda} \right|^2 r^{2\alpha+2m+n-1} \right) dr; \qquad (2.5.112)$$

in case b) when $m \leq l$,

$$|u|^2_{m,\Omega_e} = \int \psi(\theta)d\theta \int_1^\infty \left(\sum_{|\lambda|=m} \left| \frac{\partial^{|\lambda|}u}{\partial x^\lambda} \right|^2 \frac{r^{2\alpha+2m+n-1}}{(\text{Log} r)^2} \right) dr. \qquad (2.5.113)$$

Applying Hardy's inequality, with the values of the parameters $\gamma = 0$ and $\beta = 2\alpha + 2m + n - 3$, shows, in case a) that

$$|u|^2_{m-1,\Omega_e} \leq c|u|^2_{m,\Omega_e}. \qquad (2.5.114)$$

A careful check shows that the change to the parameters, $\gamma = -2\beta = -1$, occurs when m is $l+1$ in case b). When $r \geq 1$, we have $r^2 \leq 1 + r^2 \leq 2r^2$, which shows that the weights r and $(1 + r^2)^{1/2}$ are comparable. Using this property, we obtain the final result, linking up the inequalities.

Theorem 2.5.12 *If Ω_i is non-empty, then*

$$\|u\|_{W^{k,\alpha}_0(\Omega_e)} \leq c|u|_{k,\alpha,\Omega_e}. \qquad (2.5.115)$$

Proof
We extend u by zero to the interior domain Ω_i. The trace theorem shows that this extension belongs to $W^{k,\alpha}_0(B_e)$ for every ball contained in Ω_i, and the proof follows from the previous theorem.

Theorem 2.5.13 *The notation $\mathbb{P}_l$ stands for the space of polynomials of degree l, l defined in Lemma 2.5.8. We have*
   a) when $-\alpha - n/2 < 0$,

$$\|u\|_{W^{k,\alpha}(\Omega_e)} \leq c|u|_{k,\alpha,\Omega_e}; \qquad (2.5.116)$$

b) when $-\alpha - n/2 \geq 0$,

$$\inf_{p \in \mathbb{P}_l} \|u - p\|_{W^{k,\alpha}(\Omega_e)} \leq c|u|_{k,\alpha,\Omega_e}. \qquad (2.5.117)$$

Proof
Let $\mathcal{O}$ be a non-empty bounded open domain contained in Ω_e. The inequalities (2.5.116) and (2.5.117) are equivalent to showing that the semi-norm

$$[u]^2_{k,\alpha,\Omega_e} = |u|^2_{k,\alpha,\Omega_e} + \sum_{0 \leq |\lambda| \leq l} \left| \int_{\mathcal{O}} D^\lambda u dx \right|^2 \qquad (2.5.118)$$

is a norm on $W^{k,\alpha}(\Omega_e)$. In both cases, we use a contradiction argument.

Suppose the statement is false. Then, there exists a sequence of functions u_j in $W^{k,\alpha}(\Omega_e)$ which satisfies

$$
\begin{cases}
[u_j]_{k,\alpha,\Omega_e} \leq \dfrac{1}{j}, \\[2mm]
\|u_j\|_{k,\alpha,\Omega_e} = 1.
\end{cases}
\tag{2.5.119}
$$

We can choose the radius R of the ball B_R big enough so that $\Omega_i \subset B_R$. Let φ and ψ be two cut-off functions which satisfy

$$
\begin{cases}
\varphi \geq 0, \quad \psi \geq 0, \quad \varphi + \psi = 1, \text{ in } \Omega_e, \\[1mm]
\varphi = 0 \quad \text{for } |x| \geq 2R, \\[1mm]
\psi = 0 \quad \text{for } |x| \leq R,
\end{cases}
\tag{2.5.120}
$$

and consider the two sequences φu_j and ψu_j. The Leibniz' formula for differentiating products shows that

$$
|\psi(u_j - u_i)|_{k,\alpha,\Omega_e} \leq |u_j - u_i|_{k,\alpha,\Omega_e} + c\|u_j - u_i\|_{k-1,\alpha,\Omega_e \cap B_{2R}}.
\tag{2.5.121}
$$

For the part on the bounded domain, we note that $W^{k,\alpha}(\Omega_e \cap B_{2R})$ is identical to $H^k(\Omega_e \cap B_{2R})$ and is compactly imbedded in $W^{k-1,\alpha}(\Omega_e \cap B_{2R})$. We can extract from the sequence of functions u_j a subsequence, still denoted u_j, which converges to u in this space. Inequalities (2.5.119) and (2.5.121) show that ψu_j is a Cauchy sequence for the semi-norm. These functions belong to $W_0^{k,\alpha}(B_R^c)$, so Theorem 2.5.12 (with some obvious modification to take care of a ball with arbitrary radius) shows that the sequence is Cauchy in this latter space as well. It follows that

$$
\psi u_j \to v, \quad v \in W^{k,\alpha}(\Omega_e).
\tag{2.5.122}
$$

Similarly, we have

$$
|\varphi(u_j - u_i)|_{k,\alpha,\Omega_e} \leq |u_j - u_i|_{k,\alpha,\Omega_e \cap B_{2R}} + c\|u_j - u_i\|_{k-1,\alpha,\Omega_e \cap B_{2R}}.
\tag{2.5.123}
$$

Inequality (2.5.121) and the compactness property show that this subsequence is Cauchy for the semi-norm. Being also Cauchy in $W^{k-1,\alpha}(\Omega_e)$, it must be a Cauchy sequence in $W^{k,\alpha}(\Omega_e)$ and converges to some w. By addition, it follows that

$$
\begin{cases}
u = v + w \in W^{k,\alpha}(\Omega_e), \\[1mm]
\|u\|_{k,\alpha,\Omega_e} = 1.
\end{cases}
\tag{2.5.124}
$$

In case a) using (2.5.121), it follows that

$$
D^\lambda u = \lim_{j \to \infty} D^\lambda u_j = 0, \quad |\lambda| = k,
\tag{2.5.125}
$$

which implies that u is a polynomial of degree at most $k - 1$. This is contradictory since there are no polynomials in the space $W^{k,\alpha}(\Omega_e)$.

In case b) we have moreover

$$\int_{\mathcal{O}} D^{\lambda} u dx = 0, \quad |\lambda| \leq l, \tag{2.5.126}$$

which eliminates the polynomials of degree l and also leads to a contradiction.   ■

Remark
The above theorem is still valid when Ω_e is the whole $\mathbb{R}^3$.

We are now in a position to solve Laplace's equation in the exterior domain Ω_e (and also in $\mathbb{R}^3$).

Theorem 2.5.14 *The exterior Dirichlet problem*

$$\begin{cases} -\Delta u(x) = f(x), & x \in \Omega_e, \\ u_{|\Gamma} = u_d \end{cases} \tag{2.5.127}$$

has a unique solution in the space $W^{1,-1}(\Omega_e)$ when u_d belongs to $H^{1/2}(\Gamma)$ and when f is in the dual space $(W_0^{1,-1}(\Omega_e))^*$. The corresponding mapping is continuous from $H^{1/2}(\Gamma) \otimes H^{-1}(\Omega_i)$ into $W_0^{1,-1}(\Omega_i)$.

Proof
Using the lifting $\mathcal{R}u_d$ of u_d in $H^1(\Omega_e)$, we introduce the new unknown $v = u - \mathcal{R}u_d$ which vanishes on the surface Γ. As the terms coming from the exterior sphere at infinity vanish, due to the behavior of the functions in $W_0^{1,-1}(\Omega_e)$, a variational formulation for this problem is

$$\begin{cases} \int_{\Omega_e} (\nabla v(x).\nabla w(x)) dx \\ \quad = \int_{\Omega_e} [fw - (\nabla \mathcal{R}u_d.\nabla w))] dx; \quad \forall w \in W_0^{1,-1}(\Omega_e). \end{cases} \tag{2.5.128}$$

It follows from Theorem 2.5.12 that the bilinear form in the left-hand side is coercive on the space $W_0^{1,-1}(\Omega_e)$ and we obtain the result using the Lax-Milgram theorem.   ■

Remarks
  – Distributions in the dual space $(W_0^{1,-1}(\Omega_e))^*$ have the form

$$f = \sum_{i=1}^{3} \frac{\partial}{\partial x_i} g_i + f_0, \quad g_i \in L^2(\Omega_e), \quad r f_0 \in L^2(\Omega_e). \tag{2.5.129}$$

  – The above theorem is also essentially valid in an **open exterior set** Ω_e **in** $\mathbb{R}^2$; but a logarithmic weight enters in the definition of the space $W_0^{1,-1}(\Omega_e)$ in that case.   ■

We can now state

Theorem 2.5.15 *The exterior Neumann problem*

$$\begin{cases} -\Delta u(x) = f(x), & x \in \Omega_e, \\ \dfrac{\partial u}{\partial n}\Big|_\Gamma = u_n \end{cases} \tag{2.5.130}$$

admits a unique solution in the space $W^{1,-1}(\Omega_e)$ when f is in the dual space $(W^{1,-1}(\Omega_e))^*$ and u_n is in $H^{-1/2}(\Gamma)$. The corresponding mapping is continuous from $H^{-1/2}(\Gamma) \otimes H^{-1}(\Omega_i)$ into $W_0^{1,-1}(\Omega_i)$.

Proof

A variational formulation of this problem is

$$\begin{cases} \displaystyle\int_{\Omega_e} (\nabla u(x).\nabla v(x))dx = -\int_\Gamma u_n v d\gamma + \int_{\Omega_e} f v dx, \\ \forall v \in W^{1,-1}(\Omega_e). \end{cases} \tag{2.5.131}$$

(The minus sign is related to the interior orientation of the normal to Ω_e!). It follows from Theorem 2.5.13 that the bilinear form in the left-hand side is coercive on the space $W^{1,-1}(\Omega_e)$, and we obtain the result using the Lax-Milgram theorem. ∎

Remark

In the case of an **open exterior set** Ω_e in $\mathbb{R}^2$, the space

$$W^{1,-1}(\Omega_e) = \left\{ u; \frac{u}{(1+r^2)^{1/2}\mathrm{Log}(2+r^2)} \in L^2(\Omega_e), \nabla u \in L^2(\Omega_e) \right\}$$

contains the constant functions.

Theorem 2.5.11 yields the existence of a solution defined up to an additive constant, when f and u_n satisfy $-\int_\Gamma u_n d\gamma + \int_{\Omega_e} f dx = 0$. (In this case, the integrals may in fact be dualities.)

2.5.5 Regularity properties of solutions in $\mathbb{R}^n$

When the data u_d, u_n and f are more regular, the solutions of Dirichlet and Neumann problems are also more regular than the variational solutions obtained in the above theorems. This section is devoted to showing some results of this type. These properties are essentially local and are related to the ellipticity of the operator Δ.

In $\mathbb{R}^3$, we can state

Theorem 2.5.16 *The operator $-\Delta$ is an isomorphism of $W^{k,-1}(\mathbb{R}^3)$ onto $W^{k-2,1}(\mathbb{R}^3)$ for $k \geq 2$.*

Proof

We proceed by induction on the index k. Let u be the variational solution in the space $W^{1,-1}(\Omega_e)$ of

$$\Delta u = f. \tag{2.5.132}$$

When f is in $W^{0,1}(\mathbb{R}^3)$, it holds in the space $S'(\mathbb{R}^3) : \Delta(\partial u/\partial x_i) = \partial f/\partial x_i$. The variational form of this identity,

$$\int_{\mathbb{R}^3} \left(\nabla \frac{\partial u}{\partial x_i} \cdot \nabla \varphi \right) dx = \int_{\mathbb{R}^3} f \frac{\partial \varphi}{\partial x_i} dx, \quad \forall \varphi \in W^{1,-1}(\mathbb{R}^3), \qquad (2.5.133)$$

shows that $\partial u/\partial x_i$ is in the space $W^{1,-1}(\mathbb{R}^3)$.

To be more precise, we need to introduce the differential quotient $\nabla_i u = (u(x + he_i) - u(x))/h$ which satisfies $\Delta \nabla_i u = \nabla_i f$. The variational form of this equation and the coercivity inequality show that the sequence of differential quotients is bounded in $W^{1,-1}(\mathbb{R}^3)$ and so it converges in this space toward $\partial u/\partial x_i$. The product $x_i(\partial u/\partial x_j)$ satisfies

$$-\Delta \left(x_i \frac{\partial u}{\partial x_j} \right) = -x_i \frac{\partial}{\partial x_j} f - 2 \frac{\partial^2 u}{\partial x_i \partial x_j}. \qquad (2.5.134)$$

The variational formulation of this equation is

$$\begin{cases} \int_{\mathbb{R}^3} (\nabla \left(x_i \frac{\partial u}{\partial x_j} \right) \cdot \nabla \varphi) dx \\ \qquad = \int_{\mathbb{R}^3} f \frac{\partial}{\partial x_j} (x_i \varphi) dx + 2 \int_{\mathbb{R}^3} \frac{\partial u}{\partial x_i} \frac{\partial \varphi}{\partial x_j} dx. \end{cases} \qquad (2.5.135)$$

When f is $W^{0,1}(\mathbb{R}^3)$ and u is $W^{1,-1}(\mathbb{R}^3)$, the right-hand side is a continuous linear form on $W^{1,-1}(\mathbb{R}^3)$. Using an approximation of $x_i(\partial u/\partial x_i)$ by $x_j \nabla_i u$, (2.5.135) shows that $x_i(\partial u/\partial x_j)$ belongs to $W^{1,-1}(\mathbb{R}^3)$.

The equality

$$x_i \frac{\partial^2 u}{\partial x_l \partial x_j} = \frac{\partial}{\partial x_l} \left(x_i \frac{\partial u}{\partial x_j} \right) - \delta_i^l \frac{\partial u}{\partial x_j} \qquad (2.5.136)$$

shows that $x_i(\partial^2 u/\partial x_l \partial x_j)$ is in $L^2(\mathbb{R}^3)$ and so is $\partial^2 u/\partial x_i \partial x_j$ as results from (2.5.135). It follows that $(1+r^2)^{1/2}(\partial^2 u/\partial x_i \partial x_j)$ is in $L^2(\mathbb{R}^3)$ and so proves the result when $k = 2$.

The proof of the general case is similar and consists in applying the above regularity result to $\partial u/\partial x_l$ and $x_j(\partial u/\partial x_l)$. The operator $-\Delta$ is continuous from $W^{k,-1}(\mathbb{R}^3)$ to $W^{k-2,1}(\mathbb{R}^3)$ and injective. So it is bijective with a continuous inverse. ∎

The case of $\mathbb{R}^2$ is slightly different due to the logarithmic weight in the definition of the space $W^{k,-1}(\mathbb{R}^2)$. We have seen that the operator $-\Delta$ is an isomorphism from $W^{1,-1}(\mathbb{R}^2)/\mathbb{P}_0$ onto the subspace of $(W^{1,-1}(\mathbb{R}^2))^*$ of functions satisfying (1 is in the space $(W^{1,-1}(\mathbb{R}^2))$) :

$$\langle f, 1 \rangle = 0. \qquad (2.5.137)$$

It is natural to think of an isomorphism from $W^{2,-1}(\mathbb{R}^2)/\mathbb{P}_0$ onto $W^{0,1}(\mathbb{R}^2)$. But this is not possible since the function 1 is not in $(W^{0,1}(\mathbb{R}^2))^*$, and so condition (2.5.137) has no meaning in the space

$W^{0,1}(\mathbb{R}^2)$. We introduce a new space

$$
\begin{cases}
X^k(\mathbb{R}^2) = \left\{ f \in \left(W^{1,-1}(\mathbb{R}^2)\right)^*, \right. \\
\\
\left. \qquad x^\alpha D^\lambda f \in W^{k-|\lambda|,0}(\mathbb{R}^2), \ 0 \le |\alpha| - 1 \le |\lambda| \le k \right\}.
\end{cases}
$$

Now the relation (2.5.137) makes sense, which shows that this space is strictly included in $W'^{k-2,-1}(\mathbb{R}^2)$.

We now have

Theorem 2.5.17 *The operator* $-\Delta$ *is an isomorphism from the space* $W^{k,-1}(\mathbb{R}^2)/\mathbb{P}_0$ *onto the subspace of functions in* $X^k(\mathbb{R}^2)$ *satisfying* (2.5.137).

Proof
We repeat step by step the proof of Theorem 2.5.16, checking at each level the needed hypothesis on f.

When $k = 2$, (2.5.133) uses $f \in L^2(\mathbb{R}^2)$. Examining the factor of f in (2.5.135), we see that we only need $x_i f \in L^2(\mathbb{R}^2)$ and $f \in \left(W^{1,-1}(\mathbb{R}^2)\right)^*$, or else $f \in X^0(\mathbb{R}^2)$. For $k = 2$, we have shown that $f \in X^0(\mathbb{R}^2)$ yields $u \in W^{2,-1}(\mathbb{R}^2)/\mathbb{P}_0$. It is easy to check the continuity of the operator $-\Delta$ from $W^{2,-1}(\mathbb{R}^2)$ into $X^0(\mathbb{R}^2)$. The result for $k = 2$ follows.

The general case is similar. The logarithmic weight appears in the product $f\varphi$, which has a meaning as a duality, not as an integral. We refer to J. Giroire [79] for further results on this subject.

2.5.6 Elementary differential geometry

We introduce in this section some differential operators on a surface Γ defined by a system of charts as described in Subsection 2.5.2. These operators will appear later when studying the regularity of the interior and exterior problems. They will also be used to obtain variational formulations for some of the integral equations associated with the Helmholtz equation. More essentially, they are needed in the study of the Maxwell system.

Our presentation is elementary and non-canonical. We avoid describing the usual tools of differential geometry and in particular the vocabulary of differential forms, although all the operators that we describe are of such nature. We hope to be forgiven by the mathematicians and hope that it will ease the access of this book to engineers.

For every point y in $\mathbb{R}^3$, we denote by $\delta(y)$ the distance of y to the surface Γ

$$
\delta(y) = \inf_{x \in \Gamma} |y - x|. \tag{2.5.138}
$$

A collection of points whose distance to the surface is less than ε is called a tubular neighbourhood of the surface Γ. For ε small enough and when the surface is regular and oriented, any point y in such a neighbourhood Γ_ε, has a unique projection $\mathcal{P}(y)$ on the surface which satisfies

$$|y - \mathcal{P}(y)| = \delta(y). \tag{2.5.139}$$

It is easy to check that for a regular surface Γ which admits a tangent plane at the point $\mathcal{P}(y)$, the line $y - \mathcal{P}(y)$ is directed along the normal to the surface Γ at this point. Remark that we have the possibility to choose all the open sets ω_i that define the surface Γ to be contained in the neighbourhood Γ_ε, which we will do from now on.

Notice that the unit normal n is the gradient of the function $\delta(y)$ oriented towards the interior or the exterior depending on the position of the point y. It holds that

$$\begin{cases} n(\mathcal{P}(y)) = \nabla\delta(y), & \text{when } y \in \Omega_e \\ n(\mathcal{P}(y)) = -\nabla\delta(y), & \text{when } y \in \Omega_i. \end{cases} \tag{2.5.140}$$

Using the unit normal, we introduce the following diffeomorphisms from the sets ω_i to open subsets of R^3, which will be used jointly with the charts ϕ_i from now on. Any point y in the tubular neighbourhood Γ_ε is located using $\mathcal{P}(y)$ and $\delta(y)$, or more precisely we write:

$$\begin{cases} y = \mathcal{P}(y) + sn(\mathcal{P}(y)), & -\varepsilon \leq s \leq \varepsilon, \\ s = \delta(y), & \text{when } y \in \Omega_e, \quad s = -\delta(y), \quad \text{when } y \in \Omega_i. \end{cases} \tag{2.5.141}$$

In restriction to each open set ω_i, the piece of surface $\Gamma_i = \omega_i \cap \Gamma$ is parametrized by the diffeomorphism ϕ_i^{-1} which maps γ on the set Γ_i. We denote by (ξ_1, ξ_2) the associated variables. We have built a parametrization of the set ω_i in the form

$$y(\xi_1, \xi_2, s) = x(\xi_1, \xi_2) + sn(\xi_1, \xi_2), \quad -\varepsilon \leq s \leq \varepsilon. \tag{2.5.142}$$

To any function u defined on the surface Γ, we associate the lifting $\tilde{u}$ defined on the tubular neighbourhood Γ_ε by

$$\tilde{u}(y) = u(\mathcal{P}(y)). \tag{2.5.143}$$

We continue to denote by $n(y)$ the vector field ∇s which is defined on Γ_ε and has the value n on the surface Γ.

We introduce the family Γ_s of parallel surfaces

$$\Gamma_s = \{y; y = x + sn(x); x \in \Gamma\}. \tag{2.5.144}$$

The vector field n is the field of normals to the surfaces Γ_s.

We introduce the first family of **differential operators** which acts **on functions defined on the surfaces Γ or Γ_s. The tangential gradient** $\nabla_\Gamma u$ is defined as

$$\nabla_\Gamma u = \nabla\tilde{u}_{|\Gamma}. \tag{2.5.145}$$

The **tangential rotational of a function** is defined as

$$\overrightarrow{\mathrm{curl}}_\Gamma u = \mathrm{curl}(\tilde{u}n)_{|_\Gamma}. \tag{2.5.146}$$

The field of normals is a gradient, which implies

$$\mathrm{curl}\, n = 0. \tag{2.5.147}$$

We use the vectorial calculus formula

$$\mathrm{curl}(u\vec{v}) = \nabla u \wedge \vec{v} + u\,\mathrm{curl}\,\vec{v} \tag{2.5.148}$$

which yields

$$\overrightarrow{\mathrm{curl}}_\Gamma u = \nabla_\Gamma u \wedge n. \tag{2.5.149}$$

The curvature operator $\mathcal{R}_s$ is

$$\mathcal{R}_s = \nabla n. \tag{2.5.150}$$

Notation

When we are on the surface Γ (i.e., when $s = 0$), we omit the index s.

Theorem 2.5.18 *The curvature operator $\mathcal{R}_s$ is a symmetric operator acting in the tangent plane. We denote by $1/R_1(s)$ and $1/R_2(s)$ its two eigenvalues, called* **principal curvatures.**

We define its **mean curvature** H_s *to be half its trace*

$$H_s = \frac{1}{2}\,\mathrm{div}\,n = \frac{1}{2}\left(\frac{1}{R_1(s)} + \frac{1}{R_2(s)}\right). \tag{2.5.151}$$

We define its **Gauss curvature** G_s *to be its determinant*

$$G_s = \frac{1}{R_1 R_2} = \det\left(\mathcal{R}_{s|\Gamma}\right). \tag{2.5.152}$$

The area element on the surface Γ_s at the point $y = x + sn(x)$ is related to the area element on the surface Γ at the point x through the relation

$$d\gamma_s(y) = \left(1 + 2sH(x) + s^2 G(x)\right)d\gamma(x). \tag{2.5.153}$$

The normal derivative of the curvature tensor $\mathcal{R}_s$ is

$$\frac{\partial}{\partial s}\mathcal{R}_s = -\mathcal{R}_s^2. \tag{2.5.154}$$

Moreover, it holds that

$$\frac{\partial}{\partial s}H_s = -\frac{1}{2}\,\mathrm{trace}\left(\mathcal{R}_s^2\right) = -\frac{1}{2}\left(\frac{1}{(R_1(s))^2} + \frac{1}{(R_2(s))^2}\right), \tag{2.5.155}$$

$$\frac{\partial}{\partial s}G_s = -2H_s G_s, \tag{2.5.156}$$

$$[\mathcal{R}_s(v \wedge n) - 2H_s(v \wedge n)] \wedge n = \mathcal{R}_s v, \text{ (for every vector } v). \tag{2.5.157}$$

Proof

The symmetry of the operator $\mathcal{R}_s$ results from the identity

$$\mathcal{R}_s = D^2 s. \tag{2.5.158}$$

Differentiating the norm of the normal n which is 1 and using the symmetry of $\mathcal{R}_s$, it follows that

$$2\nabla n \cdot n = 2\mathcal{R}_s n = \nabla(n \cdot n) = 0, \tag{2.5.159}$$

which shows that the normal is contained in the kernel of the curvature operator. So, this operator acts in the tangent plane to Γ_s. Its two eigenvalues in this plane are real and are the principal curvatures.

Formula (2.5.153) is linked to the way the jacobians are changed in integrals when modifying the variable of integration. The area element is multiplied by the jacobian of the tangent transform. The mapping $x \rightarrow x + sn(x)$ is a diffeomorphism from the surface Γ on the surface Γ_s, which tangent mapping in the tangent plane is

$$D\phi = (I + s\nabla n(x)). \tag{2.5.160}$$

The jacobian of this matrix is given by $1 + 2sH + s^2 G$.

The normal derivative $\partial/\partial s$ can be expressed in a local system of axes in $\mathbb{R}^3$ as

$$\frac{\partial}{\partial s} = \frac{d}{dn} = \sum_{i=1}^{3} n_i \frac{\partial}{\partial x_i} = n \cdot \nabla. \tag{2.5.161}$$

It follows that

$$\frac{\partial}{\partial s} n = n \cdot \nabla n = \mathcal{R}_s n = 0, \tag{2.5.162}$$

$$\begin{cases} \dfrac{\partial}{\partial s}\mathcal{R}_{ij} = \sum_{k=1}^{3} n_k \dfrac{\partial}{\partial x_k}\left(\dfrac{\partial n_i}{\partial x_j}\right) \\[4mm] = \sum_{k=1}^{3}\left[\dfrac{\partial}{\partial x_j}\left(n_k \dfrac{\partial n_i}{\partial x_k}\right) - \dfrac{\partial n_k}{\partial x_j}\dfrac{\partial n_i}{\partial x_k}\right]. \end{cases} \tag{2.5.163}$$

The first term in the right-hand side is $\nabla(\mathcal{R}n) = 0$, while the second term is nothing but $-\mathcal{R}^2$, (we use the symmetry of the tensor $\mathcal{R}$). The mean curvature is the trace of the tensor $\mathcal{R}$ and we obtain formula (2.5.155) taking the trace in the previous formula (2.5.154). It follows from (2.5.154) and from the Leibniz rule that

$$\frac{d}{ds}\mathcal{R}_s^2 = \frac{d\mathcal{R}_s}{ds}\mathcal{R}_s + \mathcal{R}_s\frac{d\mathcal{R}_s}{ds} = -2\mathcal{R}_s^3. \tag{2.5.164}$$

Moreover, it holds that

$$G_s = \frac{1}{2}\left[(\text{trace } \mathcal{R}_s)^2 - \text{trace}\left(\mathcal{R}_s^2\right)\right]. \tag{2.5.165}$$

Differentiating, we obtain

$$\begin{cases} \frac{dG_s}{ds} = \frac{1}{2}\left[2\operatorname{trace}\mathcal{R}_s \cdot \frac{d}{ds}\operatorname{trace}\mathcal{R}_s - \frac{d}{ds}\operatorname{trace}\left(\mathcal{R}_s{}^2\right)\right] \\ = -\left[\operatorname{trace}(\mathcal{R}_s)\operatorname{trace}\mathcal{R}_s^2 - \operatorname{trace}\mathcal{R}_s^3\right] = -2H_sG_s. \end{cases} \tag{2.5.166}$$

Formula (2.5.157) can be easily checked using the eigenvector basis of the operator $\mathcal{R}_s$. ■

In order to introduce a new family of operators acting on tangent vector fields v, we need to define a lifting $\tilde{v}$ of this field in the neigbourhood Γ_ε. It is natural to ask it to be tangent to the surfaces Γ_s. But this property leaves some degrees of freedom in the tangent planes. We introduce the moving frame to understand what the different possible choices are.
Going back to the parametrization (2.5.142), it is **natural** to choose as a **vector basis in the tangent plane** to Γ_s,

$$\begin{cases} e_1(x,s) = \frac{\partial y}{\partial \xi_1} = \frac{\partial x}{\partial \xi_1} + s\frac{\partial n}{\partial \xi_1} = e_1(x) + s\mathcal{R}(x)e_1(x), \\ e_2(x,s) = \frac{\partial x}{\partial \xi_2} + s\frac{\partial n}{\partial \xi_2} = e_2(x) + s\mathcal{R}(x)e_2(x). \end{cases} \tag{2.5.167}$$

A vector in the tangent plane has the expression

$$v = v^1 e_1 + v^2 e_2. \tag{2.5.168}$$

We introduce the **metric tensor** g which acts from the tangent plane into the cotangent plane and which matrix is:

$$g_{ij} = (e_i \cdot e_j). \tag{2.5.169}$$

The inverse matrix is denoted g^{-1} and its coordinates are g^{ij}. The image of the vector basis (e_1, e_2) by this inverse matrix are the vectors $\left(e^1, e^2\right)$, that we choose as **vector basis in the cotangent plane**. It holds that

$$\left(e^i \cdot e_j\right) = \delta_j^i. \tag{2.5.170}$$

A vector in the cotangent plane (or covariant vector) takes the form

$$v = v_1 e^1 + v_2 e^2. \tag{2.5.171}$$

In this system of coordinates the surface area element is

$$d\gamma = |e_1 \wedge e_2|\, d\xi_1 d\xi_2 = \sqrt{\det g}\, d\xi_1 d\xi_2. \tag{2.5.172}$$

When the orientation of the surface satisfies

$$n = \frac{e_1 \wedge e_2}{|e_1 \wedge e_2|}, \tag{2.5.173}$$

then

$$\begin{cases} n \wedge e^1 = \dfrac{e_2}{\sqrt{\det g}}, \\[2mm] n \wedge e^2 = -\dfrac{e_1}{\sqrt{\det g}}. \end{cases} \tag{2.5.174}$$

The differential of the function $\widetilde{u}$ is

$$d\widetilde{u} = \frac{\partial u}{\partial \xi_1} d\xi_1 + \frac{\partial u}{\partial \xi_2} d\xi_2 = (\nabla \widetilde{u} \cdot dx). \qquad (2.5.175)$$

Another expression is $(dx = e_1 d\xi_1 + e_2 d\xi_2 + nds)$

$$\nabla_\Gamma u = \frac{\partial u}{\partial \xi_1} e^1 + \frac{\partial u}{\partial \xi_2} e^2, \qquad \left((\nabla \widetilde{u} \cdot e_i) = \frac{\partial u}{\partial \xi_i}\right). \qquad (2.5.176)$$

The vector $\overrightarrow{\mathrm{curl}}_\Gamma u$ can be expressed as a covariant vector and then (2.5.174) shows that it takes the form :

$$\overrightarrow{\mathrm{curl}}_\Gamma u = \frac{1}{\sqrt{\det g}} \left(\frac{\partial u}{\partial \xi_2} e_1 - \frac{\partial u}{\partial \xi_1} e_2\right). \qquad (2.5.177)$$

The matrix g changes the contravariant coordinates into the covariant coordinates, and the matrix g^{-1} changes the covariant coordinates into the contravariant coordinates.

We want to built a lifting $\widetilde{v}$ of a tangent vector field v (a covariant or a contravariant vector) defined in the neigbourhood Γ_ε. When this field is a gradient, **a natural formula** results from the expression (2.5.176) choosing

$$\widetilde{\nabla_\Gamma u} = \frac{\partial u}{\partial \xi_1} e^1(s) + \frac{\partial u}{\partial \xi_2} e^2(s) = \nabla_{\Gamma_s} \widetilde{u}. \qquad (2.5.178)$$

We will now use the formulas (2.5.167) to rewrite the previous expression independently of the system of coordinates.

The vector $e^i(s)$ satisfies

$$\left(e^i(s) \cdot (e_j + s\mathcal{R}e_j)\right) = \delta^i_j \qquad (2.5.179)$$

which in view of the symmetry of $\mathcal{R}$ takes the form

$$\left((I + s\mathcal{R})e^i(s) \cdot e_j(x)\right) = \delta^i_j \qquad (2.5.180)$$

or equivalently

$$e^i(s) = (I + s\mathcal{R})^{-1} e^i(x). \qquad (2.5.181)$$

A natural transport formula for a gradient field is

$$\widetilde{\nabla_{\Gamma_s} u} = (I + s\mathcal{R}(x))^{-1} \nabla_\Gamma u. \qquad (2.5.182)$$

We obtain a different expression when computing directly $\nabla \widetilde{u}$. The diffeomorphism which maps Γ_s on Γ is

$$x = y - sn(y). \qquad (2.5.183)$$

It follows that

$$\nabla_y \widetilde{u} = (I - s\nabla_y n)\nabla_x u, \qquad (2.5.184)$$

which implies

$$\nabla_{\Gamma_s} \widetilde{u} = \widetilde{\nabla_{\Gamma_s} u} = \nabla_\Gamma u - s\mathcal{R}_s \nabla_\Gamma u. \qquad (2.5.185)$$

This expression, which looks different from (2.5.182) is in fact identical. This is related to a very specific property of this system of local coordinates which is the object of the following lemma:

Lemma 2.5.9 *The diffeomorphism from Γ onto Γ_s defined by $y = x + sn(x)$ has $x = y - sn(y)$ for inverse. It satisfies*

$$\begin{cases} \mathcal{R}(x) - \mathcal{R}_s(y) = s\mathcal{R}_s\mathcal{R} = s\mathcal{R}\mathcal{R}_s, \\ \\ (I + s\mathcal{R}(x))^{-1} = I - s\mathcal{R}_s(y). \end{cases} \tag{2.5.186}$$

Proof
We differentiate the identity

$$n(x) = n(x + sn(x)) \tag{2.5.187}$$

to obtain the first form of (2.5.186). The second form is easily seen to be equivalent. ∎

From now on, we choose to **transport the tangential vector fields** in the neighbourhood Γ_ε using (2.5.185), which is also

$$\tilde{v}(y) = v(x) - s\mathcal{R}_s(y)v(x). \tag{2.5.188}$$

We introduce the following first order **differential operators:**
The **surfacic divergence** of a tangent vector field v is

$$\operatorname{div}_\Gamma v = \operatorname{div} \tilde{v}_{|\Gamma}. \tag{2.5.189}$$

The **surfacic rotational** of a vector field v is

$$\operatorname{curl}_\Gamma v = (\operatorname{curl} \tilde{v} \cdot n(x))_{|\Gamma}. \tag{2.5.190}$$

We introduce the following second order **differential operators:**
The **scalar Laplacian** or **Laplace-Beltrami operator** acting on a function u is

$$\Delta_\Gamma u = \operatorname{div}_\Gamma \nabla_\Gamma u = -\operatorname{curl}_\Gamma \overrightarrow{\operatorname{curl}}_\Gamma u. \tag{2.5.191}$$

The **vectorial Laplacian** or **Hodge operator** acting on a tangent vector field v is

$$\Delta_\Gamma v = \nabla_\Gamma \operatorname{div}_\Gamma v - \overrightarrow{\operatorname{curl}}_\Gamma \operatorname{curl}_\Gamma v. \tag{2.5.192}$$

Theorem 2.5.19 *Let $u \in C^1(\Gamma)$ be a function and $v \in \left(C^1(\Gamma)\right)^2$ a tangent vector field defined on the surface Γ. The following **Stokes identities** hold:*

$$\int_\Gamma (\nabla_\Gamma u \cdot v) \, d\gamma + \int_\Gamma u \operatorname{div}_\Gamma v d\gamma = 0, \tag{2.5.193}$$

$$\int_\Gamma \left(\overrightarrow{\operatorname{curl}}_\Gamma u \cdot v\right) d\gamma - \int_\Gamma u \operatorname{curl}_\Gamma v d\gamma = 0, \tag{2.5.194}$$

$$\operatorname{div}_\Gamma \overrightarrow{\operatorname{curl}}_\Gamma u = 0, \tag{2.5.195}$$

$$\operatorname{curl}_\Gamma \nabla_\Gamma u = 0, \tag{2.5.196}$$

$$\operatorname{div}_\Gamma(v \wedge n) = \operatorname{curl}_\Gamma v. \tag{2.5.197}$$

Moreover, if w is a function $C^2(\Gamma)$, we have:

$$-\int_\Gamma \Delta_\Gamma w u \, d\gamma = \int_\Gamma (\nabla_\Gamma w \cdot \nabla_\Gamma u) \, d\gamma = \int_\Gamma \left(\overrightarrow{\operatorname{curl}}_\Gamma w \cdot \overrightarrow{\operatorname{curl}}_\Gamma u\right) d\gamma. \tag{2.5.198}$$

If w is a tangent vector field $\left(C^2(\Gamma)\right)^2$, it holds that

$$-\int_\Gamma (\Delta_\Gamma w \cdot v) \, d\gamma = \int_\Gamma \operatorname{div}_\Gamma w \operatorname{div}_\Gamma v \, d\gamma + \int_\Gamma \operatorname{curl}_\Gamma w \operatorname{curl}_\Gamma v \, d\gamma. \tag{2.5.199}$$

Proof

The two identities (2.5.195) and (2.5.196) follow directly from the definitions of the operators and from the analogous formulas in $\mathbb{R}^3$, which are $\operatorname{curl} \nabla u = 0$ and $\operatorname{div} \operatorname{curl} v = 0$. Introduce the field $\widetilde{u}\widetilde{v}\chi(s)$, where χ is a cutoff function vanishing outside Γ_ε and whose value is 1 in a neighbourhood of Γ. It holds that

$$\int_\Gamma (\operatorname{curl}(\widetilde{u}\widetilde{v}) \cdot n) \, d\gamma = \int_{\Omega_i} \operatorname{div} \operatorname{curl}(\widetilde{u}\widetilde{v}\chi(s)) \, dx = 0. \tag{2.5.200}$$

The vectorial calculus formula (φ is a function and a a vector)

$$\operatorname{curl}(\varphi a) = \nabla\varphi \wedge a + \varphi \operatorname{curl} a \tag{2.5.201}$$

shows that

$$(\operatorname{curl}(\widetilde{u}\widetilde{v}) \cdot n) = u \operatorname{curl}_\Gamma v - \left(\overrightarrow{\operatorname{curl}}_\Gamma u \cdot v\right) \tag{2.5.202}$$

which yields (2.5.194). The vectorial calculus formula (a, b are vectors)

$$\operatorname{div}(a \wedge b) = (\operatorname{curl} a \cdot b) - (\operatorname{curl} b \cdot a) \tag{2.5.203}$$

shows (2.5.197) using $\operatorname{curl} n = 0$. From formulas (2.5.197) and (2.5.149), it follows that the identity (2.5.193) is equivalent to (2.5.194) just changing v in $v \wedge n$. The identities (2.5.198) and (2.5.199) are direct consequences of the previous ones using the definitions of the Laplace-Beltrami and the Hodge operators. ∎

We will compute the expressions of these operators in the moving frame coordinates. We already know the expression of the gradient of a function u given by (2.5.176). The formula (2.5.193) shows that the operator $\operatorname{div}_\Gamma$ is in duality with the operator ∇_Γ. Let v be a vector in the tangent plane, then formula (2.5.193) takes the form:

$$\begin{cases} \int_\Gamma \left(\dfrac{\partial u}{\partial \xi_1} v^1 + \dfrac{\partial u}{\partial \xi_2} v^2\right) \sqrt{\det g} \, d\xi_1 d\xi_2 \\[4mm] \qquad = -\int_\Gamma u \operatorname{div}_\Gamma v \sqrt{\det g} \, d\xi_1 d\xi_2. \end{cases} \tag{2.5.204}$$

It follows that the expression of the **surfacic divergence of a contravariant vector** is

$$\text{div}_\Gamma \, v = \frac{1}{\sqrt{\det g}} \left(\frac{\partial}{\partial \xi_1} \sqrt{\det g} v^1 + \frac{\partial}{\partial \xi_2} \sqrt{\det g} v^2 \right). \qquad (2.5.205)$$

From identities (2.5.197) and (2.5.174), we infer the expression of **the rotational of a covariant vector**

$$\text{curl}_\Gamma \, v = \frac{1}{\sqrt{\det g}} \left(\frac{\partial}{\partial \xi_1} v_2 - \frac{\partial}{\partial \xi_2} v_1 \right). \qquad (2.5.206)$$

From these expressions, we compute now the expression of the **scalar Laplacian**

$$\Delta_\Gamma u = \frac{1}{\sqrt{\det g}} \left(\sum_{i,j=1}^{2} \frac{\partial}{\partial \xi_i} \sqrt{\det g} \, g^{ij} \frac{\partial u}{\partial \xi_j} \right). \qquad (2.5.207)$$

The expression of the vectorial Laplacian of a tangent vector is more complex in both the covariant and contravariant frames. We can compute them from the above expressions, but we do not give the results here as we are not going to use them.

We will now give some of the links between the surfacic operators in the domain Γ_ε and the usual three dimensional operators in $\mathbb{R}^3$. These expressions will be very usefull later in the study of the Maxwell equations. We use the parametrization (2.5.142).

Theorem 2.5.20 *Let u be a function defined on Γ_ε and v a field of vectors in R^3 defined on Γ_ε. The following decompositions hold:*

$$\nabla u = \nabla_{\Gamma_s} u + \frac{\partial u}{\partial s} n, \qquad (2.5.208)$$

$$v = (v \cdot n)n + v_{\Gamma_s}, \quad v_{\Gamma_s} = n \wedge (v \wedge n), \qquad (2.5.209)$$

$$\text{div} \, v = \text{div}_{\Gamma_s} \, v_{\Gamma_s} + 2H_s(v \cdot n) + \frac{\partial}{\partial s}(v \cdot n), \qquad (2.5.210)$$

$$\left\{ \begin{aligned} &\text{curl} \, v = (\text{curl}_{\Gamma_s} \, v_{\Gamma_s})n + \overrightarrow{\text{curl}}_{\Gamma_s}(v \cdot n) + \mathcal{R}_s(v \wedge n) \\ &\qquad - 2H_s(v \wedge n) - \frac{\partial}{\partial s}(v \wedge n), \end{aligned} \right. \qquad (2.5.211)$$

$$\Delta u = \Delta_{\Gamma_s} u + 2H_s \frac{\partial u}{\partial s} + \frac{\partial^2}{\partial s^2} u, \qquad (2.5.212)$$

$$\begin{cases} \Delta v = \Delta_{\Gamma_s} v_{\Gamma_s} + 2H_s \frac{\partial}{\partial s} v_{\Gamma_s} + \frac{\partial^2}{\partial s^2} v_{\Gamma_s} \\[2mm] \quad + \left[\Delta_{\Gamma_s}(v \cdot n) + 2H_s \frac{\partial}{\partial s}(v \cdot n) + \frac{\partial^2}{\partial s^2}(v \cdot n) \right] n \\[2mm] \quad + \left[2(v \cdot \nabla_{\Gamma_s} H_s) - 2 \operatorname{div}_{\Gamma_s}(\mathcal{R}_s v) + 2(v \cdot n)\frac{\partial}{\partial s} H_s \right] n \\[2mm] \quad + (2H_s I - 2\mathcal{R}_s)\mathcal{R}_s v + 2\mathcal{R}_s \nabla_{\Gamma}(v \cdot n) + 2(v \cdot n)\nabla_{\Gamma} H_s. \end{cases} \tag{2.5.213}$$

Proof

Formulas (2.5.208) and (2.5.209) are obvious. To prove identity (2.5.209), we start from (2.5.209) and the vectorial calculus formula

$$\operatorname{div}(\varphi a) = (\nabla \varphi \cdot a) + \varphi \operatorname{div} a. \tag{2.5.214}$$

We get

$$\operatorname{div} v = \operatorname{div} v_{\Gamma_s} + (v \cdot n) \operatorname{div} n + (n \cdot \nabla(v \cdot n)). \tag{2.5.215}$$

The two last terms are also the two last ones in the expression (2.5.210), so that (2.5.209) follows from

$$\operatorname{div} v_{\Gamma_s} = \operatorname{div} \tilde{v}_{|\Gamma} \ (\text{when } s = 0), \tag{2.5.216}$$

with the decomposition

$$v_{\Gamma_s} - \tilde{v} = v_{\Gamma_s}(y) - v_{\Gamma}(x) + s\mathcal{R}_s(y)v_{\Gamma}(x). \tag{2.5.217}$$

Identity (2.5.214) shows that the part $s\mathcal{R}_s(y)v_{\Gamma}(x)$ is tangential to Γ and has a zero divergence when $s = 0$. The remainder $v_{\Gamma_s}(y) - v_{\Gamma}(x)$ vanishes like s on the surface Γ and hence takes the form

$$v_{\Gamma_s}(y) - v_{\Gamma}(x) = sw(s, x) \tag{2.5.218}$$

where w is a regular tangent vector to Γ_s. Identity (2.5.216) follows now from (2.5.214).

The decomposition (2.5.208) and the formulas (2.5.210) and (2.5.214) imply (2.5.212). To prove the identity (2.5.211), we start from the following vectorial calculus formulas,

$$\operatorname{curl}(a \wedge b) = a \operatorname{div} b - b \operatorname{div} a + (b \cdot \nabla)a - (a \cdot \nabla)b, \tag{2.5.219}$$

$$\operatorname{curl}(\varphi a) = \varphi \operatorname{curl} a + \nabla \varphi \wedge a. \tag{2.5.220}$$

Starting from the decomposition (2.5.209), we get on one hand

$$\operatorname{curl}((v \cdot n)n) = \nabla(v \cdot n) \wedge n = \overrightarrow{\operatorname{curl}}_{\Gamma_s}(v \cdot n) \tag{2.5.221}$$

and on the other hand

$$\begin{cases} \operatorname{curl}(n \wedge (v \wedge n)) = \operatorname{div}(v \wedge n)n - (v \wedge n) \operatorname{div} n \\[2mm] \qquad + ((v \wedge n) \cdot \nabla)n - (n \cdot \nabla)(v \wedge n). \end{cases} \tag{2.5.222}$$

But $\mathrm{div}(v \wedge n)$ is nothing but $\mathrm{div}_{\Gamma_x}(v \wedge n)$ which is also $\mathrm{curl}_{\Gamma_x} v_{\Gamma_x}$. This shows (2.5.211). Formula (2.5.213) is related to the following vectorial calculus formula:

$$\Delta v = \nabla \, \mathrm{div}\, v - \mathrm{curl}\, \mathrm{curl}\, v. \tag{2.5.223}$$

Computing from the already established formulas, it holds that

$$\left\{ \begin{aligned} \nabla \, \mathrm{div}\, v &= \nabla_\Gamma \, \mathrm{div}_\Gamma \, v_\Gamma + \frac{\partial}{\partial n}\left(\mathrm{div}_\Gamma \, v_\Gamma\right) n + 2H \frac{\partial}{\partial n}(v \cdot n) n \\[2mm] &\quad + 2H \nabla_\Gamma(v \cdot n) + 2(v \cdot n)\nabla_\Gamma H + 2(v \cdot n)\frac{\partial}{\partial n} H \cdot n \\[2mm] &\quad + \nabla_\Gamma \frac{\partial}{\partial n}(v \cdot n) + \frac{\partial^2}{\partial n^2}(v \cdot n)n. \end{aligned} \right. \tag{2.5.224}$$

Similarly, from (2.5.149) and (2.5.157) we obtain

$$\mathrm{curl}\, v \wedge n = -\nabla_\Gamma(v \cdot n) + \mathcal{R}v + \frac{\partial}{\partial n} v_\Gamma \tag{2.5.225}$$

and besides using (2.5.197) :

$$\left\{ \begin{aligned} \mathrm{curl}\, \mathrm{curl}\, v &= \left(\mathrm{curl}_\Gamma \, \overrightarrow{\mathrm{curl}}_\Gamma (v \cdot n)\right) n + \mathrm{div}_\Gamma(\mathcal{R}v)n \\[2mm] &\quad + \mathrm{div}_\Gamma\left(\frac{\partial}{\partial n} v_\Gamma\right) n + \overrightarrow{\mathrm{curl}}_\Gamma \, \mathrm{curl}_\Gamma \, v_\Gamma \\[2mm] &\quad + \left[\mathcal{R} - 2H - \frac{\partial}{\partial n}\right]\left[-\nabla_\Gamma(v \cdot n) + \mathcal{R}v_\Gamma + \frac{\partial}{\partial n} v_\Gamma\right]. \end{aligned} \right. \tag{2.5.226}$$

Adding these two expressions, it follows that

$$\left\{ \begin{aligned} \Delta v &= \Delta_\Gamma v_\Gamma + 2H \frac{\partial}{\partial n} v_\Gamma + \frac{\partial^2}{\partial n^2} v_\Gamma \\[2mm] &\quad + \left[\Delta_\Gamma(v \cdot n) + 2H \frac{\partial}{\partial n}(v \cdot n) + \frac{\partial^2}{\partial n^2}(v \cdot n)\right] n \\[2mm] &\quad + \left[\frac{\partial}{\partial n}\left(\mathrm{div}_\Gamma \, v_\Gamma\right) - \mathrm{div}_\Gamma\left(\frac{\partial}{\partial n} v_\Gamma\right)\right] n \\[2mm] &\quad + \nabla_\Gamma \frac{\partial}{\partial n}(v \cdot n) - \frac{\partial}{\partial n}\nabla_\Gamma(v \cdot n) \\[2mm] &\quad + \frac{\partial}{\partial n}\left(\mathcal{R}v_\Gamma\right) - \mathcal{R}\frac{\partial}{\partial n} v_\Gamma \\[2mm] &\quad + \left[2\frac{\partial}{\partial n} H \cdot (v \cdot n) - \mathrm{div}_\Gamma(\mathcal{R}v)\right] n \\[2mm] &\quad + 2\nabla_\Gamma H(v \cdot n) + \mathcal{R}\nabla_\Gamma(v \cdot n) - (\mathcal{R} - 2HI)\mathcal{R}v. \end{aligned} \right. \tag{2.5.227}$$

In the above expression, two commutators appear that we need to compute to obtain the final expression of (2.5.213). The expressions of these two commutators are the object of the following lemma:

Lemma 2.5.10 *It holds that*

$$
\left\{
\begin{aligned}
&\frac{\partial}{\partial s}\left(\operatorname{div}_{\Gamma_s} v_{\Gamma_s}\right) - \operatorname{div}_{\Gamma_s}\left(\frac{\partial}{\partial s} v_{\Gamma_s}\right) \\
&= -\operatorname{div}_{\Gamma_s}\left(\mathcal{R}_s v\right) + 2\left(v \cdot \nabla_{\Gamma_s} H_s\right),
\end{aligned}
\right.
\tag{2.5.228}
$$

$$
\nabla_{\Gamma_s}\frac{\partial}{\partial s}(v \cdot n) - \frac{\partial}{\partial s}\nabla_{\Gamma_s}(v \cdot n) = \mathcal{R}_s \nabla_{\Gamma_s}(v \cdot n),
\tag{2.5.229}
$$

$$
\frac{\partial}{\partial s}\left(\mathcal{R}_s v_{\Gamma_s}\right) - \mathcal{R}_s\frac{\partial}{\partial s}v_{\Gamma_s} = \left(\frac{\partial}{\partial s}\mathcal{R}_s\right) v_{\Gamma_s}.
\tag{2.5.230}
$$

Proof

Let us first examine (2.5.228). From (2.5.210) follows

$$
\frac{\partial}{\partial s}\left(\operatorname{div}_{\Gamma_s} v_{\Gamma_s}\right) = \frac{\partial}{\partial s}\operatorname{div} v - \frac{\partial^2}{\partial s^2}(v \cdot n) - 2\frac{\partial}{\partial s}\left(H_s(v \cdot n)\right),
\tag{2.5.231}
$$

$$
\frac{\partial v_{\Gamma_s}}{\partial s} = \frac{\partial v}{\partial s} - \left(\frac{\partial v}{\partial s} \cdot n\right) n,
\tag{2.5.232}
$$

$$
\operatorname{div}_{\Gamma_s}\frac{\partial v_{\Gamma_s}}{\partial s} = \operatorname{div}\frac{\partial v}{\partial s} - 2H_s\left(\frac{\partial v}{\partial s} \cdot n\right) - \frac{\partial^2}{\partial s^2}(v \cdot n).
\tag{2.5.233}
$$

Subtracting, it follows that

$$
\left\{
\begin{aligned}
&\frac{\partial}{\partial s}\left(\operatorname{div}_{\Gamma_s} v_{\Gamma_s}\right) - \operatorname{div}_{\Gamma_s}\left(\frac{\partial}{\partial s} v_{\Gamma_s}\right) \\
&= \frac{\partial}{\partial s}\operatorname{div} v - \operatorname{div}\frac{\partial v}{\partial s} - 2(v \cdot n)\frac{\partial}{\partial s}H_s.
\end{aligned}
\right.
\tag{2.5.234}
$$

A new commutator appears which can be computed in a fixed system of coordinates (using here the Einstein convention!)

$$
\left\{
\begin{aligned}
\frac{\partial}{\partial n}\operatorname{div} v - \operatorname{div}\frac{\partial v}{\partial n} &= n_i\frac{\partial}{\partial x_i}\frac{\partial v_k}{\partial x_k} - \frac{\partial}{\partial x_i}\left(n_k\frac{\partial v_i}{\partial x_k}\right) \\
&= -\frac{\partial n_k}{\partial x_i}\frac{\partial v_i}{\partial x_k} \\
&= -\frac{\partial}{\partial x_k}\left(\frac{\partial n_k}{\partial x_i}v_i\right) + \frac{\partial}{\partial x_i}\left(\frac{\partial n_k}{\partial x_k}\right)v_i \\
&= -\operatorname{div}(\mathcal{R}v) + 2\left(\nabla_{\Gamma_s}H_s \cdot v\right) + 2(v \cdot n)\frac{\partial}{\partial n}H_s.
\end{aligned}
\right.
\tag{2.5.235}
$$

Let us examine (2.5.229). It holds that

$$
\begin{cases}
\nabla_{\Gamma_*}\dfrac{\partial}{\partial n}(v \cdot n) = \nabla\dfrac{\partial}{\partial n}(v \cdot n) - \dfrac{\partial^2}{\partial n^2}(v \cdot n)n \\[3mm]
\dfrac{\partial}{\partial n}\nabla_{\Gamma_*}(v \cdot n) = \dfrac{\partial}{\partial n}\nabla(v \cdot n) - \dfrac{\partial}{\partial n}((\nabla v \cdot n)n)
\end{cases}
\tag{2.5.236}
$$

from which follows

$$
\nabla_{\Gamma_*}\frac{\partial}{\partial n}(v \cdot n) - \frac{\partial}{\partial n}\nabla_{\Gamma_*}(v \cdot n) = \nabla\frac{\partial}{\partial n}(v \cdot n) - \frac{\partial}{\partial n}\nabla(v \cdot n).
\tag{2.5.237}
$$

This last commutator can be computed in a fixed system of coordinates and leads to (2.5.229). Similarly, from the previous results follows

$$
\frac{\partial}{\partial s}(\mathcal{R}_s v_{\Gamma_*}) - \mathcal{R}_s\frac{\partial}{\partial s}v_{\Gamma_*} = \frac{\partial}{\partial s}(\mathcal{R}_s v) - \mathcal{R}_s\frac{\partial}{\partial s}v
\tag{2.5.238}
$$

which is easily computed in a fixed system of coordinates.   ∎

2.5.7 Regularity properties

We prove some regularity results for solutions of the interior and exterior Dirichlet and Neumann problems. They are based on one side on the corresponding regularity properties of solutions in $\mathbb{R}^3$ given in Section 2.5.5, and on the other side on the use of localization and the tools of differential geometry that we introduced in the last section.

We consider the interior (resp. exterior) Dirichlet problem

$$
\begin{cases}
-\Delta u(x) = f(x), & x \in \Omega_i \text{ (resp. } x \in \Omega_e), \\[2mm]
u_{|\Gamma} = u_d,
\end{cases}
\tag{2.5.239}
$$

and the interior (resp. exterior) Neumann problem

$$
\begin{cases}
-\Delta u(x) = f(x), & x \in \Omega_i \text{ (resp. } x \in \Omega_e), \\[2mm]
\dfrac{\partial u}{\partial n}\Big|_{\Gamma} = u_n.
\end{cases}
\tag{2.5.240}
$$

We have introduced variational formulations which admit a unique solution in the Hilbert spaces $H^1(\Omega_i)$ for the interior problems and $W^{1,-1}(\Omega_e)$ also for the interior problems (and an extra condition linking f and u_n for the Neumann problem).

Theorem 2.5.21 *For any integer $k \geq 2$, the variational solutions of the interior (resp. exterior) Dirichlet problem (2.5.239) and the interior (resp. exterior) Neumann problem (2.5.240) are in the Hilbert space $H^k(\Omega_i)$ (resp. $W^{k,-1}(\Omega_e)$), when $u_d \in H^{k-1/2}(\Gamma)$ and $f \in H^{k-2}(\Omega_i)$ (resp. $W^{k-2,1}(\Omega_e)$) and when $u_n \in H^{k-3/2}(\Gamma)$. The corresponding mappings are continuous.*

Proof

Let χ be a regular positive cut-off function whose value is 1 on the surface Γ and whose support is contained in the tubular neighbourhood Γ_ε. The function $(1-\chi)u$ has a support strictly included in Ω_i (resp. Ω_e). It satisfies

$$-\Delta((1-\chi)u) = (1-\chi)f + 2(\nabla u \cdot \nabla\chi) + u\Delta\chi. \qquad (2.5.241)$$

Extending this function by zero, equation (2.5.241) holds in $\mathbb{R}^3$ and so the regularity of the solution is obtained upon applying Theorem 2.5.16. The support of the function χu is included in $\Omega_i \cap \Gamma_\varepsilon$ (resp. $\Omega_e \cap \Gamma_\varepsilon$). It satisfies

$$-\Delta(\chi u) = \chi f - 2(\nabla u \cdot \nabla\chi) - u\Delta\chi \qquad (2.5.242)$$

and the boundary condition associated with Dirichlet or with Neumann. The support of the right-hand side is also included in the tubular neighbourhood Γ_ε.

The two interior and exterior problems are now set up in a similar way and the regularity properties are related to the regularity of problem (2.5.242). We proceed by induction on the integer k, separately on $(1-\chi)u$ and χu. When we have proved the regularity H_{loc}^k of the function u, the equation (2.5.241) shows that $(1-\chi)u$ has one more index of regularity and belongs to $W^{k+1,-1}(\mathbb{R}^3)$. Using then equation (2.5.242), it follows that χu is in $H^{k+1}(\mathbb{R}^3)$. Thus u belongs to $H^{k+1}(\Omega_i)$ (resp. $W^{k+1,-1}(\Omega_e)$) and the induction can be continued.

So, we study problem (2.5.242) in the domain $\Omega_i \cap \Gamma_\varepsilon$ (resp. $\Omega_e \cap \Gamma_\varepsilon$). The lifting theorem (2.5.4) shows that we only have to study the associated homogeneous problem:

$$\left\{ \begin{array}{ll} -\Delta w = h, & h \in H^k(\Omega_e), \quad \text{supp}(h) \subset \Gamma_\varepsilon; \\[2mm] w_{|\Gamma} = 0, & \text{supp}(w) \subset \Gamma_\varepsilon. \end{array} \right. \qquad (2.5.243)$$

$$\left\{ \begin{array}{ll} -\Delta w = h, & h \in H^k(\Omega_e), \quad \text{supp}(h) \subset \Gamma_\varepsilon; \\[2mm] \dfrac{\partial w}{\partial n}_{|\Gamma} = 0, & \text{supp}(w) \subset \Gamma_\varepsilon. \end{array} \right. \qquad (2.5.244)$$

The proof of Theorem 2.5.21 is equivalent to that of

Theorem 2.5.22 *For any integer $k \geq 2$, the variational solution of the interior and exterior Dirichlet and Neumann problems (2.5.243) and (2.5.244), are in the space $H^k(\Omega_e \cap \Gamma_\varepsilon)$ when $h \in H^{k-2}(\Omega_e \cap \Gamma_\varepsilon)$.*

Proof

The general structure of the proof consists in obtaining an equation for the gradient of the function w, just deriving the equation. We then use the expression of the vectorial Laplacian (2.5.213) given in Theorem 2.5.20, looking separately at the tangential derivatives and at the normal derivative. In each case, we derive boundary conditions for these derivatives. From the variational formulation, we obtain estimates which increase by one the regularity. We then use an induction to end the proof.

Let us consider first the **Dirichlet problem** with a right-hand side $h \in L^2\left(\Omega_e \cap \Gamma_\varepsilon\right)$, i.e., $k = 2$. We show the regularity of the **tangential derivatives**. First introduce a regular positive function ρ, whose support is contained in the unit ball and whose mean value is 1. Consider the associated **regularization kernel**

$$\rho_\eta(x) = \frac{1}{\eta^3} \rho\left(\frac{x}{\eta}\right), \quad \eta > 0. \tag{2.5.245}$$

The convolute function w $(w = 0$ in $\Omega_i)$

$$w_\eta(y) = \int_{\mathbb{R}^3} \rho_\eta(y - x) w(x - \eta n(x)) dx \tag{2.5.246}$$

vanishes on the surface Γ and has the same regularity as ρ. Besides, it holds that

$$\begin{cases} -\Delta w_\eta = h_\eta, \\ w_{\eta|\Gamma} = 0, \end{cases} \tag{2.5.247}$$

where h_η is the convolute of the function h using (2.5.246). It holds that

$$-\Delta \nabla_{\Gamma_s} w_\eta = -\nabla \Delta w_\eta + \Delta\left(\frac{\partial w_\eta}{\partial n} n\right) \tag{2.5.248}$$

and multiplying by $\nabla_{\Gamma_s} w_\eta$, then integrating by parts

$$\begin{cases} \int_{\Omega_e} |\nabla \nabla_{\Gamma_s} w_\eta|^2 dx = \int_{\Omega_e} (\nabla_{\Gamma_s} h_\eta \cdot \nabla_{\Gamma_s} w_\eta) dx \\ \qquad + \int_{\Omega_e} \left(\Delta\left(\frac{\partial w_\eta}{\partial n} n\right)\right) \cdot \nabla_{\Gamma_s} w_\eta\right) dx. \end{cases} \tag{2.5.249}$$

The above identities result from the properties of the tangential differential operators and the theorem (2.5.20) (the operator ∇n acts in the tangent plane)

$$\begin{cases} \Delta\left(\frac{\partial w_\eta}{\partial n} n\right) = \Delta\left(\frac{\partial w_\eta}{\partial n}\right) n + 2\left(\nabla\frac{\partial w_\eta}{\partial n} \cdot \nabla n\right) + \frac{\partial w_\eta}{\partial n}\Delta n, \\ \left(\Delta\left(\frac{\partial w_\eta}{\partial n} n\right)\right) \cdot \nabla_{\Gamma_s} w_\eta = 2\left(\nabla_{\Gamma_s}\frac{\partial w_\eta}{\partial n} \cdot \nabla n \nabla_{\Gamma_s} w_\eta\right) \\ \qquad + \left(\frac{\partial w_\eta}{\partial n}\Delta n \cdot \nabla_{\Gamma_s} w_\eta\right). \end{cases} \tag{2.5.250}$$

It follows from Stokes formula (2.5.193) on the surface Γ_s, and a limit as η tends to zero that

$$
\begin{cases}
\displaystyle\int_{\Omega_e} |\nabla\nabla_{\Gamma_s} w|^2\, dx = -\int_{\Omega_e} h\Delta_{\Gamma_s} w\, dx \\[3mm]
\displaystyle\qquad\qquad - 2\int_{\Omega_e} \frac{\partial w}{\partial n}\, \mathrm{div}_{\Gamma_s}\, (\nabla n\nabla_{\Gamma_s} w)\, dx \qquad (2.5.251) \\[3mm]
\displaystyle\qquad\qquad + \int_{\Omega_e} \frac{\partial w}{\partial n}\, (\Delta n \cdot \nabla_{\Gamma_s} w)\, dx.
\end{cases}
$$

This identity implies that $\nabla_{\Gamma_s} w$ is in $H^1(\Omega_e \cap \Gamma_\varepsilon)$.

To continue the proof, we need a similar estimate on the **normal derivative** $\partial w / \partial n$. We will omit to mention at each step the use of regularization, although it is technically necessary. We write directly an equation for $\partial w / \partial n$ in order to simplify the presentation.

Differentiating, we get

$$
\nabla\frac{\partial w}{\partial n} = \nabla(\nabla w \cdot n) = \frac{\partial}{\partial n}\nabla w + \nabla n\nabla w, \qquad (2.5.252)
$$

$$
\Delta\frac{\partial w}{\partial n} = \Delta(\nabla w \cdot n) = \frac{\partial}{\partial n}\Delta w + 2\,\mathrm{div}(\nabla n\nabla w) - (\nabla w \cdot \Delta n). \qquad (2.5.253)
$$

We now use the equalities

$$
\begin{cases}
\Delta n = \nabla\,\mathrm{div}\, n, \\[2mm]
\nabla_\Gamma w_{|\Gamma} = 0,
\end{cases}
\qquad (2.5.254)
$$

which we multiply by φ and integrate by parts to obtain

$$
\begin{cases}
\displaystyle\int_{\Omega_e} \left(\nabla\frac{\partial w}{\partial n} \cdot \nabla\varphi\right) dx = -\int_\Gamma \frac{\partial^2 w}{\partial n^2}\varphi\, d\gamma - \int_{\Omega_e} \frac{\partial}{\partial n}(\Delta w)\varphi\, dx \\[3mm]
\displaystyle\qquad + 2\int_{\Omega_e} (\nabla n\nabla w \cdot \nabla\varphi)\, dx + \int_{\Omega_e} (\nabla\,\mathrm{div}\, n \cdot \nabla w)\varphi\, dx.
\end{cases}
\qquad (2.5.255)
$$

When looking at this equality, we see that we have good estimates on all terms in the right-hand side except for $\partial^2 w / \partial n^2$. It follows from relation (2.5.212) that this term can be written (using $w_{|\Gamma} = 0$) in the form

$$
\int_\Gamma \frac{\partial^2 w}{\partial n^2}\varphi\, dx = \int_\Gamma \Delta w\varphi\, d\gamma - 2\int_\Gamma H\frac{\partial w}{\partial n}\varphi\, d\gamma. \qquad (2.5.256)
$$

Besides,

$$
\mathrm{div}(\Delta w\varphi n) = \frac{\partial}{\partial n}(\Delta w)\varphi + \Delta w\frac{\partial\varphi}{\partial n} + \Delta w\varphi\,\mathrm{div}\, n \qquad (2.5.257)
$$

which leads to

$$
\begin{cases}
\displaystyle \int_\Gamma \Delta w \varphi \, d\gamma + \int_{\Omega_\varepsilon} \frac{\partial}{\partial n}(\Delta w)\varphi \, dx \\
\displaystyle \qquad = -\int_{\Omega_\varepsilon} \Delta w \frac{\partial \varphi}{\partial n} dx - \int_{\Omega_\varepsilon} \Delta w \varphi \operatorname{div} n \, dx.
\end{cases}
\tag{2.5.258}
$$

There remains one term on Γ which hopefully vanishes due to

$$
\operatorname{div}(H \varphi \nabla w) = H \varphi \Delta w + (\nabla H \cdot \nabla w)\varphi + H(\nabla \varphi \cdot \nabla w),
\tag{2.5.259}
$$

which yields

$$
\begin{cases}
\displaystyle \int_\Gamma H \varphi \frac{\partial w}{\partial n} d\gamma \\
\displaystyle \qquad = -\int_{\Omega_\varepsilon} [H\varphi \Delta w + (\nabla H \cdot \nabla w)\varphi + H(\nabla \varphi \cdot \nabla w)] \, dx.
\end{cases}
\tag{2.5.260}
$$

Adding all the terms, it follows that

$$
\begin{cases}
\displaystyle \int_{\Omega_\varepsilon} \left(\nabla \frac{\partial w}{\partial n} \cdot \nabla \varphi \right) dx \\
\displaystyle \qquad = \int_{\Omega_\varepsilon} \Delta w \frac{\partial \varphi}{\partial n} dx + 2 \int_{\Omega_\varepsilon} ((\mathcal{R} - HI)\nabla w \cdot \nabla \varphi) dx.
\end{cases}
\tag{2.5.261}
$$

Choosing $\varphi = \partial w / \partial n$ in this equality yields an estimate which shows that $\partial w / \partial n$ is in the space $H_{\text{loc}}^1(\Omega_e \cap \Gamma_\varepsilon)$. It follows from the trace theorem that $\partial w / \partial n \in H^{1/2}(\Gamma)$.

Consider now the **Neumann problem** when $h \in L^2(\Omega_e)$ i.e., $k = 2$. In order to simplify the presentation, we omit here also to mention the use of regularization. It follows from (2.5.253) that the **normal derivative** satisfies

$$
\begin{cases}
\displaystyle \Delta \frac{\partial w}{\partial n} = \frac{\partial}{\partial n} h + 2 \operatorname{div}(\nabla n \cdot \nabla w) - (\nabla w \cdot \Delta n), \\
\displaystyle \frac{\partial w}{\partial n}\Big|_\Gamma = 0.
\end{cases}
\tag{2.5.262}
$$

It solves a Dirichlet problem, which shows that $\partial w / \partial n$ is in $H^1(\Omega_e \cap \Gamma_\varepsilon)$.

The **tangential derivatives** satisfy the equation (2.5.248), or

$$
-\Delta \nabla_{\Gamma_*} w = -\nabla h + \Delta \left(\frac{\partial w}{\partial n} n \right),
\tag{2.5.263}
$$

and from (2.5.229) they admit the boundary condition

$$
\frac{\partial}{\partial n} \nabla_\Gamma w = -\mathcal{R} \nabla w.
\tag{2.5.264}
$$

We multiply (2.5.263) by $\nabla_{\Gamma_s} w$, then integrate by parts and use (2.5.250) and the Stokes formula to obtain the equality

$$
\begin{cases}
\displaystyle \int_{\Omega_e} |\nabla \nabla_{\Gamma_s} w|^2 \, dx = \int_{\Omega_e} h \Delta_{\Gamma_s} w \, dx \\[2ex]
\qquad \displaystyle -2 \int_{\Omega_e} \frac{\partial w}{\partial n} \operatorname{div}_{\Gamma_s} (\nabla n \nabla_{\Gamma_s} w) \, dx \\[2ex]
\qquad \displaystyle + \int_{\Omega_e} \frac{\partial w}{\partial n} (\Delta n \cdot \nabla_{\Gamma_s} w) \, dx + \int_{\Gamma} (\mathcal{R} \nabla_{\Gamma} w \cdot \nabla_{\Gamma} w) \, d\gamma.
\end{cases}
\tag{2.5.265}
$$

The only term in the right-hand side not already known to be bounded is the last one. It can be written in the form

$$
\int_{\Gamma} (\mathcal{R} \nabla_{\Gamma} w \cdot \nabla_{\Gamma} w) \, d\gamma = - \int_{\Gamma} \int_0^\varepsilon \frac{\partial}{\partial s} (\mathcal{R}_s \nabla_{\Gamma_s} w \cdot \nabla_{\Gamma_s} w) \, d\gamma ds \tag{2.5.266}
$$

which leads to the expected estimate. Expression (2.5.265) shows that $\nabla_{\Gamma_s} w$ is in the space $H^1 (\Omega_e \cap \Gamma_\varepsilon)$.

The **general case** $(k > 2)$ is now proved by induction. We know that the Dirichlet problem is such that the tangential derivatives solve a Laplace equation with Dirichlet boundary data, while the normal derivative solves a Laplace equation with Neumann boundary data. The regularity H^k for the Neumann problem implies the regularity H^{k+1} of the normal derivative of the Dirichlet problem, which implies then, looking at the right-hand side of (2.5.251), the regularity H^{k+1} of the tangential derivatives.

Similarly, the normal derivative of the Neumann problem satisfies a Dirichlet problem. The regularity H^k for the Dirichlet problem implies the regularity H^{k+1} of the normal derivative of the Neumann problem. The tangential derivatives satisfy a Neumann problem whose right-hand side given in (2.5.265), has the expected regularity. ∎

Remarks

- Theorem 2.5.16 can be extended without essential modification to the case of domains in $\mathbb{R}^2$. We only need to replace the spaces $W^{k,-1}$ by the "ad hoc" spaces X^k.

- The expressions that appear in the previous proof show the necessary regularity of the surface Γ. We use the curvature tensor to obtain the regularity H^2 in the normal derivative expression and its derivatives in the expression of the tangential derivatives. It follows that we need a surface with regularity C^3 (or at least $W^{3,\infty}$). In the general case $(k > 2)$, we need a surface with regularity C^{k+1}.

2.6   The Helmholtz Equation in $\mathbb{R}^3$

We have seen in Section 2.2 that the interior and exterior Helmholtz problems are of quite different nature. In the case of an **interior problem** we can restrict our attention to real solutions without lost of generality. The two model equations are the **Dirichlet and the Neumann problems**

$$\begin{cases} \Delta u + k^2 u = 0, & x \in \Omega_i, \\ u_{|\Gamma} = u_d. \end{cases} \tag{2.6.1}$$

$$\begin{cases} \Delta u + k^2 u = 0, & x \in \Omega_i, \\ \left. \dfrac{\partial u}{\partial n} \right|_\Gamma = u_n. \end{cases} \tag{2.6.2}$$

Fredholm's alternative shows that they have a unique solution except when k^2 is an eigenvalue of the associated operator for a Dirichlet or for a Neumann boundary condition. The regularity properties are analoguous to that of the Laplace equation.

In the case of **exterior problems**, we seek complex-valued solutions. We need to specify the behavior at infinity and in particular to impose the radiation condition.

The two model equations are the Dirichlet and the Neumann problems

$$\begin{cases} \Delta u + k^2 u = 0, & x \in \Omega_e, \\ u_{|\Gamma} = u_d. \end{cases} \tag{2.6.3}$$

$$\begin{cases} \Delta u + k^2 u = 0, & x \in \Omega_e, \\ \left. \dfrac{\partial u}{\partial n} \right|_\Gamma = u_n. \end{cases} \tag{2.6.4}$$

In both cases, the conditions at infinity are (c is a fixed constant)

$$\begin{cases} |u| \leq \frac{c}{r}, & r \text{ large}, \\ |\nabla u| \leq \frac{c}{r}, \end{cases} \tag{2.6.5}$$

and the Sommerfeld condition

$$\left| \frac{\partial u}{\partial r} - iku \right| \leq \frac{c}{r^2}. \tag{2.6.6}$$

Notice that this last condition implies that the solutions are complex-valued, although k^2 is real. The associated operator is not symmetric.

We study now, with care, the case where the domain is the interior or the exterior of the unit sphere.

2.6.1 The spherical Bessel functions

In the case of a sphere, exactly as for the Laplace equation, we can solve quite explicitly the interior and exterior Helmholtz problems. The solution is expressed as a sum of separated variable solutions in the variable r on one side, and (θ, φ) on the other side. We seek solutions in the form

$$v_\ell^m(r, \theta, \varphi) = h_\ell(kr) Y_\ell^m(\theta, \varphi) \tag{2.6.7}$$

where $Y_\ell^m(\theta, \varphi)$ is a spherical harmonic function. We use the expression of the Laplacian in spherical coordinates (identity (2.4.2)) and the properties of the spherical harmonics to obtain an equation for the function $h_\ell(r)$ which is independent of the index m. It is the **spherical Bessel equation**

$$\frac{1}{r^2}\frac{d}{dr}r^2\frac{dh_\ell(r)}{dr} + \left(1 - \frac{\ell(\ell+1)}{r^2}\right)h_\ell(r) = 0, \quad r > 0. \tag{2.6.8}$$

The spherical Bessel equation takes the form

$$\frac{d^2}{dr^2}h_\ell(r) + \frac{2}{r}\frac{d}{dr}h_\ell(r) + \left(1 - \frac{\ell(\ell+1)}{r^2}\right)h_\ell(r) = 0, \quad r > 0. \tag{2.6.9}$$

It is a second-order homogeneous equation. The coefficients are singular at $r = 0$, and at infinity. It admits two families of solutions defined up to a multiplicative factor.

Theorem 2.6.1 *The spherical Bessel equation (2.6.8) admits two families of solutions, called **spherical Hankel functions** or also **half-integer Bessel functions**, which satisfy the recursion formulas*

$$\frac{d}{dr}h_\ell = \frac{\ell}{r}h_\ell - h_{\ell+1} = -\frac{\ell+1}{r}h_\ell + h_{\ell-1}, \tag{2.6.10}$$

$$h_{\ell+1} + h_{\ell-1} = \frac{2\ell+1}{r}h_\ell. \tag{2.6.11}$$

They are given by the expressions

$$h_\ell^{(1)}(r) = (-r)^\ell \left(\frac{1}{r}\frac{d}{dr}\right)^\ell \left(\frac{e^{ir}}{r}\right), \tag{2.6.12}$$

$$h_\ell^{(2)}(r) = (-r)^\ell \left(\frac{1}{r}\frac{d}{dr}\right)^\ell \left(\frac{e^{-ir}}{r}\right); \tag{2.6.13}$$

or else

$$\begin{cases} h_\ell^{(1)}(r) = (-i)^\ell \frac{e^{ir}}{r}\left(\beta_0^\ell + i\beta_1^\ell \frac{1}{r} + \cdots \right. \\ \\ \left. + (i)^m \beta_m^\ell \left(\frac{1}{r}\right)^m + \cdots + (i)^\ell \beta_\ell^\ell \left(\frac{1}{r}\right)^\ell\right); \end{cases} \tag{2.6.14}$$

$$h_\ell^{(2)}(r) = \overline{h}_\ell^{(1)}(r); \tag{2.6.15}$$

$$\beta_m^\ell = \frac{(m+\ell)!}{m!(\ell-m)!\,2^m}. \tag{2.6.16}$$

The function

$$z_\ell(r) = r \frac{\dfrac{d}{dr} h_\ell^{(1)}(r)}{h_\ell^{(1)}(r)} \tag{2.6.17}$$

satisfies the recursion formula

$$(z_{\ell-1} - (\ell - 1))(z_\ell + \ell + 1) = -r^2. \tag{2.6.18}$$

It is given by the expression

$$z_\ell(r) = -\frac{p_\ell}{q_\ell} + i\frac{r}{q_\ell} \tag{2.6.19}$$

where p_ℓ and q_ℓ are polynomials of degré ℓ in the variable $\frac{1}{r^2}$ and

$$q_\ell = 1 + \alpha_1^\ell \frac{1}{r^2} + \cdots + \alpha_\ell^\ell \frac{1}{r^{2\ell}} = r^2 \left| h_\ell^{(1)}(r) \right|^2, \tag{2.6.20}$$

$$p_\ell = 1 + 2\alpha_1^\ell \frac{1}{r^2} + \cdots + (\ell + 1)\alpha_\ell^\ell \frac{1}{r^{2\ell}}, \tag{2.6.21}$$

$$\alpha_m^\ell = \beta_m^\ell \beta_m^m. \tag{2.6.22}$$

Besides,

$$1 \le -\Re(z_\ell(r)) = \frac{p_\ell}{q_\ell} \le \ell + 1, \tag{2.6.23}$$

$$0 \le \Im z_\ell(r) \le r. \tag{2.6.24}$$

Proof

We can check that $(dh_\ell/dr) - (\ell/r)h_\ell$ satisfies the Bessel equation with index $\ell + 1$, which shows the recursion formula (2.6.10), except for the choice of a multiplicative constant. The choice of this constant is linked to the recursion formula in the following way: $h_{\ell+1}$ is defined through the recursion $h_{\ell+1} = (\ell/r)h_\ell - dh_\ell/dr$ and we check that with this choice, h_ℓ satisfies the other recursion formula $h_{\ell-1} = ((\ell + 1)/r)h_\ell + dh_\ell/dr$. Adding the two expressions from (2.6.10) gives (2.6.11).

When $\ell = 0$, the equation simplifies and its solutions are

$$\begin{cases} h_0^{(1)}(r) = \dfrac{e^{ir}}{r}, \\[2mm] h_0^{(2)}(r) = \dfrac{e^{-ir}}{r}. \end{cases} \tag{2.6.25}$$

The recursion formula (2.6.10) can be written in the form

$$h_{\ell+1}(r) = -r^{\ell+1} \frac{1}{r} \frac{d}{dr} \left(\frac{1}{r^\ell} h_\ell(r) \right) \tag{2.6.26}$$

from which we infer (2.6.12) and (2.6.13).

Starting from the expression (2.6.14), the recursion formula (2.6.10) yields

$$\beta_m^{\ell+1} = \beta_m^\ell + (m + \ell)\beta_{m-1}^\ell \qquad (2.6.27)$$

with

$$\beta_0^0 = 1. \qquad (2.6.28)$$

We can check that this double recursion formula yields (2.6.16).

The recursion formula (2.6.18) is a consequence of (2.6.10) dividing by h_ℓ and multiplying the terms corresponding to indexes ℓ and $\ell - 1$. The recursion formula (2.6.11) implies

$$h_{\ell+1}\,\overline{h}_\ell + h_{\ell-1}\overline{h}_\ell = \frac{2\ell + 1}{r}\,h_\ell\overline{h}_\ell, \qquad (2.6.29)$$

and thus

$$\Im\left(rh_{\ell+1}\overline{h}_\ell\right) = \Im\left(rh_1\overline{h}_0\right) = -\frac{1}{r}. \qquad (2.6.30)$$

Moreover,

$$\Re\left(\frac{r\dfrac{d}{dr}h_\ell}{h_\ell}\right) = \frac{r}{2}\frac{\dfrac{d}{dr}\left(h_\ell\overline{h}_\ell\right)}{h_\ell\overline{h}_\ell}. \qquad (2.6.31)$$

From (2.6.30) and (2.6.10), it follows that

$$\Im\left(\frac{r\dfrac{d}{dr}h_\ell}{h_\ell}\right) = \frac{1}{rh_\ell\overline{h}_\ell}. \qquad (2.6.32)$$

We see that all the quantities depend on the squared modulus of h_ℓ. It follows from (2.6.14) that it is a polynomial in the variable $1/r^2$ which takes the form

$$h_\ell\overline{h}_\ell(r) = \frac{1}{r^2}\left(\alpha_0^\ell + \alpha_1^\ell\frac{1}{r^2} + \cdots + \alpha_\ell^\ell\frac{1}{r^{2\ell}}\right). \qquad (2.6.33)$$

The recursion (2.6.14) does not give anything simple.

We introduce

$$q_\ell(r) = \alpha_0^\ell + \alpha_1^\ell\frac{1}{r^2} + \cdots + \alpha_\ell^\ell\frac{1}{r^{2\ell}}, \qquad (2.6.34)$$

$$p_\ell(r) = \alpha_0^\ell + 2\alpha_1^\ell\frac{1}{r^2} + \cdots + (\ell + 1)\alpha_\ell^\ell\frac{1}{r^{2\ell}}, \qquad (2.6.35)$$

which, from (2.6.31) and (2.6.34), satisfy

$$z_\ell(r) = -\frac{p_\ell(r)}{q_\ell(r)} + \frac{ir}{q_\ell(r)}. \qquad (2.6.36)$$

It follows from the recursion relation (2.6.18) that

$$q_\ell q_{\ell+1} = 1 + \frac{1}{r^2}(p_\ell + \ell q_\ell)^2, \tag{2.6.37}$$

$$p_{\ell+1} = (\ell+2)q_{\ell+1} - (p_\ell + \ell q_\ell). \tag{2.6.38}$$

This last recursion is linear and thus is equivalent to a linear recursion formula on the coefficients α_m^ℓ which is

$$(\ell - m)\alpha_m^\ell = (\ell + m)\alpha_m^{\ell-1}. \tag{2.6.39}$$

Using the equality

$$\alpha_m^m = (\beta_m^m)^2 \tag{2.6.40}$$

the recursion (2.6.39) then leads to (2.6.22).

The inequalities (2.6.23) and (2.6.24) are then obvious from the expression of p and q. ∎

Notice that the imaginary part of h_0 is the function

$$j_0 = \frac{\sin r}{r} \tag{2.6.41}$$

which is analytical and so has no singularity in a neighbourhood of the origin. It is the spherical Bessel function of order zero. We introduce the usual **spherical Bessel functions**

$$j_\ell(r) = (-r)^\ell \left(\frac{1}{r}\frac{d}{dr}\right)^\ell \left(\frac{\sin r}{r}\right). \tag{2.6.42}$$

They are analytical in a neighbourhood of the origin since the series expansion of j_0 has only even terms.

The above expressions show that the function h_ℓ behaves at infinity as

$$h_\ell(r) \sim (-i)^\ell \frac{e^{ir}}{r}, \quad r \text{ large}, \tag{2.6.43}$$

and at the origin as

$$h_\ell(r) \sim \frac{(2\ell)!}{\ell\,!2^\ell}\frac{1}{r^{\ell+1}}, \quad r \text{ small}. \tag{2.6.44}$$

It is clear in view of the expression (2.6.20) that the modulus $|h_\ell|$ is a decreasing function of r and so is $r|h_\ell|$, while $r^{\ell+1}|h_\ell|$ is an increasing function. The following lemma gives a more precise behavior.

Lemma 2.6.1 *The function* $r^{\ell+1}|h_\ell(r)|\left(r^2 + 2\ell(\ell - 1/2)\right)^{-\ell/2}$ *increases from* γ_ℓ *to* 1,

$$\gamma_\ell = \frac{2\ell!}{\ell\,!2^\ell}\frac{1}{[2\ell(\ell - 1/2)]^{\ell/2}}. \tag{2.6.45}$$

The function $r^{\ell+1} |h_\ell(r)| \left(r^2 + \frac{\ell+1}{2}\right)^{-\ell/2}$ *decreases from* δ_ℓ *to 1,*

$$\delta_\ell = \frac{2\ell!}{\ell!2^\ell} \frac{2^{\ell/2}}{(\ell+1)^{\ell/2}}. \tag{2.6.46}$$

Proof

In view of the expression (2.6.20) of q_ℓ,

$$r^{\ell+1} |h_\ell(r)| \left(r^2 + \gamma\right)^{-\ell/2} = \sqrt{\frac{\alpha_\ell^\ell + \cdots + \alpha_1^\ell r^{2\ell-2} + r^{2\ell}}{(r^2 + \gamma)^\ell}}. \tag{2.6.47}$$

We only need to show that the rational function under the square symbol in (2.6.47) is increasing or decreasing depending on the choice of γ. Its logarithmic derivative is

$$\frac{2\alpha_{\ell-1}^\ell r + \cdots + (2\ell-2)\alpha_1^\ell r^{2\ell-3} + 2\ell r^{2\ell-1}}{\alpha_\ell^\ell + \cdots + \cdots + r^{2\ell}} - \frac{2\ell r}{r^2 + \gamma} \tag{2.6.48}$$

where the numerator has the same sign as

$$\left\{ \begin{aligned} & \left(\alpha_{\ell-1}^\ell + \cdots + (\ell-1)\alpha_1^\ell r^{2\ell-4} + \ell r^{2\ell-2}\right)\left(r^2 + \gamma\right) \\ & \quad -\ell\left(\alpha_\ell^\ell + \cdots + \alpha_1^\ell r^{2\ell-2} + r^{2\ell}\right) \\ & = \gamma\alpha_{\ell-1}^\ell - \ell\alpha_\ell^\ell + \left(\alpha_{\ell-1}^\ell + 2\gamma\alpha_{\ell-2}^\ell - \ell\alpha_{\ell-1}^\ell\right)r^2 + \cdots \\ & \quad + \left[(\ell-m)\alpha_m^\ell + (\ell+1-m)\gamma\alpha_{m-1}^\ell - \ell\alpha_m^\ell\right]r^{2\ell-2m} \\ & \quad + \cdots + \left[(\ell-1)\alpha_1^\ell + \gamma\ell - \ell\alpha_1^\ell\right]r^{2\ell-2}. \end{aligned} \right. \tag{2.6.49}$$

This sign depends only on the quantity $(1 \leq m \leq \ell)$

$$\frac{m\alpha_m^\ell}{(\ell+1-m)\alpha_{m-1}^\ell} - \gamma = (m+\ell)(m-1/2) - \gamma. \tag{2.6.50}$$

The bounds are realized respectively when $m = 1$ or when $m = \ell$ which gives the values of γ : $\gamma = (\ell+1)/2$ and $\gamma = 2\ell(\ell - 1/2)$.   ∎

2.6.2 Dirichlet and Neumann problems for a sphere

In order to exhibit the solutions of the interior and exterior problems, we expand u_d and u_n in spherical harmonics. It holds that

$$\left\{ \begin{aligned} & u_d(\theta, \varphi) = \sum_{\ell=0}^\infty \sum_{m=-\ell}^\ell u_\ell^m Y_\ell^m(\theta, \varphi), \\ & u_\ell^m = \int_S u_d(\theta, \varphi)\overline{Y_\ell^m}(\theta, \varphi)d\sigma, \end{aligned} \right. \tag{2.6.51}$$

$$\begin{cases} u_n(\theta, \varphi) = \sum_{\ell=0}^{\infty} \sum_{m=-\ell}^{\ell} v_\ell^m Y_\ell^m(\theta, \varphi), \\ v_\ell^m = \int_S u_n(\theta, \varphi) \overline{Y_\ell^m}(\theta, \varphi) d\sigma. \end{cases} \qquad (2.6.52)$$

The solution of the **interior Dirichlet problem** (2.6.1) is given by

$$u(r, \theta, \varphi) = \sum_{\ell=0}^{\infty} \sum_{m=-\ell}^{\ell} u_\ell^m \frac{j_\ell(kr)}{j_\ell(k)} Y_\ell^m(\theta, \varphi). \qquad (2.6.53)$$

We see that this expression has no meaning when k is a zero of one of the spherical Bessel functions, except when the corresponding coefficient of the data is also zero. This is typical of the Fredholm alternative, and the set of all corresponding values of $-k^2$ is exactly the set of eigenvalues of the Laplacian on the unit sphere with a Dirichlet boundary condition. The associated eigenfunctions are $j_\ell(kr) Y_\ell^m(\theta, \varphi), m = -\ell$ to ℓ, and the multiplicity is $2\ell + 1$.

The solution of the **interior Neumann problem** (2.6.2) is given by

$$u(r, \theta, \varphi) = \sum_{\ell=0}^{\infty} \sum_{m=-\ell}^{\ell} v_\ell^m \frac{j_\ell(kr)}{k\frac{d}{dr}j_\ell(k)} Y_\ell^m(\theta, \varphi). \qquad (2.6.54)$$

Similarly, this expression has no meaning when k is a zero of the derivative of one of the spherical Bessel functions, except when the corresponding coefficient of the data is also zero. Through the change of k in k^2, the set of zeros of the functions dj_ℓ/dr is exactly the set of eigenvalues of the Laplacian in the unit sphere with a Neumann boundary condition. The associated eigenspace is spanned by the functions $j_\ell(kr) Y_\ell^m(\theta, \varphi)$ and the multiplicity is $2\ell + 1$.

When k^2 is not an eigenvalue, these interior and exterior problems have a unique solution, whose regularity is exactly the same as the one of the associated Laplace problem. The term $k^2 u$ is two orders of derivation less than the main term and changes only the constants in the regularity estimates. This is also true for the exterior problems when we are interested only in the local regularity for a bounded value of k.

The solution of the **exterior Dirichlet problem** (2.6.3) is given by

$$u(r, \theta, \varphi) = \sum_{\ell=0}^{\infty} \sum_{m=-\ell}^{\ell} u_\ell^m \frac{h_\ell^{(1)}(kr)}{h_\ell^{(1)}(k)} Y_\ell^m(\theta, \varphi). \qquad (2.6.55)$$

The solution of the **exterior Neumann problem** (2.6.4) is given by

$$u(r, \theta, \varphi) = \sum_{\ell=0}^{\infty} \sum_{m=-\ell}^{\ell} v_\ell^m \frac{h_\ell^{(1)}(kr)}{k\frac{d}{dr}h_\ell^{(1)}(k)} Y_\ell^m(\theta, \varphi). \qquad (2.6.56)$$

The following theorem exhibits some continuity bounds of these solutions in Sobolev spaces:

Theorem 2.6.2 *The solution of the exterior Dirichlet problem given by (2.6.55), satisfies the estimates*

$$
\left\{
\begin{aligned}
\left\|\frac{u}{r}\right\|^2_{L^2(B_c)} &= \sum_{\ell=0}^{\infty} \sum_{m=-\ell}^{\ell} \gamma(k,\ell) \, |u_\ell^m|^2 \\
&\leq \|u_d\|^2_{H^{-1/2}(S)} + 3k^2 \|u_d\|^2_{H^{-1}(S)} ;
\end{aligned}
\right.
\tag{2.6.57}
$$

$$
\gamma(k,\ell) = \frac{1 + \cdots + \dfrac{1}{2m+1}\alpha_m^\ell \dfrac{1}{k^{2m}} + \cdots + \dfrac{1}{2\ell+1}\alpha_\ell^\ell \dfrac{1}{k^{2\ell}}}{1 + \cdots + \alpha_m^\ell \dfrac{1}{k^{2m}} + \cdots + \alpha_\ell^\ell \dfrac{1}{k^{2\ell}}} ;
\tag{2.6.58}
$$

$$
\gamma(k,\ell) \leq \frac{k^2 + \dfrac{\ell+1}{6}}{k^2 + \dfrac{(2\ell+1)(\ell+1)}{6}} .
\tag{2.6.59}
$$

Let $\nabla_T u$ be the tangential derivative on the sphere of radius r:

$$
\left\{
\begin{aligned}
\|\nabla_T u\|^2_{L^2(B_c)} &= \sum_{\ell=0}^{\infty} \sum_{m=-\ell}^{\ell} \ell(\ell+1)\gamma(k,\ell) \, |u_\ell^m|^2 \\
&\leq \frac{1}{2}\|u_d\|^2_{H^{1/2}(S)} + 3k^2 \|u_d\|^2_{L^2(S)} ;
\end{aligned}
\right.
\tag{2.6.60}
$$

$$
\left\{
\begin{aligned}
\left\|\frac{1}{r}\frac{\partial u}{\partial r}\right\|^2_{L^2(B_c)} &\leq \sum_{\ell=0}^{\infty} \sum_{m=-\ell}^{\ell} k^2 \delta(k,\ell) \, |u_\ell^m|^2 \\
&\leq \frac{1}{2}\|u_d\|^2_{H^{1/2}(S)} + k^2 \|u_d\|^2_{L^2(S)} ;
\end{aligned}
\right.
\tag{2.6.61}
$$

$$
\left\{
\begin{aligned}
\delta(k,\ell) &\leq \frac{k^2 + \frac{\ell+1}{6}}{k^2 + \frac{(2\ell+1)(\ell+1)}{6}} + \frac{(\ell+1)^2}{3k^2}\frac{k^2 + 3\frac{\ell+1}{10}}{k^2 + \frac{(2\ell+3)(\ell+1)}{10}} \\
&\leq 1 + \frac{\ell+1}{2k^2} ;
\end{aligned}
\right.
\tag{2.6.62}
$$

$$
\left\{
\begin{aligned}
\left\|\frac{\partial u}{\partial r} - iku\right\|^2_{L^2(B_c)} &\leq \sum_{\ell=0}^{\infty} \sum_{m=-\ell}^{\ell} 2(\ell+1)^2\gamma(k,\ell) \, |u_\ell^m|^2 \\
&\leq \|u_d\|^2_{H^{1/2}(S)} + 6k^2 \|u_d\|^2_{L^2(S)} .
\end{aligned}
\right.
\tag{2.6.63}
$$

Proof

It follows from the orthogonality of the spherical harmonics that

$$\int_{B_r} \frac{|u(x)|^2}{r^2} dx = \sum_{\ell=0}^{\infty} \sum_{m=-\ell}^{\ell} |u_\ell^m|^2 \int_1^\infty \frac{\left|h_\ell^{(1)}(kr)\right|^2}{\left|h_\ell^{(1)}(k)\right|^2} dr. \tag{2.6.64}$$

Formula (2.6.20) helps to compute the expression inside the integral

$$\left\{ \frac{\left|h_\ell^{(1)}(kr)\right|^2}{\left|h_\ell^{(1)}(k)\right|^2} = \frac{1 + \alpha_1^\ell \frac{1}{(kr)^2} + \cdots + \alpha_\ell^\ell \frac{1}{(kr)^{2\ell}}}{1 + \alpha_1^\ell \frac{1}{k^2} + \cdots + \alpha_\ell^\ell \frac{1}{k^{2\ell}}} \frac{1}{r^2}, \right. \tag{2.6.65}$$

and the change of variable $t = (1/r)$ leads to the value of this integral which is

$$\left\{ \begin{aligned} \gamma(k,\ell) &= \int_1^\infty \frac{\left|h_\ell^{(1)}(kr)\right|^2}{\left|h_\ell^{(1)}(k)\right|^2} dr \\[2mm] &= \frac{1 + \frac{1}{3}\alpha_1^\ell \frac{1}{k^2} + \cdots + \frac{1}{2m+1}\alpha_m^\ell \frac{1}{k^{2m}} + \cdots + \frac{1}{2\ell+1}\alpha_\ell^\ell \frac{1}{k^{2\ell}}}{1 + \alpha_1^\ell \frac{1}{k^2} + \cdots + \alpha_\ell^\ell \frac{1}{k^{2\ell}}}. \end{aligned} \right. \tag{2.6.66}$$

The value of the quantity $\gamma(k,\ell)$ is $1/(2\ell + 1)$ when $k = 0$, and 1 when k is infinite. It is bounded by 1. In order to prove the estimate (2.6.59), we seek a bound of the form $(k^2 + \beta)/(k^2 + \beta(2\ell + 1))$. Identifying the coefficients of $1/k^{2m}$, it follows that

$$\beta \le \frac{(2m+2)}{(2m+3)} \frac{(2m+1)}{2(\ell-m)} \frac{\alpha_{m+1}^\ell}{\alpha_m^\ell} = \frac{(2m+1)^2}{(2m+3)} \frac{(m+\ell+1)}{2}. \tag{2.6.67}$$

The right-hand side is an increasing function of m and so the lowest bound occurs when $m = 0$ with $\beta = (\ell+1)/6$. The estimate (2.6.59) implies

$$\gamma(k,\ell) \le \frac{1}{2\ell+1} + \frac{6k^2}{(2\ell+1)(\ell+1)} \tag{2.6.68}$$

which is the bound (2.6.57).

From the upper bound of β follows

$$\frac{k^2 + \beta}{k^2 + (2\ell+1)\beta} \le \gamma(k,\ell), \quad \beta = \frac{\ell(2\ell-1)^2}{2\ell+1}. \tag{2.6.69}$$

The gradient of the function u can be decomposed in the form

$$\nabla u = \frac{\partial u}{\partial r}\overrightarrow{e_r} + \frac{1}{r}\nabla_S u = \frac{\partial u}{\partial r}\overrightarrow{e_r} + \nabla_T u, \tag{2.6.70}$$

$$\nabla_S u = \sum_{\ell=0}^{\infty} \sum_{m=-\ell}^{\ell} u_\ell^m \frac{h_\ell^{(1)}(kr)}{h_\ell^{(1)}(k)} \nabla_S Y_\ell^m(\theta, \varphi), \tag{2.6.71}$$

$$\int_S |\nabla_S Y_\ell^m|^2 \, d\sigma = \ell(\ell+1). \tag{2.6.72}$$

The associated part of the norm is

$$\int_{B_r} \frac{|\nabla_S u|^2}{r^2} dx = \sum_{\ell=0}^{\infty} \sum_{m=-\ell}^{\ell} \ell(\ell+1) |u_\ell^m|^2 \gamma(k,\ell) \tag{2.6.73}$$

from which we obtain the bound (2.6.60). It holds that

$$\frac{\partial u}{\partial r} = \sum_{\ell=0}^{\infty} \sum_{m=-\ell}^{\ell} u_\ell^m \frac{k\frac{d}{dr}h_\ell^{(1)}(kr)}{h_\ell^{(1)}(kr)} \frac{h_\ell^{(1)}(kr)}{h_\ell^{(1)}(k)} Y_\ell^m(\theta,\varphi). \tag{2.6.74}$$

Estimates (2.6.23) and (2.6.24) yield

$$\frac{\left|\frac{d}{dr}h_\ell^{(1)}(kr)\right|^2}{\left|h_\ell^{(1)}(kr)\right|^2} \le 1 + \frac{(\ell+1)^2}{k^2 r^2}, \tag{2.6.75}$$

from which it follows that

$$\left\{ \begin{aligned} &\int_{B_\epsilon} \frac{1}{r^2}\left|\frac{\partial u}{\partial r}\right|^2 dx \\ &\le \sum_{\ell=0}^{\infty} \sum_{m=-\ell}^{\ell} |u_\ell^m|^2 \int_1^\infty \left(k^2 + \frac{(\ell+1)^2}{r^2}\right) \frac{\left|h_\ell^{(1)}(kr)\right|^2}{\left|h_\ell^{(1)}(k)\right|^2} dr. \end{aligned} \right. \tag{2.6.76}$$

The change of variable $t = 1/r$ allows an easy computation of the value of this integral which is the quantity $\delta(k,\ell)$:

$$\delta(k,\ell) = \gamma(k,\ell) + \frac{(\ell+1)^2}{k^2} \frac{\frac{1}{3} + \frac{1}{5}\alpha_1^\ell \frac{1}{k^2} + \cdots + \frac{1}{2\ell+3}\alpha_\ell^\ell \frac{1}{k^{2\ell}}}{1 + \alpha_1^\ell \frac{1}{k^2} + \cdots + \alpha_\ell^\ell \frac{1}{k^{2\ell}}}. \tag{2.6.77}$$

Using the same technique as for the case of γ, it follows that

$$\delta(k,\ell) \le \frac{k^2 + \frac{\ell+1}{6}}{k^2 + \frac{(2\ell+1)(\ell+1)}{6}} + \frac{(\ell+1)^2}{3k^2} \frac{k^2 + 3\frac{\ell+1}{10}}{k^2 + \frac{(2\ell+3)(\ell+1)}{10}}. \tag{2.6.78}$$

Inequality (2.6.61) results from the previous estimates and from

$$\delta(k,\ell) \le 1 + \frac{(\ell+1)^2}{k^2}\left(\frac{1}{2\ell+3} + \frac{20k^2\ell}{3(2\ell+3)^2(\ell+1)}\right). \tag{2.6.79}$$

The quantity $\dfrac{\partial u}{\partial r} - iku$ admits the expansion

$$\frac{\partial u}{\partial r} - iku = \sum_{\ell=0}^{\infty} \sum_{m=-\ell}^{\ell} k u_\ell^m \frac{\dfrac{d}{dr} h_\ell^{(1)}(kr) - i h_\ell^{(1)}(kr)}{h_\ell^{(1)}(k)} Y_\ell^m(\theta, \varphi). \qquad (2.6.80)$$

There appears the quantity

$$\frac{1}{r}\left(z_\ell(kr) - ikr\right) = -\frac{p_\ell(kr)}{r q_\ell(kr)} + i\frac{k\left(1 - q_\ell(kr)\right)}{q_\ell(kr)}. \qquad (2.6.81)$$

Starting from the expressions of p_ℓ and q_ℓ, we obtain the bounds

$$\frac{k^2 + \frac{(\ell+1)^2}{2}}{k^2 + \frac{\ell+1}{2}} \le \frac{p_\ell(k)}{q_\ell(k)} \le \frac{k^2 + 4\ell(\ell - 1/2)(\ell + 1)}{k^2 + 4\ell(\ell - 1/2)}, \qquad (2.6.82)$$

$$|z_\ell(kr) - ikr|^2 \le \frac{k^2 r^2 + 5(\ell + 1)^4}{k^2 r^2 + 4(\ell + 1)^2} \le 1 + \frac{5}{4}(\ell + 1)^2, \qquad (2.6.83)$$

which implies

$$\begin{cases} \displaystyle\int_{B_r} \left|\frac{\partial u}{\partial r} - iku\right|^2 dx \\[3mm] \displaystyle \le \sum_{\ell=0}^{\infty} \sum_{m=-\ell}^{\ell} |u_\ell^m|^2 \left(1 + \frac{5}{4}(\ell + 1)^2\right) \int_1^{\infty} \frac{\left|h_\ell^{(1)}(kr)\right|^2}{\left|h_\ell^{(1)}(k)\right|^2} dr, \end{cases} \qquad (2.6.84)$$

and then from the estimate (2.6.68) follows (2.6.63). ∎

Similarly, the following theorem gives more precise information on the continuity bounds of the solution of the exterior Neumann problem in some Sobolev spaces.

Theorem 2.6.3 *The solution of the exterior Neumann problem* (2.6.56) *satisfies the estimates*

$$\begin{cases} \displaystyle \left\|\frac{u}{r}\right\|_{L^2(B_r)}^2 \le \sum_{\ell=0}^{\infty} \sum_{m=-\ell}^{\ell} \left(\frac{k^2 + \frac{\ell+1}{2}}{k^2 + \frac{(\ell+1)^2}{2}}\right)^2 \gamma(k, \ell)\, |v_\ell^m|^2 \\[5mm] \qquad\qquad \begin{cases} \|u_n\|_{L^2(S)}^2 ; \\[3mm] \le \begin{cases} 8(\|u_n\|_{H^{-3/2}(S)}^2 + 3k^2 \|u_n\|_{H^{-2}(S)}^2 \\[3mm] \quad + 3k^4 \|u_n\|_{H^{-5/2}(S)}^2 + k^6 \|u_n\|_{H^{-3}(S)}^2); \end{cases} \end{cases} \end{cases} \qquad (2.6.85)$$

$$
\left\{
\begin{aligned}
&\|\nabla_T u\|^2_{L^2(B_r)} \\[2mm]
&\le \sum_{\ell=0}^{\infty} \sum_{m=-\ell}^{\ell} (\ell^2 + \ell) \left(\frac{k^2 + \frac{\ell+1}{2}}{k^2 + \frac{(\ell+1)^2}{2}} \right)^2 \gamma(k,\ell) |v_\ell^m|^2 \\[2mm]
&\le \begin{cases} \|u_n\|^2_{H^1(S)}; \\[2mm] 8(\|u_n\|^2_{H^{-1/2}(S)} + 3k^2 \|u_n\|^2_{H^{-1}(S)} \\[2mm] \quad +3k^4 \|u_n\|^2_{H^{-3/2}(S)} + k^6 \|u_n\|^2_{H^{-2}(S)}); \end{cases}
\end{aligned}
\right.
\tag{2.6.86}
$$

$$
\left\{
\begin{aligned}
&\left\| \frac{1}{r} \frac{\partial u}{\partial r} \right\|^2_{L^2(B_c)} \le \sum_{\ell=0}^{\infty} \sum_{m=-\ell}^{\ell} k^2 \left(\frac{k^2 + \frac{\ell+1}{2}}{k^2 + \frac{(\ell+1)^2}{2}} \right)^2 \delta(k,\ell) |v_\ell^m|^2 \\[2mm]
&\le \begin{cases} 8(\|u_n\|^2_{H^{-1/2}(S)} + 3k^2 \|u_n\|^2_{H^{-1}(S)} \\[2mm] \quad +3k^4 \|u_n\|^2_{H^{-3/2}(S)} + k^6 \|u_n\|^2_{H^{-2}(S)}); \end{cases}
\end{aligned}
\right.
\tag{2.6.87}
$$

$$
\left\{
\begin{aligned}
&\left\| \frac{\partial u}{\partial r} - iku \right\|_{L^2(B_c)} \\[2mm]
&\le \sum_{\ell=0}^{\infty} \sum_{m=-\ell}^{\ell} (\ell+1)^2 \left(\frac{k^2 + \frac{\ell+1}{2}}{k^2 + \frac{(\ell+1)^2}{2}} \right)^2 \gamma(k,\ell) |v_\ell^m|^2 \\[2mm]
&\le \begin{cases} 8(\|u_n\|^2_{H^{-1/2}(S)} + 3k^2 \|u_n\|^2_{H^{-1}(S)} \\[2mm] \quad +3k^4 \|u_n\|^2_{H^{-3/2}(S)} + k^6 \|u_n\|^2_{H^{-2}(S)}). \end{cases}
\end{aligned}
\right.
\tag{2.6.88}
$$

Proof

It follows from (2.6.56) that

$$
u(r,\theta,\varphi) = \sum_{\ell=0}^{\infty} \sum_{m=-\ell}^{\ell} \frac{h_\ell^{(1)}(k)}{k \frac{d}{dr} h_\ell^{(1)}(k)} \frac{h_\ell^{(1)}(kr)}{h_\ell^{(1)}(k)} v_\ell^m Y_\ell^m(\theta,\varphi).
\tag{2.6.89}
$$

This expression is similar to the one already used in the case of the Dirichlet problem, except for division by the function $z_\ell(k)$. It follows from (2.6.1) and the previous bounds that

$$
\frac{1}{|z_\ell|^2} = \frac{q_\ell^2(k)}{k^2 + |p_\ell(k)|^2} \le \left(\frac{k^2 + \frac{\ell+1}{2}}{k^2 + \frac{(\ell+1)^2}{2}} \right)^2.
\tag{2.6.90}
$$

We then infer the above estimates using

$$\frac{k^2 + \frac{\ell+1}{2}}{k^2 + \frac{(\ell+1)^2}{2}} \leq \frac{2}{\ell+1} + \frac{2k^2}{(\ell+1)^2}. \tag{2.6.91}$$

2.6.3 The capacity operator T

We denoted by T the capacity operator which associates with a function u on the unit sphere S, the normal derivative of the solution of the exterior Dirichlet problem whose boundary data is u.

It holds that

$$\begin{cases} u = \displaystyle\sum_{\ell=0}^{\infty} \sum_{m=-\ell}^{\ell} u_\ell^m Y_\ell^m(\theta, \varphi), \\[2mm] Tu = \displaystyle\sum_{\ell=0}^{\infty} \sum_{m=-\ell}^{\ell} z_\ell(k) u_\ell^m Y_\ell^m(\theta, \varphi). \end{cases} \tag{2.6.92}$$

The capacity operator satisfies the following properties:

Theorem 2.6.4 *The operator T is continuous from $H^{s+1}(S)$ into $H^s(S)$. It satisfies the inequalities of coercivity and continuity*

$$\|u\|_{L^2(S)}^2 \leq -\Re\,(Tu, u)_{L^2(S)} \leq \|u\|_{H^{1/2}(S)}^2, \tag{2.6.93}$$

$$0 \leq \Im\,(Tu, u)_{L^2(S)} \leq k\,\|u\|_{L^2(S)}^2. \tag{2.6.94}$$

∎

These are direct consequences of Theorem 2.6.1 and inequalities (2.6.23) and (2.6.24).

2.6.4 The case of a plane wave

A plane wave is a solution of a Helmholtz equation of the form

$$u_{\text{inc}} = e^{i(\overrightarrow{k} \cdot \overrightarrow{x})}; \ |\overrightarrow{k}| = k. \tag{2.6.95}$$

A plane wave does not satisfy the radiation condition. When the sphere is soft, the associated acoustic problem satisfies the zero Dirichlet boundary condition. We seek a solution u of the exterior problem in the form

$$u = u_{\text{inc}} + u_{\text{sc}}, \tag{2.6.96}$$

where **the scattered field** u_{sc} **satisfies the outgoing radiation condition.** Thus, the boundary condition takes the form

$$u_{\text{sc}} = -u_{\text{inc}}, \quad \text{on } S. \tag{2.6.97}$$

The solution of the associated interior problem is then $-u_{\text{inc}}$. The problem is invariant by rotation around the axis oriented along $\overline{k}$, which we can choose to coincide with the axis ox_3. We use the spherical coordinates defined in section (2.4). It holds that

$$u_{\text{diff}} = -e^{ik\cos\theta}. \tag{2.6.98}$$

This function is invariant by rotation around the axis ox_3 and thus Theorem 2.4.3 shows that its expansion contains only spherical harmonics whose index m is zero. It is given by

Lemma 2.6.2 *The following expansion holds:*

$$e^{ikx} = \sum_{\ell=0}^{\infty}(i)^{\ell}(2\ell+1)j_{\ell}(k)\mathbb{P}_{\ell}(x), \tag{2.6.99}$$

where $\mathbb{P}_{\ell}$ is the Legendre polynomial of order ℓ. An equivalent expression is

$$e^{ik\cos\theta} = \sum_{\ell=0}^{\infty}(i)^{\ell}\,\sqrt{8\pi}\,\sqrt{\ell+1/2}j_{\ell}(k)Y_{\ell}^{0}(\theta). \tag{2.6.100}$$

Proof
We define

$$v_{\ell}(k) = (-i)^{\ell}\int_{-1}^{+1}e^{ikx}\mathbb{P}_{\ell}(x)dx \tag{2.6.101}$$

which is real since $\mathbb{P}_{\ell}$ has the same parity as ℓ. Theorem 2.6.2 implies

$$\begin{cases} v_{\ell}(k) &= -\dfrac{(-i)^{\ell}}{\ell(\ell+1)}\displaystyle\int_{-1}^{+1}e^{ikx}\dfrac{d}{dx}\left((1-x^2)\dfrac{d}{dx}\mathbb{P}_{\ell}(x)\right)dx \\[2mm] &= \dfrac{(-i)^{\ell}ik}{\ell(\ell+1)}\displaystyle\int_{-1}^{+1}e^{ikx}(1-x^2)\dfrac{d}{dx}\mathbb{P}_{\ell}(x)dx. \end{cases} \tag{2.6.102}$$

From the recursion formula (2.4.46) follows

$$v_{\ell}(k) = -\frac{(-i)^{\ell+1}k}{2\ell+1}\int_{-1}^{+1}e^{ikx}\left(\mathbb{P}_{\ell-1}(x)-\mathbb{P}_{\ell+1}(x)\right)dx \tag{2.6.103}$$

which is the recursion formula

$$\frac{(2\ell+1)}{k}v_{\ell}(k) = v_{\ell+1}(k) + v_{\ell-1}(k). \tag{2.6.104}$$

We recognize one of the recursion formulas satisfied by the spherical Bessel functions. Identifying v_0 and v_1, it follows that

$$v_{\ell}(k) = 2j_{\ell}(k). \tag{2.6.105}$$

The orthogonality of the basis $\sqrt{\ell + 1/2} \; \mathbb{P}_\ell$ leads to (2.6.99). The expression of $Y_\ell^0(\theta)$ given in Theorem 2.4.4,

$$Y_\ell^0(\theta) = \sqrt{\frac{\ell + 1/2}{2\pi}} \; \mathbb{P}_\ell(\cos \theta), \qquad (2.6.106)$$

implies (2.6.100). ∎

Besides, we infer from these expressions some usual properties and estimates of the Bessel function $j_\ell(k)$.

Lemma 2.6.3 *It holds that*

$$j_\ell(k) = (-i)^\ell \frac{1}{2} \int_{-1}^{+1} e^{ikx} \mathbb{P}_\ell(x) dx; \qquad (2.6.107)$$

$$j_\ell(k) = \frac{k^\ell}{2^{\ell+1}\ell!} \int_{-1}^{+1} e^{ikx} \left(1 - x^2\right)^\ell ds; \qquad (2.6.108)$$

$$ixe^{ikx} = \sum_{\ell=0}^{\infty} (i)^\ell (2\ell + 1) \frac{d}{dk} j_\ell(k) \mathbb{P}_\ell(x); \qquad (2.6.109)$$

$$|j_\ell(k)| \leq \frac{k^\ell}{2^{\ell+1}\ell!} \int_{-1}^{+1} \left(1 - x^2\right)^\ell dx = \frac{k^\ell 2^\ell \ell!}{(2\ell + 1)!}; \qquad (2.6.110)$$

$$\sum_{\ell=0}^{\infty} (\ell + 1/2) \left(j_\ell(k)\right)^2 = \frac{1}{2}; \qquad (2.6.111)$$

$$\sum_{\ell=0}^{\infty} (\ell + 1/2) \left(\frac{d}{dk} j_\ell(k)\right)^2 = \frac{1}{6}. \qquad (2.6.112)$$

Proof

Formula (2.6.107) is (2.6.101). Expression (2.6.108) results from the Rodrigues formula (2.4.31) (which is the definition of $\mathbb{P}_\ell$) and an integration by parts. It follows the estimate (2.6.110). The squared $L^2(S)$ norm of $e^{ik \cos \theta}$ is 4π, which is also from (2.6.100)

$$4\pi = \sum_{\ell=0}^{\infty} 8\pi (\ell + 1/2) \left(j_\ell(k)\right)^2.$$

This proves (2.6.111).

Differentiating under the integral gives the expression of dj_ℓ/dr. Differentiating identity (2.6.99) with respect to the variable k leads to (2.6.109), from which we infer (2.6.112) by computing the $L^2(S)$ norm of the left-hand side. ∎

The solution of the interior problem is $-u_{\text{inc}}$, whose normal derivative is

$$\frac{\partial u}{\partial n}\bigg|_{\text{int}} = -i\,k\cos\theta e^{ik\cos\theta}. \tag{2.6.113}$$

The spherical harmonics expansion of this function is given by

$$\frac{\partial u}{\partial n}\bigg|_{\text{int}} = -\sum_{\ell=0}^{\infty} k\,(i)^{\ell}\,\sqrt{8\pi}\,\sqrt{\ell+1/2}\,\frac{d}{dk}j_{\ell}(k)Y_{\ell}^{0}(\theta). \tag{2.6.114}$$

We have already computed the normal derivative of the scattered part of the exterior solution which is given by

$$\frac{\partial u}{\partial n}\bigg|_{\text{ext}} = -Tu_{\text{inc}} = -\sum_{\ell=0}^{\infty} (i)^{\ell}\,\sqrt{8\pi}\sqrt{\ell+1/2}\,z_{\ell}(k)j_{\ell}(k)Y_{\ell}^{0}(\theta). \tag{2.6.115}$$

The difference of these two normal derivatives takes a quite simple expression when using the following **property of the wronskian,**

$$\frac{d}{dk}\left(h_{\ell}^{(1)}(k)\right)j_{\ell}(k) - \frac{d}{dk}\left(j_{\ell}(k)\right)h_{\ell}^{(1)}(k) = -\frac{1}{k^{2}}, \tag{2.6.116}$$

from which we obtain

$$\frac{\partial u}{\partial n}\bigg|_{\text{ext}} - \frac{\partial u}{\partial n}\bigg|_{\text{int}} = \sum_{\ell=0}^{\infty}(i)^{\ell}\,\sqrt{8\pi}\,\frac{\sqrt{\ell+1/2}}{kh_{\ell}^{(1)}(k)}Y_{\ell}^{0}(\theta). \tag{2.6.117}$$

Formally we can obtain the "limit" of this expression **when k tends to infinity.** We have

$$h_{\ell}^{(1)}(k) \sim (-i)^{\ell}\frac{e^{ik}}{k}$$

and thus

$$\frac{\partial u}{\partial n}\bigg|_{\text{ext}} - \frac{\partial u}{\partial n}\bigg|_{\text{int}} \sim e^{-ik}\sum_{\ell=0}^{\infty}(-1)^{\ell}\sqrt{8\pi}\,\sqrt{\ell+1/2}Y_{\ell}^{0}(\theta). \tag{2.6.118}$$

The distribution in the right-hand side appears to be the Dirac mass at the south pole of the sphere S.

Lemma 2.6.4 *The Dirac mass located at the point $(0,0,-1)$ of the sphere S is expressed in the form*

$$\delta = \sum_{\ell=0}^{\infty}(-1)^{\ell}\sqrt{\frac{\ell+1/2}{2\pi}}Y_{\ell}^{0}(\theta). \tag{2.6.119}$$

Proof
The Dirac mass at a point x of the sphere S applies to regular functions and its value on a function φ is

$$\langle\delta,\varphi\rangle = \sum_{\ell=0}^{\infty}\sum_{m=-\ell}^{\ell}\overline{\delta_{\ell}^{m}}\varphi_{\ell}^{m} = \varphi(x). \tag{2.6.120}$$

It is invariant by rotation around the axis ox, and when the point x is the south pole its expansion contains only spherical harmonics with an index $m = 0$. Thus, we only need to consider functions φ depending on the variable θ. When the point x is the south pole, it holds that

$$\varphi(x) = \sum_{\ell=0}^{\infty} \varphi_\ell Y_\ell^0(-\pi) = \sum_{\ell=0}^{\infty} \sqrt{\frac{\ell + 1/2}{2\pi}} \mathbb{P}_\ell(-1)\, \varphi_\ell. \tag{2.6.121}$$

Using $\mathbb{P}_\ell(-1) = (-1)^\ell$, then identifying, it follows (2.6.119).   ∎

We find in a **formal way** that

$$\left.\frac{\partial u}{\partial n}\right|_{\text{ext}} - \left.\frac{\partial u}{\partial n}\right|_{\text{int}} \sim 4\pi e^{-ik}\delta(0, 0, -1). \tag{2.6.122}$$

The physical meaning is the following: in the high-frequency approximation, the far field scattered by a sphere is exactly that of a bright point situated at the south pole, with an amplitude proportional to the area of this sphere.

The **exterior Neumann problem** will be solved in the same way. We seek a solution of the form (2.6.96), sum of an incident and a scattered wave. The boundary condition is

$$\frac{\partial u_{\text{sc}}}{\partial n} = -\frac{\partial u_{\text{inc}}}{\partial n}, \quad \text{on } S. \tag{2.6.123}$$

The expansion in spherical harmonics of $\partial u_{\text{inc}}/\partial n$ is given by (2.6.114). We apply the inverse of the capacity operator T which gives the value of u_{sc} on the sphere S,

$$u_{\text{sc}} = -\sum_{\ell=0}^{\infty} (i)^\ell \sqrt{8\pi} \sqrt{\ell + 1/2} \frac{\frac{d}{dk} j_\ell(k) h_\ell^{(1)}(k)}{\frac{d}{dk} h_\ell^{(1)}(k)} Y_\ell^0(\theta). \tag{2.6.124}$$

It is convenient to examine the quantity $u_{\text{inc}} + u_{\text{sc}}$.

$$\begin{cases} u_{\text{sc}} + u_{\text{inc}} = -\sum_{\ell=0}^{\infty} (i)^\ell \frac{\sqrt{8\pi(\ell + 1/2)}}{\frac{d}{dk} h_\ell^{(1)}(k)} \\ \qquad\qquad \times \left(\frac{d}{dk} j_\ell(k) h_\ell^{(1)}(k) - \frac{d}{dk} h_\ell^{(1)}(k) j_\ell(k) \right) Y_\ell^0(\theta), \end{cases} \tag{2.6.125}$$

which, using the property of the wronskian can be written

$$u_{\text{sc}} + u_{\text{inc}} = -\frac{1}{k^2} \sum_{\ell=0}^{\infty} (i)^\ell \sqrt{8\pi} \frac{\sqrt{\ell + 1/2}}{\frac{d}{dk} h_\ell^{(1)}(k)} Y_\ell^0(\theta). \tag{2.6.126}$$

Formally, using the asymptotics of the function $\frac{d}{dk} h_\ell(k)$,

$$\frac{d}{dk} h_\ell(k) \sim (-i)^{\ell-1} \frac{e^{ik}}{k},$$

it follows that

$$u_{sc} + u_{inc} \sim -\frac{1}{k}e^{-ik} \sum_{\ell=0}^{\infty} (i)^{2\ell-1} \sqrt{8\pi} \sqrt{\ell+1/2} \, Y_\ell^0(\theta); \qquad (2.6.127)$$

$$u_{sc} + u_{inc} \sim \frac{i}{k}e^{-ik} 4\pi\delta(0, 0, -1). \qquad (2.6.128)$$

2.6.5 The exterior problem for the Helmholtz equation

We gave in Section 2.2 a quite formal formulation of the exterior problem

$$\begin{cases} \Delta u + k^2 u = 0, & \text{in } \Omega_e, \\[2mm] u_{|\Gamma} = u_d, & \left(\text{or } \left.\dfrac{\partial u}{\partial n}\right|_\Gamma = u_n\right), \end{cases} \qquad (2.6.129)$$

$$\left|\frac{\partial u}{\partial r} - iku\right| \le \frac{c}{r^2}, \quad \text{for distant } r. \qquad (2.6.130)$$

It is natural to look for a solution such that

$$\frac{u}{(1+r^2)^{1/2}} \in L^2(\Omega_e), \qquad \frac{\nabla u}{(1+r^2)^{1/2}} \in L^2(\Omega_e).$$

Indeed, these estimates were the ones that appeared in the case of the sphere. The solution of (2.6.129) is not unique in this space, as we have seen in the case of the sphere. We need to add the radiation condition. A form of this condition is (2.6.130). A weaker form is

$$\int_{\Omega_e} \left|\frac{\partial u}{\partial r} - iku\right|^2 dx \le c. \qquad (2.6.131)$$

Let us **denote by** H the associated Hilbert space

$$H = \left\{ u, \; \frac{u}{(1+r^2)^{1/2}} \in L^2(\Omega_e), \; \frac{\nabla u}{(1+r^2)^{1/2}} \in L^2(\Omega_e), \right.$$

$$\left. \frac{\partial u}{\partial r} - iku \in L^2(\Omega_e) \right\}.$$

We will also need a ball B_R whose boundary is the sphere S_R. Its radius R will be chosen large enough to enclose the interior domain Ω_i.

Theorem 2.6.5 (Uniqueness theorem) *The exterior Dirichlet and Neumann problems (2.6.129) admit at most one solution in the Hilbert space H.*

Proof

The difference u between two solutions u_1 and u_2 has a zero boundary condition, i.e., satisfies $u_{|\Gamma} = 0$, or $(\partial u / \partial n)|_\Gamma = 0$. We multiply this equation by $\bar{u}$ and integrate by parts to obtain

$$\int_{\Omega_r \cap B_R} \left(|\nabla u|^2 - k^2 |u|^2 \right) dx - \int_{S_R} \partial u / \partial r \bar{u} d\sigma = 0. \qquad (2.6.132)$$

At the exterior of the ball B_R, we expand the solution in the spherical harmonics. It holds that

$$u(r, \theta, \varphi) = \sum_{\ell=0}^{\infty} \sum_{m=-\ell}^{\ell} \left(\alpha_\ell^m \frac{h_\ell^{(1)}(kr)}{h_\ell^{(1)}(kR)} + \beta_\ell^m \frac{h_\ell^{(2)}(kr)}{h_\ell^{(2)}(kR)} \right) Y_\ell^m(\theta, \varphi), \quad (2.6.133)$$

and $\frac{\partial u}{\partial r} - iku$ has the expansion

$$\begin{cases} \frac{\partial u}{\partial r} - iku = \sum_{\ell=0}^{\infty} \sum_{m=-\ell}^{\ell} \left[\frac{k\alpha_\ell^m}{h_\ell^{(1)}(kR)} \left(\frac{d}{dr} h_\ell^{(1)}(kr) - i h_\ell^{(1)}(kr) \right) \right. \\ \\ \left. + \frac{k\beta_\ell^m}{h_\ell^{(2)}(kR)} \left(\frac{d}{dr} h_\ell^{(2)}(kr) - i h_\ell^{(2)}(kr) \right) \right] Y_\ell^m(\theta, \varphi). \end{cases} \qquad (2.6.134)$$

The orthogonality of the spherical harmonics implies that the integral on the complement of B_R of a function of the form

$$\begin{cases} v = \sum_{\ell=0}^{\infty} \sum_{m=-\ell}^{\ell} \varphi_\ell^m(r) Y_\ell^m(\theta, \varphi), \qquad \text{for r > R, is} \\ \\ \int_{B_R^c} |v|^2 dx = \sum_{\ell=0}^{\infty} \sum_{m=-\ell}^{\ell} \int_R^{\infty} |\varphi_\ell^m(r)|^2 r^2 dr. \end{cases} \qquad (2.6.135)$$

The dominant term in the expression $\partial u / \partial r - iku$ is the one coming from $h_\ell^{(2)}$. It behaves as $-2i \, (e^{-ikr}/r)(\beta_\ell^m / h^{(2)}(kR))$. Thus, the convergence of the associated series is possible only when all the coefficients

$$\beta_\ell^m = 0. \qquad (2.6.136)$$

We have proved that the solution contains only the terms $h_\ell^{(1)}$. The imaginary part in the expression (2.6.132) reduces to

$$\Im \int_{S_R} (T_R u, \bar{u}) \, d\sigma = 0 \qquad (2.6.137)$$

where T_R is the capacity operator on the ball with radius R. From the expression of this operator follows

$$\alpha_\ell^m = 0, \qquad (2.6.138)$$

from which we infer that

$$u_{|S_R} = 0, \tag{2.6.139}$$

$$\frac{\partial u}{\partial n}\Big|_{S_R} = 0, \tag{2.6.140}$$

and thus the solution vanishes on the exterior of the ball B_R. Moreover, the Helmholtz equation is an elliptic operator, from which it follows that the solution is **analytic** in the domain Ω_e. Thus, it is zero.

Remarks
There exists a weaker form for the radiation condition which is

$$\lim_{R\to\infty} \int_{S_R} \left| \frac{\partial u}{\partial r} - iku \right|^2 d\sigma = 0. \tag{2.6.141}$$

In the same way as above, it implies $\beta_\ell^m = 0$ and so the uniqueness of the solution. Another weaker form for the radiation condition is

$$\int_{\Omega_e} \frac{1}{r} \left| \frac{\partial u}{\partial r} - iku \right|^2 dx \le c. \tag{2.6.142}$$

∎

The existence proof for solutions is based on the introduction of an equivalent formulation. Let us denote by T_R **the capacity operator** on the sphere S_R whose expression is

$$\begin{cases} u(R,\theta,\varphi) = \displaystyle\sum_{\ell=0}^{\infty} \sum_{m=-\ell}^{\ell} \alpha_\ell^m Y_\ell^m(\theta,\varphi), \\[2mm] T_R u = \displaystyle\sum_{\ell=0}^{\infty} \sum_{m=-\ell}^{\ell} \frac{1}{R} z_\ell(kR)\alpha_\ell^m Y_\ell^m(\theta,\varphi). \end{cases} \tag{2.6.143}$$

Lemma 2.6.5 *Any restriction to $\Omega_e \cap B_R$ of the solution of the Dirichlet or the Neumann problem given by Theorem 2.6.5 is a solution of the problem: find u in the space $H^1(\Omega_e \cap B_R)$ which satisfies*

$$\begin{cases} \Delta u + k^2 u = 0, & \text{in } \Omega_e \cap B_R, \\[2mm] \dfrac{\partial u}{\partial n}\Big|_{S_R} = T_R u, \\[2mm] u_{|\Gamma} = u_d \quad (\text{or } \dfrac{\partial u}{\partial n}_{|\Gamma} = u_n). \end{cases} \tag{2.6.144}$$

Proof
It is a direct consequence of the above proof: as it belongs to the space H, it admits an expansion outside B_R in the form (2.6.133) with vanishing coefficients β_ℓ^m, from which follows the lemma.   ∎

From now on, we suppose that $u_d \in H^{1/2}(\Gamma)$(resp. $u_n \in H^{-1/2}(\Gamma)$). The trace Theorem 2.5.3, shows that u_d admits a lifting in $H^1(\Omega_e \cap B_R)$ whose support is contained in B_R. Thus, we can replace the Dirichlet problem (2.6.144) by a problem with a vanishing boundary condition and a right-hand side:

$$\begin{cases} \Delta u + k^2 u = g(x), \\[2mm] \dfrac{\partial u}{\partial n}\Big|_{S_R} = T_R u, \\[2mm] u_{|\Gamma} = 0. \end{cases} \qquad (2.6.145)$$

Theorem 2.6.6 (Existence theorem) *The Dirichlet problem (2.6.145) admits a unique variational solution in the space $H^1(\Omega_e \cap B_R)$, when $g \in H^{-1}(\Omega_e \cap B_R)$, which satisfies*

$$\begin{cases} \displaystyle\int_{\Omega_e \cap B_R} (\nabla u \cdot \nabla v)\, dx - k^2 \int_{\Omega_e \cap B_R} uv\, dx - \int_{S_R} T_R uv\, d\sigma \\[4mm] \displaystyle = -\int_{\Omega_e} gv\, dx, \quad \forall v \in \left\{ H^1(\Omega_e \cap B_R)\,;\, v_{|\Gamma} = 0 \right\}. \end{cases} \qquad (2.6.146)$$

The Dirichlet problem (2.6.129) admits a unique solution in the space H when $u_d \in H^{1/2}(\Gamma)$. The associated mapping is continuous from $H^{1/2}(\Gamma)$ into the space H.

The Neumann problem (2.6.144) admits, when $u_n \in H^{-1/2}(\Gamma)$, a unique variational solution in the space $H^1(\Omega_e \cap B_R)$, which satisfies

$$\begin{cases} \displaystyle\int_{\Omega_e \cap B_R} (\nabla u \cdot \nabla v)\, dx - k^2 \int_{\Omega_e \cap B_R} uv\, dx - \int_{S_R} T_R uv\, d\sigma \\[4mm] \displaystyle = -\int_{\Gamma} u_n v\, d\gamma \quad \forall v \in H^1(\Omega_e \cap B_R). \end{cases} \qquad (2.6.147)$$

The Neumann problem (2.6.144) admits a unique solution in the space H when $u_n \in H^{-1/2}(\Gamma)$. The associated mapping is continuous from $H^{-1/2}(\Gamma)$ into the space H.

Proof

We give only the proof in the case of the Dirichlet problem, the proof in the case of the Neumann problem being almost exactly the same. The bilinear form (2.6.146) is split in three parts. The first one is the one associated with the Laplacian and is coercive. The second one has the wrong sign, but is continuous into the space $L^2(\Omega_e \cap B_R)$ and thus is compact compared to the first one. It follows from Theorem 2.6.4 that the third part has a real part which is coercive on $L^2(\Gamma)$ and a negative imaginary part. Hence, we can use the Fredholm alternative, from which we know that uniqueness implies existence. Uniqueness was the object of theorem (2.6.5).

Let us be more precise. We consider the case where k^2 is a complex number denoted z. We introduce a perturbation problem depending on the parameter z, which admits a unique solution (from the Lax-Milgram theorem).

Lemma 2.6.6 *The variational problem*

$$
\begin{cases}
\displaystyle \int_{\Omega_e} (\nabla u_z \cdot \nabla v)\, dx - z \int_{\Omega_e} u_z v dx = - \int_{\Omega_e} gv dx, \\[4mm]
\forall v \in H^1(\Omega_e), \ v_{|\Gamma} = 0
\end{cases}
\tag{2.6.148}
$$

admits, when z satisfies $\Im z > 0$, a unique solution in $H^1(\Omega_e)$ vanishing on Γ. It depends analytically on z in this half complex plane and is a solution of equation (2.6.146) for $z = k^2$.

Proof
The choice $v = \bar{u}_z$ yields

$$
\int_{\Omega_e} |\nabla u_z|^2 \, dx - \Re z \int_{\Omega_e} |u_z|^2 \, dx = -\Re \int_{\Omega_e} g\bar{u}_z dx,
\tag{2.6.149}
$$

$$
\Im z \int_{\Omega_e} |u_z|^2 \, dx = \Im \int_{\Omega_e} g\bar{u}_z dx.
\tag{2.6.150}
$$

Let us denote by $a(u,v)$ the associated quadratic form. It holds that

$$
\begin{cases}
|a(u,\bar{u})|^2 = \\[4mm]
\left(\displaystyle\int_{\Omega_e} |\nabla u|^2 \, dx - \Re z \int_{\Omega_e} |u|^2 \, dx \right)^2 + (\Im z)^2 \left(\displaystyle\int_{\Omega_c} |u|^2 \, dx \right)^2 \\[4mm]
\geq \dfrac{(\Im z)^2 / 2}{(\Re z)^2 + (\Im z)^2} \left(\displaystyle\int_{\Omega_e} |\nabla u|^2 \, dx \right)^2 + \dfrac{(\Im z)^2}{2} \left(\displaystyle\int_{\Omega_c} |u|^2 \, dx \right)^2.
\end{cases}
\tag{2.6.151}
$$

Thus, this bilinear form is coercive in the space $H^1(\Omega_e)$ as soon as $\Im z$ is not zero. The Lax-Milgram Theorem 2.5.8 shows the existence and uniqueness of the function u_z in $H^1(\Omega_e)$.

Outside the ball B_R, the solution u_z has an expansion in spherical harmonics and spherical Bessel functions of the form

$$
u_z(r,\theta,\varphi) = \sum_{\ell=0}^{\infty} \sum_{m=-\ell}^{\ell} \left(\alpha_\ell^m \frac{h_\ell^{(1)}(zr)}{h_\ell^{(1)}(zR)} + \beta_\ell^m \frac{h_\ell^{(2)}(zr)}{h_\ell^{(2)}(zR)} \right) Y_\ell^m(\theta,\varphi).
\tag{2.6.152}
$$

It follows from the expression (2.6.14) that the spherical Bessel functions $h_\ell^{(1)}$ are exponentially decreasing for $\Im z > 0$, while the functions $h_\ell^{(2)}$ are exponentially increasing. Thus, all the coefficients β_ℓ^m in this expansion vanish, from which we infer that u_z satisfies the equation (2.6.146). ∎

Remark
The lower half-complex plane, $\Im z < 0$, also corresponds to a well-posed problem whose limit satisfies the ingoing radiation condition.

End of Theorem 2.6.6

We consider the limit of the slightly different following problem, when $\Im z$ tends to zero and $\Re z = k^2$:

$$\begin{cases} \int_{\Omega_r \cap B_R} (\nabla u_z \cdot \nabla v)\, dx - z \int_{\Omega_c \cap B_R} u_z v dx - \int_{S_R} T_R u_z v d\sigma \\ \\ = - \int_{\Omega_r \cap B_R} gv dx; \quad \forall v \in H^1\left(\Omega_e \cap B_R\right); v_{|\Gamma} = 0. \end{cases} \tag{2.6.153}$$

It follows from (2.6.143)) that

$$\begin{cases} \int_{\Omega_c \cap B_R} |\nabla u_z|^2\, dx + \dfrac{1}{R} \int_{S_R} |u_z|^2\, dx \\ \\ \leq k^2 \int_{\Omega_c \cap B_R} |u_z|^2\, dx - \Re \int_{\Omega_c \cap B_R} g\bar{u}_z dx. \end{cases} \tag{2.6.154}$$

We now proceed by contradiction:

Either the sequence u_z is bounded in $L^2\left(\Omega_e \cap B_R\right)$, and then it is also bounded in $H^1\left(\Omega_e \cap B_R\right)$. From the compact injection of H^1 into L^2, it converges strongly in L^2 and weakly in H^1 towards a function u which satisfies (2.6.146).

Or the sequence u_z is not bounded and its L^2 norm tends to infinity. We introduce

$$v_z = \frac{u_z}{\|u_z\|_{L^2}}. \tag{2.6.155}$$

Its H^1 norm is bounded and it converges in L^2. Its limit satisfies

$$\begin{cases} \Delta v + k^2 v = 0, \\ \\ v_{|\Gamma} = 0, \\ \\ \dfrac{\partial v}{\partial n}\bigg|_{S_R} = T_R v, \end{cases} \tag{2.6.156}$$

and is thus zero. This is contradictory as its L^2 norm is 1.   ∎

The following regularity results hold:

Theorem 2.6.7 *All the partial derivatives, up to order m, of the solution of the Dirichlet problem (2.6.129) are in the space H when $u_d \in H^{m-1/2}(\Gamma)(m \geq 1)$. All the partial derivatives, up to order m, of the solution of the Neuman problem (2.6.129) are in the space H when $u_n \in H^{m-3/2}(\Gamma)(m \geq 1)$.*

Proof

The regularity properties in the domain $\Omega_e \cap B_R$ are exactly those of the Laplace equation. The terms associated with $k^2 u$ are of less order of derivation and they do not modify the regularity, but the continuity constants are changed. Due to these terms, these constants are of order k^{2m}. From this local result, the trace theorem shows that the restriction to the sphere S_R of the solution belongs to $H^{m-1/2}(S_R)$. The regularity in the exterior domain then results from the study of the case of the sphere, for which we have explicit solutions in terms of spherical harmonics and spherical Bessel functions.

Comments

– A common problem corresponds to the scattering of an incident plane wave u_{inc} by the object. We look for the scattered field u_{sc} which satisfies the radiation condition, while the total field u

$$u = u_{\text{inc}} + u_{sc}$$

satisfies the boundary condition

$$u_\Gamma = 0.$$

– A quite common boundary condition is

$$\left(\frac{\partial u}{\partial n} - \lambda u\right)\Big|_\Gamma = g, \text{ for a complex parameter } \lambda.$$

The coefficient λ corresponds to an impedance term.

The existence and uniqueness results are not modified when λ is such that

$$\begin{cases} \Re\lambda \geq 0, \\ \\ \Im\lambda \leq 0. \end{cases} \tag{2.6.157}$$

When these quantities have different signs, the corresponding bilinear form has the wrong sign. Yet, the Fredholm alternative holds and there is uniqueness, except for a discrete set of values of λ, sometimes called **Steklov eigenvalues**.

– We have seen that the following homogeneous problem

$$\begin{cases} \Delta u + zu = 0, \\ \\ u_{|\Gamma} = 0 \quad (\text{or } \frac{\partial u}{\partial n}\big|_\Gamma = 0), \\ \\ \frac{\partial u}{\partial n}\Big|_{S_R} = T_R(z)u \end{cases} \tag{2.6.158}$$

admits a unique zero solution when $\Im z \geq 0$. This is no longer true in the lower half complex plane, where critical values can appear. The operator structure, sum of a coercive part and a compact perturbation, and the analyticity with respect to the parameter z show that these critical values are only **isolated poles** of the inverse operator.

These values are called **resonance values**. Their position in the complex plane depends heavily on the geometry of the domain Ω_e. For a non-convex object, there may exist an infinite sequence of these values converging to the real axis, although uniqueness shows that none of them are on this axis. The Helmholtz resonator is a classical example. It consists of a quasi-closed cavity. In that case, the eigenvalues of the interior limit cavity problem give rise by a continuation argument to resonance values, close to the real axis.

3
Integral Representations and Integral Equations

We introduce in this chapter the integral representations of the solutions of the Helmholtz equation. We study the associated integral equations and their main properties. When the wave number k is zero, these integral representations are those associated with the Laplace equation. They have very specific properties. In particular some of the associated integral equations are coercive. We therefore devote special attention to this setting. We will see that they also are the principal parts of the equation associated with the Helmholtz equation.

3.1 Integral Representations

The fundamental solution of the Helmholtz equation which satisfies the outgoing radiation condition is

$$E(x) = \frac{1}{4\pi} \frac{e^{ikr}}{r}, \quad r^2 = x_1^2 + x_2^2 + x_3^2. \tag{3.1.1}$$

Theorem 3.1.1 (Representation theorem) *Let u be a function such that (with possibly $k = 0$)*

$$\Delta u + k^2 u = 0 \quad \text{in } \Omega_i, \tag{3.1.2}$$

$$\Delta u + k^2 u = 0 \quad \text{in } \Omega_e, \tag{3.1.3}$$

u satisfies the outgoing radiation condition in the sense introduced in Theorem 2.6.5: $u \in H$; the traces of u, u_{int} and u_{ext}, $(\partial u/\partial n)|_{\text{int}}$ and $(\partial u/\partial n)|_{\text{ext}}$ belong to $C^0(\Gamma)$.
We define

$$[u] = u_{\text{int}} - u_{\text{ext}}, \tag{3.1.4}$$

$$\left[\frac{\partial u}{\partial n}\right] = \frac{\partial u}{\partial n}\bigg|_{\text{int}} - \frac{\partial u}{\partial n}\bigg|_{\text{ext}}. \tag{3.1.5}$$

*For $y \notin \Gamma$, the following **representation formula** holds:*

$$u(y) = \int_\Gamma E(x-y)\left[\frac{\partial u}{\partial n}(x)\right] d\gamma(x) - \int_\Gamma \frac{\partial}{\partial n_x}(E(x-y))[u(x)]d\gamma(x), \tag{3.1.6}$$

and for $y \in \Gamma$

$$\begin{cases} \dfrac{u_{\text{int}}(y) + u_{\text{ext}}(y)}{2} = \displaystyle\int_\Gamma E(x-y)\left[\frac{\partial u}{\partial n}(x)\right] d\gamma(x) \\[4mm] \hspace{2cm} - \displaystyle\int_\Gamma \frac{\partial}{\partial n_x}(E(x-y))[u(x)]d\gamma(x). \end{cases} \tag{3.1.7}$$

Proof
We use the Green formula with on one side u, and on the other side the function $E(\cdot - y)$. For any domain Ω, whose boundary is $\partial\Omega$, and for any $y \notin \Omega$, we have

$$\begin{cases} 0 = \displaystyle\int_\Omega \left(\Delta u(x) + k^2 u(x)\right) E(x-y)dx \\[4mm] \hspace{1cm} - \displaystyle\int_\Omega \left(\Delta E(x-y) + k^2 E(x-y)\right) u\,dx \\[4mm] = \displaystyle\int_{\partial\Omega} \left[\frac{\partial u}{\partial n}(x)E(x-y) - \frac{\partial}{\partial n_x}E(x-y)u(x)\right] d\gamma(x). \end{cases} \tag{3.1.8}$$

Let y be in the domain Ω_i and B_ε be a ball whose center is y, radius ε and boundary S_ε. We also introduce a ball B_R whose center is y, radius R, chosen large enough to contain the domain Ω_i, and whose boundary is S_R. We use the above Green formula in the two domains: $\Omega_i - B_\varepsilon, \Omega_e \cap B_R$. Adding the two corresponding contributions yields

$$\begin{cases} \displaystyle\int_\Gamma E(x-y)\left[\frac{\partial u}{\partial n}(x)\right] d\gamma(x) - \int_\Gamma \frac{\partial}{\partial n_x}E(x-y)[u(x)]d\gamma(x) \\[4mm] - \displaystyle\int_{S_\varepsilon} E(x-y)\frac{\partial u}{\partial n}(x)d\gamma(x) + \int_{S_\varepsilon}\frac{\partial}{\partial n_x}E(x-y)u(x)d\gamma(x) \\[4mm] + \displaystyle\int_{S_R} E(x-y)\frac{\partial u}{\partial n}(x)d\gamma(x) - \int_{S_R}\frac{\partial}{\partial n_x}E(x-y)u(x)d\gamma(x) \\[4mm] \hspace{5cm} = 0. \end{cases} \tag{3.1.9}$$

The function u is regular in the ball B_ε, as it results from the regularity theorems. Thus, when ε tends to zero, the first integral on S_ε is bounded by

$$\left| \int_{S_\varepsilon} E(x-y) \frac{\partial u}{\partial n}(x) d\gamma(x) \right| \le \varepsilon \sup_{x \in B_\varepsilon} \left| \frac{\partial u}{\partial n}(x) \right| \qquad (3.1.10)$$

and tends to zero. The second integral has the value

$$\left\{ \begin{aligned} \int_{S_\varepsilon} \frac{\partial}{\partial n_x} E(x-y) u(x) d\gamma(x) &= u(y) \int_{S_\varepsilon} \frac{\partial}{\partial n_x} E(x-y) d\gamma(x) \\ &+ \int_{S_\varepsilon} \frac{\partial}{\partial n_x} (E(x-y))(u(x) - u(y)) d\gamma(x), \end{aligned} \right. \qquad (3.1.11)$$

$$\frac{\partial}{\partial n_x} E(x-y) = \left(\frac{ik}{4\pi r} - \frac{1}{4\pi r^2} \right) e^{ikr}, \qquad (3.1.12)$$

$$\left\{ \begin{aligned} \left| \int_{S_\varepsilon} \frac{\partial}{\partial n_x} (E(x-y))(u(x) - u(y)) d\gamma(x) \right| \\ \le (k\varepsilon + 1) \sup_{x \in B_\varepsilon} |u(x) - u(y)|. \end{aligned} \right. \qquad (3.1.13)$$

This last term tends to zero. The first term in the right hand side of (3.1.11) admits the limit $-u(y)$.

Let us now examine the terms on the sphere S_R. They take the form

$$\left\{ \begin{aligned} &- \int_{S_R} \left(\frac{\partial}{\partial n_x} E(x-y) - ikE(x-y) \right) u(x) d\gamma(x) \\ &+ \int_{S_R} E(x-y) \left(\frac{\partial u}{\partial n} - iku(x) \right) d\gamma(x). \end{aligned} \right. \qquad (3.1.14)$$

The first of these terms is bounded by

$$\left| \int_{S_R} \left(\frac{\partial}{\partial n_x} E(x-y) - ikE(x-y) \right) u(x) d\gamma(x) \right| \le \sup_{x \in S_R} |u(x)|. \qquad (3.1.15)$$

For R large enough, the radiation condition shows that the function u admits an expansion in the form (2.6.133) where all the β_ℓ^m are zero, and thus is bounded by C/R. Hence this term tends to zero. The same expansion shows that the quantity $\partial u/\partial n - iku$ is bounded by C/R^2. The associated term is bounded by

$$\left| \int_{S_R} E(x-y) \left(\frac{\partial u}{\partial n}(x) - iku(x) \right) d\gamma(x) \right| \le \frac{C}{R}, \qquad (3.1.16)$$

and thus tends to zero when R tends to infinity.

When the point y is on the surface Γ, the ball B_ε is split into $\Omega_i \cap B_\varepsilon$ and $\Omega_e \cap B_\varepsilon$, which yields two pieces in the corresponding integrals. When ε tends to zero, these two pieces are asymptotically separated by the tangent plane. Thus the associated integrals on S_ε give rise to the term

$(1/2) \left(u_{\text{int}}(y) + u_{\text{ext}}(y) \right)$. We must notice that in this case, the integrands associated with the surface Γ admit a singularity at the point y. The one associated with the kernel $E(x-y)$ is singular as $1/|x-y|$ and thus belongs to the space $L^1(\Gamma)$ for x on Γ.

In the expression of the other integrand appears the kernel

$$\frac{\partial}{\partial n_x} E(x-y) = \frac{e^{ik|x-y|}}{4\pi|x-y|^2} \left(ik - \frac{1}{|x-y|} \right) ((x-y) \cdot n_x), \qquad (3.1.17)$$

which singularity is equivalent to $1/|x-y|$, since its numerator $((x-y) \cdot n_x)$ vanishes as $|x-y|^2$ in a neighbourhood of y, and thus this kernel belongs to the space $L^1(\Gamma)$ for $y \in \Gamma$. This is no longer true when y is not on Γ. ∎

The following integral expression is called **a single layer potential**:

$$u(y) = \int_\Gamma E(x-y)q(x)d\gamma(x); \qquad (3.1.18)$$

while the following integral expression is called **a double layer potential**:

$$u(y) = \int_\Gamma \frac{\partial E}{\partial n_x}(x-y)\varphi(x)d\gamma(x). \qquad (3.1.19)$$

These two types of potentials satisfy the Helmholtz equation in Ω_i and Ω_e and the outgoing radiation condition at infinity.

Theorem 3.1.2 *The single layer potential* (3.1.18) *is continuous with respect to the variable y in $\mathbb{R}^3$ and especially when crossing the surface Γ where its value is*

$$u(y) = \int_\Gamma E(x-y)q(x)d\gamma(x), \quad y \in \Gamma, \quad \text{when } q(x) \in C^0(\Gamma). \qquad (3.1.20)$$

Its normal derivatives have discontinuities when crossing Γ. Their limit values on both side of Γ are

$$\left. \frac{\partial u}{\partial n} \right|_{\text{int}} (y) = \frac{q(y)}{2} + \int_\Gamma \frac{\partial}{\partial n_y} E(x-y)q(x)d\gamma(x), \qquad (3.1.21)$$

$$\left. \frac{\partial u}{\partial n} \right|_{\text{ext}} (y) = -\frac{q(y)}{2} + \int_\Gamma \frac{\partial}{\partial n_y} E(x-y)q(x)d\gamma(x). \qquad (3.1.22)$$

The double layer potential (3.1.19) *has a discontinuity when crossing Γ. If $\varphi \in C^0(\Gamma)$, the corresponding limit values on both side of Γ are*

$$u_{\text{int}}(y) = -\frac{\varphi(y)}{2} + \int_\Gamma \frac{\partial}{\partial n_x} E(x-y)\varphi(x)d\gamma(x); \qquad (3.1.23)$$

$$u_{\text{ext}}(y) = \frac{\varphi(y)}{2} + \int_\Gamma \frac{\partial}{\partial n_x} E(x-y) \, \varphi(x)d\gamma(x). \qquad (3.1.24)$$

The normal derivative of a double layer potential is continuous when crossing Γ. The associated kernel admits a strong singularity equivalent to

$1/(|x - y|^3)$ *and so is not an integrable function. Yet, we write it formally as an improper integral*

$$\frac{\partial u}{\partial n}(y) = \oint_\Gamma \frac{\partial^2 E}{\partial n_x \partial n_y}(x - y)\varphi(x)d\gamma(x). \tag{3.1.25}$$

Consequently, any potential sum of a single layer with density q and a double layer with density φ is such that $q = [\partial u/\partial n]$ and $\varphi = [u]$.

Proof

The property (3.1.20) results from the integrability of the singularity of E in a tubular neighbourhood of Γ. This singularity is uniformly integrable with respect to the variable y and thus the classical Lebesgue theorem implies continuity with respect to y and (3.1.20). The representation Theorem 3.1.1 applies to a double layer potential and then a careful look at the formula (3.1.7), in the case where $[u] = \varphi$ and $[\partial u/\partial n] = 0$, shows the formulas (3.1.23) and (3.1.25). The Theorem 3.1.1 also applies to a single layer potential. It follows that

$$[u] = 0 \quad \text{and} \quad q = \left[\frac{\partial u}{\partial n}\right]. \tag{3.1.26}$$

The gradient of a single layer potential is

$$\nabla u(y) = \int_\Gamma \nabla_y E(x - y)q(x)d\gamma(x). \tag{3.1.27}$$

We use the specific moving system of coordinates introduced in the previous section. A point y_+ in a tubular neighbourhood of the surface Γ is determined by its projection y_0 and the distance ρ to Γ. We introduce the "symmetric" point y_- :

$$\begin{cases} y_+ = y_0 + \rho n_{y_0}, \\ y_- = y_0 - \rho n_{y_0}. \end{cases} \tag{3.1.28}$$

We compute the following sum:

$$\begin{cases} ((\nabla u\,(y_+) + \nabla u\,(y_-)) \cdot n_{y_0}) \\[2mm] = \int_\Gamma ((\nabla_{y_+} E\,(x - y_+) + \nabla_{y_-} E\,(x - y_-)) \cdot n_{y_0}) q(x)d\gamma(x) \\[2mm] = \int_\Gamma \left[\frac{e^{ik\,|x - y_+|}}{4\pi|x - y_+|^2}\left(ik - \frac{1}{|x - y_+|}\right)\right. \\[2mm] \left. + \frac{e^{ik\,|x-y_-|}}{4\pi\,|x-y_-|^2}\left(ik - \frac{1}{|x-y_-|}\right)\right]((y_0 - x)\cdot n_{y_0})\,q(x)d\gamma(x) \\[2mm] + \int_\Gamma \left[\frac{e^{ik\,|x - y_+|}}{4\pi\,|x - y_+|^2}\left(ik - \frac{1}{|x - y_+|}\right)\right. \\[2mm] \left. - \frac{e^{ik\,|x - y_-|}}{4\pi\,|x - y_-|^2}\left(ik - \frac{1}{|x - y_-|}\right)\right]\rho q(x)d\gamma(x). \end{cases} \tag{3.1.29}$$

The classical Lebesgue theorem applies to the first part of the right-hand side of (3.1.29) since $((x - y_0) \cdot n_{y_0})$ is equivalent to $|x - y_0|^2$. Its limit is

$$2 \int_\Gamma \frac{\partial}{\partial n_y} E(x - y) q(x) d\gamma(x), \quad \text{when } \rho \text{ tends to } 0.$$

It is more difficult to show that the second part tends to zero with ρ. Let us study the more delicate term which is

$$\left\{ \begin{array}{l} \displaystyle \int_\Gamma \left(\frac{1}{|x - y_+|^3} - \frac{1}{|x - y_-|^3} \right) \rho q(x) d\gamma(x) \\[4mm] \displaystyle = \int_\Gamma \frac{\left(|x-y_-|^2 - |x-y_+|^2\right)\left(|x-y_+|^2 + |x-y_-|^2 + |x-y_+||x-y_-|\right)}{|x - y_+|^3 |x - y_-|^3 \left(|x - y_+| + |x - y_-|\right)} \\[4mm] \hspace{8cm} \times \, \rho q(x) d\gamma(x). \end{array} \right. \tag{3.1.30}$$

We can then write

$$|x - y_-|^2 - |x - y_+|^2 = 4\rho \left((x - y_0) \cdot n_{y_0} \right). \tag{3.1.31}$$

This integral is bounded by

$$\left\{ \begin{array}{l} \displaystyle \left| \int_\Gamma \left(\frac{1}{|x - y_+|^3} - \frac{1}{|x - y_-|^3} \right) \rho q(x) d\gamma(x) \right| \\[6mm] \displaystyle \leq C \int_\Gamma \frac{\rho^2 |x - y_0|^2}{\left(|x - y_0|^2 + \rho^2 \right)^{5/2}} d\gamma(x). \end{array} \right. \tag{3.1.32}$$

We split this last expression into two parts:
– when $|x - y_0| \leq \rho$

$$\int_{|x - y_0| \leq \rho} \frac{\rho^2 |x - y_0|^2}{\left(|x - y_0|^2 + \rho^2 \right)^{5/2}} d\gamma \leq C\rho,$$

– when $|x - y_0| \geq \rho$

$$\left\{ \begin{array}{l} \displaystyle \int_{|x - y_0| \geq \rho} \frac{\rho^2 |x - y_0|^2}{\left(|x - y_0|^2 + \rho^2 \right)^{5/2}} d\gamma \\[6mm] \displaystyle \leq \rho^\varepsilon \int_\Gamma \frac{|x - y_0|^{4-\varepsilon}}{\left(|x - y_0|^2 + \rho^2 \right)^{5/2}} d\gamma \leq \rho^\varepsilon \int_\Gamma \frac{d\gamma}{|x - y_0|^{1+\varepsilon}}, \end{array} \right.$$

from which it follows that the limit is zero. The other terms can be treated similarly. Equalities (3.1.21) and (3.1.22) follow from (3.1.26) and the limit (3.1.29). ∎

There exists a direct proof of the above formulas, not using the jump properties of the potentials. This allows us to show that this theorem implies the uniqueness of the integral representation. Any potential which can be written as a sum of a single layer with density q and a double layer with density $-\varphi$, is such that q is the jump of the normal derivatives and φ is the jump of the limit values of u.

Four integral operators (S, D, D^*, N) on the surface Γ have appeared. They are

$$Su = \int_\Gamma E(x - y)u(x)d\gamma(x), \qquad (3.1.33)$$

$$Du = \int_\Gamma \frac{\partial}{\partial n_x} E(x - y)u(x)d\gamma(x), \qquad (3.1.34)$$

$$D^*u = \int_\Gamma \frac{\partial}{\partial n_y} E(x - y)u(x)d\gamma(x), \qquad (3.1.35)$$

$$Nu = \oint_\Gamma \frac{\partial^2}{\partial n_x \partial n_y} E(x - y)u(x)d\gamma(x). \qquad (3.1.36)$$

The operator D^* is the transpose of the operator D. These different operators are linked together by the **Calderon relations**.

Theorem 3.1.3 *Let us denote by H the operator*

$$H = \begin{pmatrix} -D & S \\ -N & D^* \end{pmatrix}. \qquad (3.1.37)$$

The associated operators

$$C_{\text{int}} = \frac{I}{2} + H, \qquad (3.1.38)$$

$$C_{\text{ext}} = \frac{I}{2} - H \qquad (3.1.39)$$

are projectors, i.e., they satisfy

$$C_{\text{ext}}^2 = C_{\text{ext}}, \quad C_{\text{int}}^2 = C_{\text{int}}, \quad C_{\text{int}} + C_{\text{ext}} = I. \qquad (3.1.40)$$

They are called **Calderon projectors**. *Identities (3.1.40) are equivalent to the set of relations*

$$H^2 = \frac{I}{4}, \qquad (3.1.41)$$

or more explicitly

$$DS = SD^*, \qquad (3.1.42)$$

$$ND = D^*N, \qquad (3.1.43)$$

$$D^2 - SN = \frac{I}{4},$$
(3.1.44)

$$D^{*2} - NS = \frac{I}{4}.$$
(3.1.45)

Proof

Let us consider a potential of the form

$$u = Sq - D\varphi.$$
(3.1.46)

It follows from the representation, Theorem 3.1.1, that the values of u and $\partial u/\partial n$ outside the surface are

$$u_{|\text{ext}} = -\left(\frac{I}{2} + D\right)\varphi + Sq,$$
(3.1.47)

$$\left.\frac{\partial u}{\partial n}\right|_{\text{ext}} = -N\varphi + \left(-\frac{I}{2} + D^*\right)q.$$
(3.1.48)

The representation, Theorem 3.1.1, shows that the potential v, which is given by

$$v = -S\left.\frac{\partial u}{\partial n}\right|_{\text{ext}} + Du|_{\text{ext}},$$
(3.1.49)

vanishes in Ω_i, and has the value u in Ω_e. Thus, its values outside the surface Γ are $u_{|\text{ext}}$ and $\partial u/\partial n|_{\text{ext}}$. It follows that C_{ext} is a projector. The proof in the case of the interior problem is similar. ∎

3.2   Integral Equations for Helmholtz Problems

We exhibit in this section the integral equations associated with the interior and exterior Dirichlet and Neumann problems. We give their expressions and study formally their solvability. We will introduce later the associated variational formulations.

3.2.1   Equations for the single layer potential

A single layer potential has the expression

$$u(y) = \int_\Gamma E(x - y)q(x)d\gamma(x).$$
(3.2.1)

It is continuous across the surface Γ. The function u satisfies the Helmholtz equation inside and outside the surface Γ, and the radiation condition at infinity. When it moreover satisfies the boundary condition, such a potential solves the **interior and exterior Dirichlet problems**. Thus the auxiliary unknowns q must satisfy

$$\int_\Gamma E(x - y)q(x)d\gamma(x) = u_d(y), \quad y \in \Gamma, \tag{3.2.2}$$

or else

$$Sq = u_d. \tag{3.2.3}$$

This equation gives the solution of the interior Dirichlet problem and thus **is not invertible when** $-k^2$ **is an eigenvalue of the interior Dirichlet problem for the Laplacian**. It also gives the unique solution of the exterior Dirichlet problem. The operator S is in fact invertible except for these critical values.

The interior and exterior Neumann problems can be solved using Theorem 3.1.2 and the expressions of the normal derivatives. The associated equations are

$$\frac{q(y)}{2} + \int_\Gamma \frac{\partial}{\partial n_y} E(x - y)q(x)d\gamma(x) = u_n(y), \quad y \in \Gamma \tag{3.2.4}$$

$$-\frac{q(y)}{2} + \int_\Gamma \frac{\partial}{\partial n_y} E(x - y)q(x)d\gamma(x) = u_n(y), \quad y \in \Gamma \tag{3.2.5}$$

or equivalently

$$\left(\frac{I}{2} + D^*\right) q = u_n, \quad \text{interior problem}, \tag{3.2.6}$$

$$\left(-\frac{I}{2} + D^*\right) q = u_n, \quad \text{exterior problem}. \tag{3.2.7}$$

Let w be an eigenvector of the interior Dirichlet problem for the Laplacian. The function whose value is w inside Ω_i and zero outside, satisfies all the hypotheses of the representation, Theorem 3.1.1. Thus, it holds that

$$0 = S\frac{\partial w}{\partial n}, \quad \text{in } \Omega_e, \tag{3.2.8}$$

and

$$\left(-\frac{I}{2} + D^*\right) \frac{\partial w}{\partial n} = 0. \tag{3.2.9}$$

Hence, **the integral equation (3.2.7) has no solution when** $-k^2$ **is an eigenvalue of the interior Dirichlet problem for the Laplacian.** The associated operator is in fact invertible except for these critical values of k^2. We will check later that **the integral equation (3.2.6) is not invertible when** $-k^2$ **is an eigenvalue of the interior Neumann problem for the Laplacian.** The associated operator is in fact invertible except for these critical values of k^2.

3.2.2 Equations for the double layer potential

A double layer potential has the expression

$$u(y) = \int_\Gamma \frac{\partial}{\partial n_x} E(x - y)\varphi(x)d\gamma(x). \tag{3.2.10}$$

Its normal derivative is continuous across the surface Γ. The function u satisfies the Helmholtz equation inside and outside the surface Γ, and the radiation condition at infinity. This potential solves the interior and exterior Dirichlet problems when its values u_{int} and u_{ext} are those given. From the formulas (3.1.23) and (3.1.24), we obtain the two integral equations

$$-\frac{\varphi(y)}{2} + \int_\Gamma \frac{\partial}{\partial n_x} E(x - y)\varphi(x)d\gamma(x) = u_d(y), \quad y \in \Gamma, \tag{3.2.11}$$

$$\frac{\varphi(y)}{2} + \int_\Gamma \frac{\partial}{\partial n_x} E(x - y)\varphi(x)d\gamma(x) = u_d(y), \quad y \in \Gamma, \tag{3.2.12}$$

or in a more condensed notation

$$\left(-\frac{I}{2} + D\right)\varphi = u_d, \quad \text{interior problem}, \tag{3.2.13}$$

$$\left(\frac{I}{2} + D\right)\varphi = u_d, \quad \text{exterior problem}. \tag{3.2.14}$$

The operators D and D^* are mutually transposed with respect to the $L^2(\Gamma)$ scalar product. The real associated eigenvalues and the corresponding real eigenvectors are the same, since in that case eigenvectors are in the kernel of the imaginary part of this operator. From these properties, it results that the **integral equation (3.2.13) is not invertible when $-k^2$ is an eigenvalue of the interior Dirichlet problem for the Laplacian.**

Let w be an eigenvector of the interior Neumann problem for the Laplacian. The function whose value is w inside Ω_i, and zero outside, satisfies all the hypothesis of the representation, Theorem 3.1.1. Its value is

$$v = -Dw_{|\Gamma}. \tag{3.2.15}$$

Its exterior value vanishes and so does its exterior normal derivative. Hence, the **integral equation (3.2.14) is not invertible when $-k^2$ is an eigenvalue of the interior Neumann problem for the Laplacian.** The associated operator is in fact invertible except for these critical values of k^2. The transposed equation associated with the operator $I/2 + D^*$, which is (3.2.14), satisfies the same properties.

The **operator N** is not clearly defined as an integral, but only as an improper integral. Yet, it follows from Representation Theorem 3.1.1 that the **operator N solves both interior and exterior Neumann problems.**

The associated integral equation can be formally written as

$$\oint_\Gamma \frac{\partial^2}{\partial n_x \partial n_y} E(x-y)\varphi(x)d\gamma(x) = -u_n(y), \quad y \in \Gamma, \qquad (3.2.16)$$

or else

$$N\varphi = -u_n. \qquad (3.2.17)$$

The operator N is not invertible when $-k^2$ is an eigenvalue of the interior Neumann problem for the Laplacian. The associated operator, which will be defined precisely later, is in fact invertible except for these critical values of k^2.

Remark *The Brakhage and Werner trick [51]*
All the above representations lead to operators which are not invertible for some critical values of the wave number k. The following trick avoids this difficulty.

Let us consider a representation, sum of a single layer and a double layer potential. It takes the form

$$u(y) = \frac{1}{4\pi}\int_\Gamma \left[i\alpha\frac{e^{ik|x-y|}}{|x-y|} - \frac{\partial}{\partial n_x}\frac{e^{ik|x-y|}}{|x-y|}\right]q(x)d\gamma(x). \qquad (3.2.18)$$

It follows from Representation Theorem 3.1.1 that this potential satisfies

$$\left[\frac{\partial u}{\partial n} - i\alpha u\right]_\Gamma = 0. \qquad (3.2.19)$$

The associated interior problem is

$$\begin{cases} \Delta u + k^2 u = 0, \\ \left(\frac{\partial u}{\partial n} - i\alpha u\right)\Big|_\Gamma = 0, \end{cases} \qquad (3.2.20)$$

which has no eigenvalues when α is real. This potential leads to an invertible operator for every value of k.

3.2.3 The spherical case

We exhibit, in the case where the domain is a sphere, the links between the above operators and their expressions in terms of spherical harmonics expansions.

The operator S^{-1} is associated with $(T_{\text{int}} - T_{\text{ext}})$, which, using the wronskian property (2.6.116) gives

$$S^{-1}u = \sum_{\ell=0}^{\infty}\sum_{m=-\ell}^{\ell}\frac{1}{kj_\ell(k)h_\ell^{(1)}(k)}u_\ell^m Y_\ell^m(\theta,\varphi), \qquad (3.2.21)$$

and thus,

$$Sq = \sum_{\ell=0}^{\infty} \sum_{m=-\ell}^{\ell} k j_\ell(k) h_\ell^{(1)}(k) q_\ell^m Y_\ell^m(\theta, \varphi). \qquad (3.2.22)$$

The operator N^{-1} is associated with $(T_{\text{ext}}^{-1} - T_{\text{int}}^{-1})$, which, using the wronskian property (2.6.116) gives

$$N^{-1}v = \sum_{\ell=0}^{\infty} \sum_{m=-\ell}^{\ell} \frac{1}{k^3 \dfrac{d}{dk} j_\ell(k) \dfrac{d}{dk} h_\ell^{(1)}(k)} v_\ell^m Y_\ell^m(\theta, \varphi), \qquad (3.2.23)$$

or else

$$N\varphi = \sum_{\ell=0}^{\infty} \sum_{m=-\ell}^{\ell} k^3 \frac{d}{dk} j_\ell(k) \frac{d}{dk} h_\ell^{(1)}(k) \varphi_\ell^m Y_\ell^m(\theta, \varphi). \qquad (3.2.24)$$

From (3.1.22), it follows that the operator D^* takes the form

$$D^*q = (T_{\text{ext}} S + 1/2I)\, q \qquad (3.2.25)$$

or else

$$D^*q = \sum_{\ell=0}^{\infty} \sum_{m=-\ell}^{\ell} \left(k^2 j_\ell(k) \frac{d}{dk} h_\ell^{(1)}(k) + 1/2 \right) q_\ell^m Y_\ell^m(\theta, \varphi), \qquad (3.2.26)$$

which has also the expression

$$\begin{cases} D^*q = \dfrac{k^2}{2} \times \\[2mm] \displaystyle\sum_{\ell=0}^{\infty} \sum_{m=-\ell}^{\ell} \left[j_\ell(k) \frac{d}{dk} h_\ell^{(1)}(k) + h_\ell^{(1)}(k) \frac{d}{dk} (j_\ell(k)) \right] q_\ell^m Y_\ell^m(\theta, \varphi). \end{cases} \qquad (3.2.27)$$

Similarly, the operator D takes the form

$$D\varphi = (T_{\text{int}}^{-1} N + 1/2I)\, \varphi \qquad (3.2.28)$$

or

$$\begin{cases} D\varphi = \dfrac{k^2}{2} \times \\[2mm] \displaystyle\sum_{\ell=0}^{\infty} \sum_{m=-\ell}^{\ell} \left[j_\ell(k) \frac{d}{dk} h_\ell^{(1)}(k) + h_\ell^{(1)}(k) \frac{d}{dk} (j_\ell(k)) \right] \varphi_\ell^m Y_\ell^m(\theta, \varphi). \end{cases} \qquad (3.2.29)$$

which is exactly D^*. The Calderon relations expressed in Theorem 3.1.3 are directly deduced from the wronskian property (2.6.116) in this case.

An easy computation leads to the expression of the **corresponding operators** associated with the **Laplace equation** which are

$$Sq = \sum_{\ell=0}^{\infty} \sum_{m=-\ell}^{\ell} \frac{1}{2\ell+1} q_\ell^m Y_\ell^m(\theta, \varphi), \tag{3.2.30}$$

$$N\varphi = -\sum_{\ell=0}^{\infty} \sum_{m=-\ell}^{\ell} \frac{\ell(\ell+1)}{2\ell+1} \varphi_\ell^m Y_\ell^m(\theta, \varphi), \tag{3.2.31}$$

$$D^* q = Dq = -\frac{1}{2} Sq. \tag{3.2.32}$$

The Calderon relations in Theorem 3.1.3 reduce in that case to the unique non-trivial relation

$$-SN + \frac{1}{4} S^2 = \frac{I}{4}. \tag{3.2.33}$$

Besides,

$$-S^{-1} N\varphi = \sum_{\ell=0}^{\infty} \sum_{m=-\ell}^{\ell} \ell(\ell+1) \varphi_\ell^m Y_\ell^m(\theta, \varphi) = -\Delta_S \varphi \tag{3.2.34}$$

from which we obtain

$$S = \frac{1}{2} \left(-\Delta_S + \frac{I}{4} \right)^{-1/2}, \tag{3.2.35}$$

$$N = \frac{1}{2} \Delta_S \left(-\Delta_S + \frac{I}{4} \right)^{-1/2}. \tag{3.2.36}$$

3.2.4   The far field

The asymptotic behavior of the solution at infinity is described by the far field. Its expression can be deduced quite easily from the knowledge of any integral representation, e.g., a single layer potential. Let there be given such a potential

$$u(y) = \frac{1}{4\pi} \int_\Gamma \frac{e^{ik|x-y|}}{|x-y|} q(x) d\gamma(x). \tag{3.2.37}$$

We expand the integral kernel with respect to the variable y. $\mathcal{O}(\varepsilon)$ denotes a function of the small parameter ε bounded by $C\varepsilon$. We can write

$$|x-y|^2 = |y|^2 \left(1 - \frac{2(x \cdot y)}{|y|^2} + \frac{|x|^2}{|y|^2} \right), \tag{3.2.38}$$

$$|x - y| = |y| \left(1 - \frac{(x \cdot y)}{|y|^2} + \mathcal{O} \left(\frac{1}{|y|^2} \right) \right), \tag{3.2.39}$$

$$\frac{1}{|x - y|} = \frac{1}{|y|} \left(1 + \frac{(x \cdot y)}{|y|^2} + \mathcal{O} \left(\frac{1}{|y|^2} \right) \right), \tag{3.2.40}$$

$$e^{ik|x - y|} = e^{ik|y|} e^{-ik(x \cdot y)/|y|} \left(1 + \mathcal{O} \left(\frac{1}{|y|} \right) \right). \tag{3.2.41}$$

It follows that

$$u(y) \sim \frac{e^{ik|y|}}{4\pi|y|} \int_\Gamma e^{-ik(x \cdot y)/|y|} q(x) d\gamma(x). \tag{3.2.42}$$

The quantity

$$A(\theta, \varphi) = \frac{1}{4\pi} \int_\Gamma e^{-ik(x \cdot y)/|y|} q(x) d\gamma(x) \tag{3.2.43}$$

is called the **scattering amplitude**. When Γ is **the unit sphere** S, the scattering amplitude takes the form

$$A(\theta, \varphi) = \sum_{\ell=0}^{\infty} \sum_{m=-\ell}^{\ell} \frac{q_\ell^m}{4\pi} \int_S e^{-ik(x \cdot y)/|y|} Y_\ell^m(\theta', \varphi') d\sigma. \tag{3.2.44}$$

We get the function $e^{-ik(x \cdot y)/|y|}$, which is a plane wave with wave vector $\vec{k} = -k\vec{y}/|y|$. Let $Q(y/|y|)$ be the rotation matrix which maps the vector $\vec{e_3}$ to the vector $\vec{y}/|y|$. From the formula (2.6.99) and Lemma 2.6.2, it follows that

$$e^{-ik(x \cdot y)/|y|} = \sum_{\ell=0}^{\infty} \sqrt{8\pi}(-i)^\ell \sqrt{\ell + 1/2} \, j_\ell(k) \overline{Y}_\ell^0 (Q^{-1}x). \tag{3.2.45}$$

Besides, from formula (2.4.107), Theorem 2.4.5 we know that

$$Y_\ell^m (Q^{-1}\vec{\tau}) = \sum_{j=-\ell}^{\ell} \alpha_j^m \left(\frac{\vec{y}}{|y|} \right) Y_\ell^j(\vec{\tau}) \tag{3.2.46}$$

from which it follows that $(\vec{\tau} = \vec{e_3})$

$$\begin{cases} Y_\ell^0 (\vec{e_3}) = \sqrt{\dfrac{\ell + 1/2}{2\pi}}, \\ Y_\ell^m (\vec{e_3}) = 0 \qquad m \neq 0, \end{cases} \tag{3.2.47}$$

and thus,

$$\alpha_0^m \left(\frac{\vec{y}}{|y|} \right) = \frac{\sqrt{2\pi}}{\sqrt{\ell + 1/2}} Y_\ell^m (Q^{-1}\vec{e_3}). \tag{3.2.48}$$

The unit vector $\vec{y}/|y|$ has spherical coordinates (θ, φ), while the vector $Q^{-1}e_3$ has spherical coordinates $(\theta, \varphi + \pi)$. It results that

$$\alpha_0^m \left(\frac{\vec{y}}{|y|} \right) = \left(\frac{(\ell - m)!}{(\ell + m)!} \right)^{1/2} e^{im\varphi} P_\ell^m(\cos\theta). \tag{3.2.49}$$

But it follows from (2.4.109) that the matrix with coefficients α_j^m is unitary, and hence, the following relation holds:

$$Y_\ell^j(\vec{\tau}) = \sum_{m=-\ell}^{\ell} \alpha_j^m \left(\frac{\vec{y}}{|y|} \right) Y_\ell^m \left(Q^{-1} \vec{\tau} \right). \tag{3.2.50}$$

The choice $Q^{-1}\vec{\tau} = \vec{e_3}$ yields

$$\alpha_j^0 \left(\frac{\vec{y}}{|y|} \right) = \frac{\sqrt{2\pi}}{\sqrt{\ell + 1/2}} Y_\ell^j(Q\vec{e_3}), \tag{3.2.51}$$

$$\alpha_j^0 \left(\frac{\vec{y}}{|y|} \right) = (-1)^m \left(\frac{(\ell - m)!}{(\ell + m)!} \right)^{1/2} e^{im\varphi} P_\ell^m(\cos\theta). \tag{3.2.52}$$

From (3.2.45) and the previous equality, we see that the **scattering amplitude for the sphere** has the expression

$$\begin{cases} A(\theta, \varphi) = \sum_{\ell=0}^{\infty} \sum_{m=-\ell}^{\ell} (-i)^\ell (-1)^m \sqrt{\frac{(\ell + 1/2)(\ell - m)!}{2\pi(\ell + m)!}} \\ \qquad\qquad \times j_\ell(k) q_\ell^m e^{im\varphi} P_\ell^m(\cos\theta), \end{cases} \tag{3.2.53}$$

$$A(\theta, \varphi) = \sum_{\ell=0}^{\infty} \sum_{m=-\ell}^{\ell} (-i)^\ell j_\ell(k) q_\ell^m Y_\ell^m(\theta, \varphi). \tag{3.2.54}$$

Moreover, if we consider a Dirichlet problem created by an incident plane wave with a wave vector $\vec{k} = k\vec{e_3}$, (2.6.117) gives the values of the coefficients q_ℓ^m. It follows that

$$\begin{cases} A(\theta, \varphi) = \sum_{\ell=0}^{\infty} 2(\ell + 1/2) \dfrac{j_\ell(k)}{k h_\ell^{(1)}(k)} P_\ell(\cos\theta) \\ \qquad\quad = \sum_{\ell=0}^{\infty} \sqrt{8\pi} \sqrt{\ell + 1/2} \dfrac{j_\ell(k)}{k h_\ell^{(1)}(k)} Y_\ell^0(\theta). \end{cases} \tag{3.2.55}$$

It is now possible in this expression to examine the high-frequency asymptotic with respect to the parameter k. Replacing the function $1/(k h_\ell^{(1)}(k))$

by $(i)^\ell e^{-ik}$, it follows that

$$
\begin{cases}
A(\theta, \varphi) \sim e^{-ik} \displaystyle\sum_{\ell=0}^{\infty} (i)^\ell (2\ell + 1) j_\ell(k) P_\ell(\cos\theta) \\[2mm]
= e^{-ik} e^{ik\cos\theta} = e^{-2ik\sin^2\frac{\theta}{2}}.
\end{cases}
\tag{3.2.56}
$$

This is exactly the scattering amplitude created by $e^{-ik}\delta(0,0,-1)$, which is a Dirac mass located at the south pole. The southern hemisphere is lit while the northern hemisphere is in the shade. The physical meaning is quite obvious: in the far field, the field scattered by the sphere is approximately that of a bright point located at the south pole with a power proportional to the apparent surface of this sphere. The scattering amplitude possesses more properties arising from the following fundamental lemma, which corresponds physically to the expression of the scattered energy:

Lemma 3.2.1 *let u be the outgoing solution of the Helmholtz equation in an exterior domain Ω_e. It holds that*

$$
\int_S |A(\theta, \varphi)|^2 \, d\sigma = \frac{1}{k} \Im \int_\Gamma \frac{\partial u}{\partial n} \bar{u} \, d\gamma.
\tag{3.2.57}
$$

Proof

Let S_R be the sphere with radius R for a large enough R. Consider u and v, two outgoing solutions of the Helmholtz equation. From the Green formula follows

$$
\int_{S_R} \left(\frac{\partial u}{\partial n} v - \frac{\partial v}{\partial n} u \right) d\sigma = \int_\Gamma \left(\frac{\partial u}{\partial n} v - \frac{\partial v}{\partial n} u \right) d\gamma.
\tag{3.2.58}
$$

Using the radiation condition with the choice $v = \bar{u}$ in this expression, it follows that the right-hand side has a limit when R tends to infinity which is

$$
\lim_{R \to \infty} 2ik \int_{S_R} |u|^2 \, d\sigma = 2ik \int_S |A(\theta, \varphi)|^2 \, d\sigma.
\tag{3.2.59}
$$

∎

It is possible to use this property in specific situations where the field is created by an incident field which satisfies the Helmholtz equation in the domain Ω_e (but not the radiation condition).

We examine two particular cases. The incident field is denoted u_{inc}. The scattered field which satisfies the outgoing radiation condition is denoted u_S. The total field is

$$
u = u_{\text{inc}} + u_S
\tag{3.2.60}
$$

and satisfies on the surface Γ the boundary condition. In the case of a Dirichlet problem, it is

$$u|_\Gamma = 0. \tag{3.2.61}$$

From the Green formula (3.2.58) follows

$$\Im \int_{S_R} \frac{\partial u}{\partial n} \bar{u} d\gamma = 0. \tag{3.2.62}$$

We begin with the case where the incident u_{inc} is an **incoming wave** which admits the expansion

$$u_{\text{inc}}(r,\theta,\varphi) = \sum_{\ell=0}^{\infty} \sum_{m=-\ell}^{\ell} \alpha_\ell^m h_\ell^{(2)}(kr) Y_\ell^m(\theta,\varphi). \tag{3.2.63}$$

(The origin is inside the domain Ω_i and thus u_{inc} satisfies a Helmholtz equation outside the domain Ω_i.) The scattered field u_S takes the form

$$u_S(r,\theta,\varphi) = \sum_{\ell=0}^{\infty} \sum_{m=-\ell}^{\ell} \beta_\ell^m h_\ell^{(1)}(kr) Y_\ell^m(\theta,\varphi). \tag{3.2.64}$$

Lemma 3.2.2 *When the trace of the incident field on the unit sphere belongs to $H^s(S)$, for a real s, it holds that*

$$\lim_{R\to\infty} \int_{S_R} |u_{\text{inc}}|^2 \, d\sigma = \lim_{R\to\infty} \int_{S_R} |u_S|^2 \, d\sigma = \int_S |A(\theta,\varphi)|^2 \, d\sigma. \tag{3.2.65}$$

Proof
In this case, equality (3.2.62) is equivalent to

$$\begin{cases} \Im \left[\int_{S_R} \left(\frac{\partial u_{\text{inc}}}{\partial n} + ik \, u_{\text{inc}} + \frac{\partial u_S}{\partial n} - iku_S \right) (\bar{u}_{\text{inc}} + \bar{u}_S) \right. \\ \qquad \left. - ik \left(|u_{\text{inc}}|^2 - |u_S|^2 \right) d\sigma \right] = 0. \end{cases} \tag{3.2.66}$$

H^s regularity of the trace implies that u_{inc} and u_S decrease as r increases in the same way as $1/r$. It follows from the radiation conditions that the first part of (3.2.66) tends to zero with R. ■

The operator which, to the scattering amplitude of an ingoing wave, associates the scattering amplitude of the outgoing scattered wave is unitary. A more interesting case is the one where the incident wave is a **plane wave** whose wave vector is y.

$$u_{\text{inc}} = e^{ik(y \cdot x)}, \quad |y| = 1. \tag{3.2.67}$$

The scattering amplitude associated with this plane wave is a function defined on the unit sphere that is denoted $A(x,y)$, $|x| = 1$. We associate with this function the integral operator

$$A\varphi = \frac{1}{4\pi} \int_S A(y,x)\varphi(x)d\sigma(x). \qquad (3.2.68)$$

Theorem 3.2.1 *The scattering amplitude $A(x,y)$ satisfies the identities*

$$\frac{k}{4\pi} \int_S |A(x,y)|^2 \, d\sigma(x) = \Im A(y,y), \qquad (3.2.69)$$

$$|(I + 2ikA)\,\varphi|^2_{L^2(S)} = |\varphi|^2_{L^2(S)}. \qquad (3.2.70)$$

Moreover, it satisfies the reciprocity principle, which is the identity

$$A(x,y) = A(-y,-x). \qquad (3.2.71)$$

Proof

Let u and v be two solutions associated with the two incident plane waves u_{inc} (with wave vector y) and v_{inc} (with wave vector z). The Green formula (3.2.58) yields

$$\int_{S_R} \left(\frac{\partial u}{\partial n}\overline{v} - \frac{\partial \overline{v}}{\partial n}u \right) d\sigma = 0. \qquad (3.2.72)$$

The Green formula in the ball B_R with the functions u_{inc} and v_{inc} shows that

$$\int_{S_R} \left(\frac{\partial u_{\text{inc}}}{\partial n}\overline{v}_{\text{inc}} - \frac{\partial \overline{v}_{\text{inc}}}{\partial n}u_{\text{inc}} \right) d\sigma = 0. \qquad (3.2.73)$$

Formula (3.2.59) leads to the limit

$$\lim_{R\to\infty} \int_{S_R} \left(\frac{\partial u_S}{\partial n}\overline{v}_S - \frac{\partial \overline{v}_S}{\partial n}u_S \right) d\sigma = 2ik \int_S A(x,y)\overline{A}(x,z)d\sigma(x). \qquad (3.2.74)$$

Thus, identity (3.2.72) takes the form

$$\left\{ \begin{array}{l} \displaystyle\int_{S_R} \left[\frac{\partial u_{\text{inc}}}{\partial n}\overline{v}_S + \frac{\partial u_S}{\partial n}\overline{v}_{\text{inc}} + \frac{\partial u_S}{\partial n}\overline{v}_S \right. \\[4mm] \displaystyle\hspace{1cm} \left. - \frac{\partial \overline{v}_{\text{inc}}}{\partial n}u_S - \frac{\partial \overline{v}_S}{\partial n}u_{\text{inc}} - \frac{\partial \overline{v}_S}{\partial n}u_S \right] d\sigma = 0, \end{array} \right. \qquad (3.2.75)$$

and we already know the limit of all the terms which contain only scattered fields. We now examine the limit of all the terms which contain both incident and scattered fields. Let us consider

$$\left\{ \begin{array}{l} \displaystyle\int_{S_R} \left[\frac{\partial u_{\text{inc}}}{\partial n}\overline{v}_S - \frac{\partial \overline{v}_S}{\partial n}u_{\text{inc}} \right] d\sigma \\[4mm] \displaystyle= \int_{S_R} \left[(\frac{\partial u_{\text{inc}}}{\partial n} + iku_{\text{inc}})\overline{v}_S - (\frac{\partial \overline{v}_S}{\partial n} + ik\overline{v}_S)u_{\text{inc}} \right] d\sigma. \end{array} \right. \qquad (3.2.76)$$

The spherical coordinates θ, φ for the sphere S_R are chosen such that the incident vector y associated with u_{inc} is the vector $\vec{e_3}$. Besides, as the fundamental solution satisfies exactly $\partial E/\partial n - ikE + E/r = 0$, it follows that

$$\left| \frac{\partial u_S}{\partial n} - iku_S + \frac{u_S}{r} \right| \leq \frac{c}{r^3}, \quad \text{for large } r, \tag{3.2.77}$$

$$\left| \frac{\partial v_S}{\partial n} - ikv_S + \frac{v_S}{r} \right| \leq \frac{c}{r^3}, \quad \text{for large } r. \tag{3.2.78}$$

Thus, it holds that

$$
\left\{
\begin{aligned}
& \lim_{R \to \infty} \frac{1}{ikR} \int_{S_R} \left(\frac{\partial u_{\mathrm{inc}}}{\partial n} + iku_{\mathrm{inc}} \right) \overline{v}_S d\sigma \\
& = \lim_{R \to \infty} \int_S e^{ikR(\cos\theta - 1)} (1 + \cos\theta) \overline{A}(\theta, \varphi, z) \sin\theta d\theta d\varphi.
\end{aligned}
\right.
\tag{3.2.79}
$$

Integrating by parts with respect to the variable θ, we obtain

$$
\left\{
\begin{aligned}
& \int_0^\pi e^{ikR\cos\theta} \varphi(\theta) \sin\theta d\theta = \frac{1}{ikR} \int_0^\pi e^{ikR\cos\theta} \frac{d\varphi}{d\theta}(\theta) d\theta \\
& \qquad + \frac{1}{ikR} \left[e^{ikR} \varphi(0) - e^{-ikR} \varphi(\pi) \right].
\end{aligned}
\right.
\tag{3.2.80}
$$

From this identity and equality (3.2.79) follows

$$
\left\{
\begin{aligned}
& Re^{-ikR} \int_S e^{ikR\cos\theta} \left(ik(1 + \cos\theta) \right) \overline{A}(\theta, \varphi, z) \sin\theta d\theta d\varphi \\
& = e^{-ikR} \int_0^\pi \int_0^{2\pi} e^{ikR\cos\theta} \Bigg[-\sin\theta \overline{A}(\theta, \varphi, z) \\
& \qquad\qquad + (1 + \cos\theta) \frac{d}{d\theta} \overline{A}(\theta, \varphi, z) \Bigg] d\theta d\varphi \\
& \qquad + 2 \int_0^{2\pi} \overline{A}(0, \varphi, z) d\varphi.
\end{aligned}
\right.
\tag{3.2.81}
$$

The **stationary phase** theorem shows that the left-hand side tends to zero in the same way as $1/\sqrt{R}$, and so does the right-hand side of identity (3.2.76) (this is clear in spherical coordinates). Using (3.2.76) and (3.2.81), a limit process gives

$$\lim_{R \to \infty} \int_{S_R} \left(\frac{\partial u_{\mathrm{inc}}}{\partial n} \overline{v}_S - \frac{\partial \overline{v}_S}{\partial n} u_{\mathrm{inc}} \right) d\sigma = 4\pi \overline{A}(y, z). \tag{3.2.82}$$

Taking into account the limits (3.2.74) and (3.2.82) and the similar formula associated with v_{inc}, it follows that

$$\int_S A(x, y) \overline{A}(x, z) d\sigma(x) + \frac{4\pi}{2ik} \left(\overline{A}(y, z) - A(z, y) \right) = 0. \tag{3.2.83}$$

When $y = z$, this is (3.2.69).

We can check that identity (3.2.83), when expressed in terms of the kernel of the operator A, is exactly the unitary property (3.2.70). Moreover, we have shown that

$$4\pi A(y, z) = \lim_{R \to \infty} \int_{S_R} \left(\frac{\partial \overline{u}_{\text{inc}}}{\partial n} v_S - \frac{\partial v_S}{\partial n} \overline{u}_{\text{inc}} \right) d\sigma. \tag{3.2.84}$$

Let us denote by w_S the solution of the Dirichlet problem associated with the incident plane wave $\overline{u}_{\text{inc}}$ with wave vector $-y$. From the Green formula (3.2.58) follows

$$\int_{S_R} \left(\frac{\partial \overline{u}_{\text{inc}}}{\partial n} v_S - \frac{\partial v_S}{\partial n} \overline{u}_{\text{inc}} \right) d\sigma = \int_\Gamma \left(\frac{\partial \overline{u}_{\text{inc}}}{\partial n} v_S - \frac{\partial v_S}{\partial n} \overline{u}_{\text{inc}} \right) d\gamma. \tag{3.2.85}$$

The boundary conditions show that this can be rewritten as

$$\int_{S_R} \left(\frac{\partial \overline{u}_{\text{inc}}}{\partial n} v_S - \frac{\partial v_S}{\partial n} \overline{u}_{\text{inc}} \right) d\sigma = - \int_\Gamma \left(\frac{\partial \overline{u}_{\text{inc}}}{\partial n} v_{\text{inc}} - \frac{\partial v_S}{\partial n} w_S \right) d\gamma. \tag{3.2.86}$$

Besides, it holds that

$$\int_\Gamma \left(\frac{\partial \overline{u}_{\text{inc}}}{\partial n} v_{\text{inc}} - \frac{\partial v_{\text{inc}}}{\partial n} \overline{u}_{\text{inc}} \right) d\gamma = 0, \tag{3.2.87}$$

which yields

$$\int_{S_R} \left(\frac{\partial \overline{u}_{\text{inc}}}{\partial n} v_S - \frac{\partial v_S}{\partial n} \overline{u}_{\text{inc}} \right) d\sigma = - \int_\Gamma \left(\frac{\partial v_{\text{inc}}}{\partial n} \overline{u}_{\text{inc}} - \frac{\partial v_S}{\partial n} w_S \right) d\gamma, \tag{3.2.88}$$

which can be rewritten as

$$\begin{cases} \displaystyle\int_\Gamma \left(\frac{\partial v_{\text{inc}}}{\partial n} \overline{u}_{\text{inc}} - \frac{\partial v_S}{\partial n} w_S \right) d\gamma \\ \\ \displaystyle = - \int_\Gamma \left(\frac{\partial v_{\text{inc}}}{\partial n} w_S - \frac{\partial w_S}{\partial n} v_{\text{inc}} + \frac{\partial v_S}{\partial n} w_S - \frac{\partial w_S}{\partial n} v_S \right) d\gamma. \end{cases} \tag{3.2.89}$$

We also have

$$\int_\Gamma \left(\frac{\partial w_S}{\partial n} v_S - \frac{\partial v_S}{\partial n} w_S \right) d\gamma = \int_{S_R} \left(\frac{\partial w_S}{\partial n} v_S - \frac{\partial v_S}{\partial n} w_S \right) d\sigma, \tag{3.2.90}$$

whose limit is zero when R tends to infinity. We obtain

$$\begin{cases} \displaystyle\lim_{R \to \infty} \int_{S_R} \left(\frac{\partial \overline{u}_{\text{inc}}}{\partial n} v_S - \frac{\partial v_S}{\partial n} \overline{u}_{\text{inc}} \right) d\sigma \\ \\ \displaystyle = \lim_{R \to \infty} \int_{S_R} \left(\frac{\partial v_{\text{inc}}}{\partial n} w_S - \frac{\partial w_S}{\partial n} v_{\text{inc}} \right) d\sigma \end{cases} \tag{3.2.91}$$

which is the reciprocity principle.   ∎

The reciprocity principle can be expressed in the following way: **The scattering amplitude created by an incoming incident plane wave with wave vector** y, **in the direction** z, **has exactly the same value as the scattering amplitude created by an incoming incident plane wave with wave vector** z, **in the direction** y.

When the domain Ω is the unit sphere, the scattering amplitude admits the expression (3.2.55), which is also equivalent to

$$
\begin{cases}
A(x,y) = 4\pi \displaystyle\sum_{\ell=0}^{\infty} \frac{j_\ell(k)}{kh_\ell^{(1)}(k)} \sum_{m=-\ell}^{\ell} \overline{Y_\ell^m}(y) Y_\ell^m(x) \\[4mm]
\quad = \displaystyle\sum_{\ell=0}^{\infty} (2\ell+1) \frac{j_\ell(k)}{kh_\ell^{(1)}(k)} P_\ell(t); \qquad t = (x \cdot y).
\end{cases}
\tag{3.2.92}
$$

Theorem 3.2.1 results directly from an easy computation in that case.

3.2.5   The physical optics approximation for the sphere

We have given a first formal approximation of the high frequency limit of the Dirichlet problem for a sphere. A more precise expression, called physical optics approximation, can be obtained starting from the nonlinear eikonal equation. We will not present this technique here, but we will exhibit this approximation in the case of a sphere, and give some estimates in terms of the small parameter $1/k$.

The physical optics approximation for an exterior Dirichlet problem consists in replacing the density q in the integral expression of the scattered field by the density

$$
q_{app} = \begin{cases}
-2\dfrac{\partial u_{\text{inc}}}{\partial n}, & \text{in the lit zone,} \\[4mm]
0, & \text{in the shade.}
\end{cases}
\tag{3.2.93}
$$

When the wave vector is $\vec{k} = k\vec{e_3}$, we have seen that the southern hemisphere is lit while the northern hemisphere is in the shade. The potential q has the expression

$$
q(\theta, \varphi) = \sum_{\ell=0}^{\infty} 2\sqrt{2\pi} \frac{\sqrt{\ell + 1/2}}{kh_\ell^{(1)}(k)} Y_\ell^0(\theta).
\tag{3.2.94}
$$

We are going to expand the approximated density q_{app} in spherical harmonics. We denote by $Y(x)$ the Heaviside function whose value is 1 when $x < 0$ and zero when $x > 0$.

Lemma 3.2.3 *It holds that*

$$
e^{ikx} Y(x) = \sum_{\ell=0}^{\infty} (i)^{\ell+1} (\ell + 1/2) v_\ell(k) P_\ell(x),
\tag{3.2.95}
$$

$$v_\ell(k) = h_\ell^{(2)}(k) - \frac{1}{k} R_\ell\left(\frac{1}{k}\right). \tag{3.2.96}$$

When ℓ is even,

$$\begin{cases} R_\ell(t) = (2\ell - 1) \times \cdots \times 5 \times 3 \times 1 \, t^\ell \\[2mm] \quad + \dfrac{(2\ell - 3) \times \cdots \times 3 \times 1}{2} t^{\ell-2} + \cdots \\[3mm] \quad + \dfrac{(\ell - 1 + 2m) \times \cdots \times 3 \times 1}{2 \times 4 \times \cdots \times (\ell - 2m)} t^{2m} + \cdots \\[3mm] \quad + \dfrac{(\ell - 1) \times \cdots \times 5 \times 3 \times 1}{2 \times 4 \times \cdots \times \ell}. \end{cases} \tag{3.2.97}$$

When ℓ is odd,

$$\begin{cases} R_\ell(t) = (2\ell - 1) \times \cdots \times 5 \times 3 \times 1 \, t^\ell + \cdots \\[2mm] \quad + \dfrac{(\ell + 2m) \times \cdots \times 3 \times 1}{2 \times 4 \times \cdots \times (\ell - 1 - 2m)} t^{2m+1} + \cdots \\[3mm] \quad + \dfrac{\ell \times \cdots \times 5 \times 3 \times 1}{2 \times 4 \times \cdots \times (\ell - 1)} t. \end{cases} \tag{3.2.98}$$

Proof

We proceed in a similar way as in Lemma 2.6.2. By construction

$$v_\ell(k) = (-i)^{\ell+1} \int_{-1}^{0} e^{ikx} P_\ell(x) dx \tag{3.2.99}$$

and in particular

$$v_0(k) = \frac{e^{-ik} - 1}{k} = h_0^{(2)}(k) - \frac{1}{k}, \tag{3.2.100}$$

$$v_1(k) = -\int_{-1}^{0} e^{ikx} x \, dx = h_1^{(2)}(k) - \frac{1}{k^2}. \tag{3.2.101}$$

It follows from Theorem 2.4.2 and Lemma 2.4.4 that

$$\begin{cases} v_\ell = -\dfrac{(-i)^{\ell+1}}{\ell(\ell + 1)} \int_{-1}^{0} e^{ikx} \dfrac{d}{dx}(1 - x^2) \dfrac{d}{dx} P_\ell(x) dx \\[3mm] \quad = \dfrac{(-i)^{\ell+1} ik}{(2\ell + 1)} \int_{-1}^{0} e^{ikx} \left(P_{\ell-1}(x) - P_{\ell+1}(x)\right) dx \\[3mm] \qquad - \dfrac{(-i)^{\ell+1}}{2\ell + 1} \left(P_{\ell+1}(0) - P_{\ell-1}(0)\right), \end{cases} \tag{3.2.102}$$

and thus the following recursion formula holds:

$$\begin{cases} \left(\dfrac{2\ell+1}{k}\right) v_\ell(k) - v_{\ell-1}(k) - v_{\ell+1}(k) \\ \qquad = -\dfrac{(-i)^{\ell+1}}{k}\left(P_{\ell+1}(0) - P_{\ell-1}(0)\right). \end{cases} \qquad (3.2.103)$$

This recursion relation admits a special solution which is a polynomial in the variable $1/k$. The solutions of the homogeneous equation are the spherical Bessel functions, from which it follows that

$$v_\ell(k) = h_\ell^{(2)}(k) - \frac{1}{k}R_\ell\left(\frac{1}{k}\right). \qquad (3.2.104)$$

The functions R_ℓ satisfy $\left(t = \dfrac{1}{k}\right)$

$$R_0 = 1, \quad R_1 = t, \qquad (3.2.105)$$

and the recursion relation

$$\begin{cases} (2\ell+1)tR_\ell(t) - R_{\ell-1}(t) - R_{\ell+1}(t) \\ \qquad = (-i)^{\ell+1}\left(P_{\ell+1}(0) - P_{\ell-1}(0)\right). \end{cases} \qquad (3.2.106)$$

From Lemma (2.6.2), it follows that the coefficients $P_\ell(0)$ satisfy the recursion

$$(\ell+1)P_{\ell+1}(0) = -\ell P_{\ell-1}(0), \qquad (P_0(0) = 0, P_1(0) = 0). \qquad (3.2.107)$$

They vanish when ℓ is odd and have the value

$$P_\ell(0) = (-1)^{\frac{\ell}{2}}\frac{1\times 3\times 5\times(\ell-1)}{2\times 4\times 6\times\cdots\times\ell} = (-1)^{\ell/2}\frac{\ell!}{2^\ell\left(\frac{\ell}{2}!\right)^2}, \qquad (3.2.108)$$

for ℓ even. The values of $R_\ell(t)$ are then computed using the recursion. ∎

Lemma 3.2.4 *It holds that*

$$xe^{ikx}Y(x) = \sum_{\ell=0}^{\infty}(i)^{\ell+1}\left(\ell+1/2\right)w_\ell(k)P_\ell(x), \qquad (3.2.109)$$

$$w_\ell(k) = \frac{d}{dk}h_\ell^{(2)}(k) + \frac{1}{k}S_\ell\left(\frac{1}{k}\right). \qquad (3.2.110)$$

When ℓ is even,

$$\begin{cases} S_\ell(t) = 1\times 3\times 5\times\cdots\times(2\ell-1)(\ell+1)t^{\ell+1} + \cdots \\ \qquad + \displaystyle\sum_{m=0}^{\frac{\ell}{2}-1}\frac{1\times 3\times\cdots\times(\ell-1+2m)}{2\times 4\times\cdots\times(\ell-2m)}(2m+1)t^{2m+1}. \end{cases} \qquad (3.2.111)$$

When ℓ is odd,

$$\begin{cases} S_\ell(t) = 1 \times 3 \times 5 \times \cdots \times (2\ell - 1)(\ell + 1)t^{\ell+1} + \cdots \\ \\ \qquad + \displaystyle\sum_{m=1}^{\frac{\ell-1}{2}-1} \frac{1 \times 3 \times \cdots \times (\ell - 2 + 2m)}{2 \times 4 \times \cdots \times (\ell - 1 - 2m)} 2mt^{2m}. \end{cases} \qquad (3.2.112)$$

Proof

From the recursion relation (2.4.35)) follows that

$$w_\ell(k) = (-i)^{\ell+1} \int_{-1}^{0} e^{ikx} x P_\ell(x) dx, \qquad (3.2.113)$$

$$\begin{cases} w_\ell(k) \\ \\ = (-i)^{\ell+1} \displaystyle\int_{-1}^{0} e^{ikx} \left(\frac{\ell+1}{2\ell+1} P_{\ell+1}(x) + \frac{\ell}{2\ell+1} P_{\ell-1}(x) \right) dx, \end{cases} \qquad (3.2.114)$$

or equivalently

$$w_\ell(k) = \frac{\ell}{2\ell + 1} v_{\ell-1}(k) - \frac{\ell+1}{2\ell+1} v_{\ell+1}(k). \qquad (3.2.115)$$

Theorem 2.6.1 implies that

$$\frac{d}{dr} h_\ell^{(2)}(r) = \frac{\ell}{2\ell+1} h_{\ell-1}^{(2)}(r) - \frac{\ell+1}{2\ell+1} h_{\ell+1}^{(2)}(r), \qquad (3.2.116)$$

which shows equality (3.2.110) with

$$S_\ell(t) = \frac{\ell+1}{2\ell+1} R_{\ell+1}(t) - \frac{\ell}{2\ell+1} R_{\ell-1}(t). \qquad (3.2.117)$$

The expression of S_ℓ then results from the recursion. ∎

We have proved that the expression of the approximate potential is

$$q_{app}(\theta, \varphi) = \sum_{\ell=0}^{\infty} 2 \, ik\sqrt{2\pi} \sqrt{\ell + \frac{1}{2}} w_l(k) Y_\ell^0(\theta). \qquad (3.2.118)$$

The difference takes the form

$$\begin{cases} q_{app}(\theta) - q(\theta) \\ \\ = \displaystyle\sum_{\ell=0}^{\infty} i\sqrt{8\pi(\ell+\frac{1}{2})} \left(\frac{i}{kh_\ell^{(1)}(k)} + k\frac{d}{dk} h_\ell^{(2)}(k) + S_\ell\left(\frac{1}{k}\right) \right) Y_\ell^0(\theta) \\ \\ = \displaystyle\sum_{\ell=0}^{\infty} i\sqrt{8\pi(\ell+\frac{1}{2})} \left(S_\ell\left(\frac{1}{k}\right) - \frac{p_\ell\left(\frac{1}{k}\right)}{q_\ell\left(\frac{1}{k}\right)} h_\ell^{(2)}(k) \right) Y_\ell^0(\theta) \end{cases} \qquad (3.2.119)$$

(the polynomials p_ℓ and q_ℓ were defined by (2.6.21), Theorem 2.6.1). We hope that this quantity tends to zero with the small parameter $1/k$. In fact, all the coefficients behave at least as $1/k$.

The equivalent for ℓ fixed and k large compared to ℓ (ℓ odd) is

$$S_\ell\left(\frac{1}{k}\right) - \frac{p_\ell\left(\frac{1}{k}\right)}{k^2 h_\ell^{(1)}(k)} \sim \left(\frac{1 \times 3 \times \cdots \times (\ell-1)}{2 \times 4 \times \cdots \times \ell} - (i)^\ell e^{-ik}\right)\frac{1}{k}, \quad (3.2.120)$$

which is bounded by $1/k$. But the above expression is not well suited to obtain bounds when ℓ is large compared to k. It admits another expression:

Lemma 3.2.5 *The coefficients $v_\ell(k)$ and $w_\ell(k)$ can be written*

$$v_\ell(k) = -i\frac{k^\ell}{2^\ell \ell!} \int_{-1}^0 e^{ikx}\left(1 - x^2\right)^\ell dx + V_\ell(k); \quad (3.2.121)$$

and for ℓ even,

$$V_\ell(k) = \sum_{m=1}^{\frac{\ell}{2}} \frac{(\ell - 2m)!}{2^\ell \left(\frac{\ell-2m}{2}\right)! \left(\frac{\ell+2m}{2}\right)!} k^{2m-1}; \quad (3.2.122)$$

for ℓ odd,

$$V_\ell(k) = \sum_{m=0}^{\frac{\ell-1}{2}} \frac{(\ell - 1 - 2m)!}{2^\ell \left(\frac{\ell-1-2m}{2}\right)! \left(\frac{\ell+1+2m}{2}\right)!} k^{2m}. \quad (3.2.123)$$

In addition

$$\begin{cases} w_\ell(k) = i\left[\frac{k^{\ell+1}}{2^{\ell+1}\ell!(2\ell+1)} \int_{-1}^0 e^{ikx}(1 - x^2)^{\ell+1}dx \right. \\ \left. - \frac{k^{\ell-1}\ell^2}{2^{\ell-1}\ell!(2\ell+1)} \int_{-1}^0 e^{ikx}(1 - x^2)^{\ell+1}dx\right] + W_\ell(k); \end{cases} \quad (3.2.124)$$

for ℓ even,

$$W_\ell(k) = \sum_{m=0}^{\frac{\ell-2}{2}} \frac{(\ell - 2 - 2m)!(2m+1)}{2^\ell \left(\frac{\ell-2-2m}{2}\right)! \left(\frac{\ell+2m+2}{2}\right)!} k^{2m} - \frac{1}{2^{\ell+1}\ell!(2\ell+1)} k^\ell; \quad (3.2.125)$$

for ℓ odd,

$$W_\ell(k) = \sum_{m=1}^{\frac{\ell-1}{2}} \frac{(\ell - 1 - 2m)!\, 2m}{2^\ell \left(\frac{\ell-1-2m}{2}\right)! \left(\frac{\ell+1+2m}{2}\right)!} k^{2m-1} - \frac{1}{2^{\ell+1}\ell!(2\ell+1)} k^\ell. \quad (3.2.126)$$

Proof

Let us examine v_ℓ. From the Rodriguez formula follows

$$v_\ell(k) = -\frac{(i)^{\ell+1}}{2^\ell \ell!} \int_{-1}^0 e^{ikx} \left(\frac{d}{dx}\right)^\ell (1 - x^2)^\ell\, dx. \quad (3.2.127)$$

We integrate by parts ℓ times and use the value of $(d/dx)^{\ell} (1 - x^2)^{\ell}$ at $x = 0$, given by (2.4.123),

$$\left(\frac{d}{dx}\right)^{\ell-m} (1 - x^2)^{\ell} = \begin{cases} 0, & \text{for } (\ell - m) \text{ odd}, \\ (-C)_{\ell}^{\frac{\ell-m}{2}} (\ell - m)!, & \text{for } (\ell - m) \text{ even}, \end{cases} \quad (3.2.128)$$

from which it follows that

$$v_{\ell}(k) = -\frac{ik^{\ell}}{2^{\ell}\ell!} \int_{-1}^{0} e^{ikx} (1 - x^2)^{\ell} dx + V_{\ell}(k). \quad (3.2.129)$$

For ℓ even,

$$V_{\ell}(k) = \frac{1}{2^{\ell}\ell!} \left(\sum_{m=1}^{\frac{\ell}{2}} C_{\ell}^{\frac{\ell-2m}{2}} (\ell - 2m)! k^{2m-1} \right). \quad (3.2.130)$$

For ℓ odd,

$$V_{\ell}(k) = \frac{1}{2^{\ell}\ell!} \left(\sum_{m=0}^{\frac{\ell-1}{2}} C_{\ell}^{\frac{\ell-1-2m}{2}} (\ell - 1 - 2m)! k^{2m} \right). \quad (3.2.131)$$

From (3.2.84), we infer

$$W_{\ell}(k) = \frac{\ell}{2\ell+1} v_{\ell-1}(k) - \frac{\ell+1}{2\ell+1} v_{\ell+1}(k). \quad (3.2.132)$$

This gives the expression of W_{ℓ}.   ∎

3.3 Integral Equations for the Laplace Problem

We give and prove in this section some specific properties of the integral equations associated with the Laplace equation, that were introduced in Section 3.1 in the specific case $k = 0$. We introduce some variational formulations with the help of the tools of differential geometry. We study and prove some coercivity results for these variational formulations. They will be prove to be useful in the Helmholtz case, since the kernel associated with the Laplace operator is the principal part of the kernel of the associated integral equations.

Theorem 3.3.1 *The integral equation associated with the* **single layer potential**

$$u(y) = \frac{1}{4\pi} \int_{\Gamma} \frac{1}{|x - y|} q(x) d\gamma(x) \quad (3.3.1)$$

for **the Dirichlet Laplace problem** *is*

$$\frac{1}{4\pi} \int_{\Gamma} \frac{1}{|x - y|} q(x) d\gamma(x) = u_d(y), \quad y \in \Gamma. \quad (3.3.2)$$

The corresponding operator, denoted by S, is an **isomorphism of**
$H^{-1/2}(\Gamma)$ **onto** $H^{1/2}(\Gamma)$ *which satisfies the coercivity property*

$$\int_\Gamma (Sq)\bar{q}d\gamma \geq \alpha \|q\|^2_{H^{-1/2}(\Gamma)}, \quad \alpha > 0, \quad \forall q \in H^{-1/2}(\Gamma). \tag{3.3.3}$$

Proof

For $q \in C^0(\Gamma)$, the potential u satisfies the system of equations

$$\begin{cases} \Delta u = 0, & \text{in } \Omega_i \text{ and } \Omega_e, \\ [u]_\Gamma = 0, \\ \left[\dfrac{\partial u}{\partial n}\right]_\Gamma = q. \end{cases} \tag{3.3.4}$$

A variational formulation for this problem is

$$\int_{\mathbb{R}^3} (\nabla u \cdot \nabla v)dx = \int_\Gamma qvd\gamma, \quad \forall v \in W^{1,-1}\left(\mathbb{R}^3\right). \tag{3.3.5}$$

The space $W^{1,-1}\left(\mathbb{R}^3\right)$ was introduced in Section 2.5.4.

Theorem 2.5.13 shows that the bilinear form in the left-hand side of (3.3.5), is coercive on the space $W^{1,-1}\left(\mathbb{R}^3\right)$. Thus, the Lax-Milgram theorem proves the existence of a unique solution to this problem when q is in the space $H^{-1/2}(\Gamma)$, as this space is the dual of the trace space $H^{1/2}(\Gamma)$. Notice that the integral in (3.3.2) is a Lebesgue integral when $q \in L^\infty(\Gamma)$, but ceases to be so for a less regular q. It is not even a duality when $q \in H^{-1/2}(\Gamma)$.

From the trace Theorem 2.5.3, we know that the trace on the surface Γ of the solution of the problem (3.3.4) belongs to the space $H^{1/2}(\Gamma)$. Thus, the operator S is continuous from $H^{-1/2}(\Gamma)$ into $H^{1/2}(\Gamma)$. The inverse operator associated with $u_d \in H^{1/2}(\Gamma)$, the value of q. Both interior and exterior Dirichlet problems admit a unique solution in $H^1(\Omega_i)$ and $W^{1,-1}(\Omega_e)$. The interior normal derivative is then defined as

$$\int_\Gamma \left.\frac{\partial u}{\partial n}\right|_{\text{int}} vd\gamma = \int_{\Omega_i} (\nabla u \cdot \nabla v)\, dx, \quad \forall v \in H^1(\Omega_i), \tag{3.3.6}$$

while the exterior normal derivative is defined as

$$\int_\Gamma \left.\frac{\partial u}{\partial n}\right|_{\text{ext}} vd\gamma = -\int_{\Omega_e} (\nabla u \cdot \nabla v)\, dx, \quad \forall v \in W^{1,-1}(\Omega_i). \tag{3.3.7}$$

The trace Theorem 2.5.3 proves that both of them are in $H^{-1/2}(\Gamma)$. The function q is now

$$q = \left.\frac{\partial u}{\partial n}\right|_{\text{int}} - \left.\frac{\partial u}{\partial n}\right|_{\text{ext}} \tag{3.3.8}$$

and satisfies

$$Sq = u_d \tag{3.3.9}$$

when q is regular enough to use the Representation Theorem 3.1.1. A density argument then shows that the operator S admits a unique extension, which is a bijective application from $H^{-1/2}(\Gamma)$ onto $H^{1/2}(\Gamma)$.

Moreover, subtracting equalities (3.3.6) and (3.3.7), we obtain

$$\int_{\Gamma}(Sq)q d\gamma = \int_{\Gamma} qu d\gamma = \int_{\mathbb{R}^3} |\nabla u|^2 dx. \tag{3.3.10}$$

The continuity of the trace implies that

$$\int_{\mathbb{R}^3} |\nabla u|^2 d\gamma \geq \beta \|u\|^2_{H^{1/2}(\Gamma)}, \quad \beta > 0, \tag{3.3.11}$$

while the continuity of S^{-1} is equivalent to

$$\|u\|^2_{H^{1/2}(\Gamma)} \geq \gamma \|q\|^2_{H^{-1/2}(\Gamma)}, \quad \gamma > 0. \tag{3.3.12}$$

Combining (3.3.11) and (3.3.12) yields (3.3.3).   ∎

Remark

Another way to express the property (3.3.3) is merely to assert that

$$\int_{\Gamma}\int_{\Gamma} \frac{q(x)q(y)}{|x-y|} d\gamma(x) d\gamma(y) \tag{3.3.13}$$

is a scalar product on the space $H^{-1/2}(\Gamma)$.   ∎

Let us consider the **double layer potential** and the equation

$$-\frac{1}{4\pi} \oint_{\Gamma} \frac{\partial^2}{\partial n_x \partial n_y} \left(\frac{1}{|x-y|}\right) \varphi(x) d\gamma(x) = u_n(y), \quad \forall y \in \Gamma. \tag{3.3.14}$$

The corresponding operator is denoted by N. We can now state

Theorem 3.3.2 *The operator N associated with the integral equation (3.3.14) is an isomorphism from $H^{1/2}(\Gamma)/\mathbb{R}$ onto the subspace of elements in $H^{-1/2}(\Gamma)$ which satisfy*

$$\langle u_n, 1 \rangle = 0. \tag{3.3.15}$$

(3.3.14) admits the variational formulation

$$\left\{ \begin{array}{l} \dfrac{1}{4\pi} \displaystyle\int_{\Gamma}\int_{\Gamma} \dfrac{\left(\overrightarrow{\mathrm{curl}}_{\Gamma}\varphi(x) \cdot \overrightarrow{\mathrm{curl}}_{\Gamma}\psi(y)\right)}{|x-y|} d\gamma(x) d\gamma(y) \\[4mm] \qquad = \displaystyle\int_{\Gamma} u_n \psi d\gamma, \qquad \forall \psi \in H^{1/2}(\Gamma)/\mathbb{R}. \end{array} \right. \tag{3.3.16}$$

Moreover, this bilinear form satisfies the coercivity property

$$\left\{ \begin{array}{l} \dfrac{1}{4\pi} \displaystyle\int_{\Gamma}\int_{\Gamma} \dfrac{\left(\overrightarrow{\mathrm{curl}}_{\Gamma}\varphi(x) \cdot \overrightarrow{\mathrm{curl}}_{\Gamma}\varphi(y)\right)}{|x-y|} d\gamma(x) d\gamma(y) \\[4mm] \qquad \geq \alpha \|\varphi\|^2_{H^{1/2}(\Gamma)/\mathbb{R}}; \alpha > 0; \qquad \forall \varphi \in H^{1/2}(\Gamma)/\mathbb{R}. \end{array} \right. \tag{3.3.17}$$

An expression in terms of standard integrals, of the normal derivative of a double layer potential and so of the operator N, is

$$
\begin{cases}
\dfrac{\partial u}{\partial n}(y) = -\dfrac{1}{4\pi}\left[\int_{\Gamma}\dfrac{\Delta_{\Gamma}\varphi(x)}{|x-y|}d\gamma(x)\right. \\[3mm]
\qquad + \int_{\Gamma}\left(\nabla_{\Gamma}\varphi(x)\cdot\nabla_{\Gamma}\dfrac{1}{|x-y|}\right)(1-(n_x\cdot n_y)) \qquad\qquad (3.3.18) \\[3mm]
\qquad \left. + \int_{\Gamma}\dfrac{\partial}{\partial n_x}\left(\dfrac{1}{|x-y|}\right)(\nabla_{\Gamma}\varphi(x)\cdot n_y)\,d\gamma(x)\right].
\end{cases}
$$

Proof

The double layer potential u, given by

$$
u(y) = -\frac{1}{4\pi}\int_{\Gamma}\frac{\partial}{\partial n_x}\left(\frac{1}{|x-y|}\right)\varphi(x)d\gamma(x), \quad y\notin\Gamma, \qquad (3.3.19)
$$

satisfies the system of equations

$$
\begin{cases}
\Delta u = 0, & \text{in }\Omega_i\text{ and }\Omega_e, \\[2mm]
[u]_{\Gamma} = \varphi, & \qquad\qquad\qquad\qquad (3.3.20) \\[2mm]
\left[\dfrac{\partial u}{\partial n}\right]_{\Gamma} = 0.
\end{cases}
$$

Let us consider the Hilbert space

$$
\begin{cases}
X = \left\{v\in\left(H^1\left(\Omega_i\right)/\mathbb{R}\right)\times W^{1,-1}\left(\Omega_e\right),\right. \\[3mm]
\qquad \left.\Delta v = 0, \quad\text{in }\Omega_i\text{ and }\Omega_e, \quad \left[\dfrac{\partial v}{\partial n}\right]_{\Gamma} = 0\right\}.
\end{cases} \qquad (3.3.21)
$$

A variational formulation for the equation (3.3.20), for any $\varphi\in H^{1/2}(\Gamma)$, is

$$
\int_{\Omega_i}(\nabla u\cdot\nabla v)\,dx + \int_{\Omega_e}(\nabla u\cdot\nabla v)\,dx = \int_{\Gamma}\varphi\frac{\partial v}{\partial n}d\gamma, \quad \forall v\in X. \qquad (3.3.22)
$$

This bilinear form satisfies the coercivity property

$$
a(u,u) \geq \alpha\left[\|u\|^2_{H^1(\Omega_i)/\mathbb{R}} + \|u\|^2_{W^{1,-1}(\Omega_e)}\right]. \qquad (3.3.23)
$$

Thus, the Lax-Milgram theorem proves the existence of a unique solution to the problem (3.3.22), for any $\varphi\in H^{1/2}(\Gamma)$. The identity

$$
\left\langle\frac{\partial v}{\partial n},1\right\rangle = 0, \quad \forall v\in X, \qquad (3.3.24)
$$

shows that the functions φ and $\varphi+c$ lead to the same solution. The normal derivative of the solution belongs to the space $H^{-1/2}(\Gamma)$ and thus the operator N is continuous from $H^{1/2}(\Gamma)/\mathbb{R}$ to $H^{-1/2}(\Gamma)$ and any element of its image satisfies (3.3.15).

The inverse operator is the one that assigns to the normal derivative u_n the value of the function φ. For a given normal derivative $u_n \in H^{-1/2}(\Gamma)$, we can solve the interior and the exterior Neumann problem in $H^1(\Omega_i)/\mathbb{R}$ and $W^{1,-1}(\Omega_e)$. It results from the Representation Theorem 3.1.1 that the quantity

$$\varphi = [u] \qquad (3.3.25)$$

is a solution of equation (3.3.14). This shows that the operator N is an isomorphism from $H^{1/2}(\Gamma)/\mathbb{R}$ onto the subspace of element of $H^{-1/2}(\Gamma)$ which satisfy (3.3.15).

But the expressions of this operator that we have at hand, are all abstract. We will give now an explicit expression of the associated bilinear form and of the mapping N. Let $\varphi \in C^1(\Gamma)$. We denote by g the vector whose value is ∇u in Ω_i and Ω_e. It differs from the distributional gradient of u by the mass concentrated on the surface Γ. More precisely, from the Green formula follows

$$\nabla u = g - \varphi \vec{n} \delta_\Gamma \qquad (3.3.26)$$

where δ_Γ is the Dirac mass on the surface Γ. The Green formula and (3.3.20) imply

$$\operatorname{div} g = 0. \qquad (3.3.27)$$

It follows from the vectorial calculus formula $(-\Delta = \operatorname{curl}\operatorname{curl} - \nabla \operatorname{div})$ that

$$-\Delta g = \operatorname{curl}\operatorname{curl}(\varphi \vec{n} \delta_\Gamma). \qquad (3.3.28)$$

Thus,

$$g = E * \operatorname{curl}\operatorname{curl}(\varphi \vec{n} \delta_\Gamma). \qquad (3.3.29)$$

Let v be given in the space X. The expected bilinear form has the expression $\int_\Gamma \varphi(\partial v/\partial n)d\gamma$. Exactly as for u, we introduce the "function" part of the gradient of v, denoted by h. It holds that

$$\int_\Gamma \varphi \frac{\partial v}{\partial n} d\gamma = \int_{\mathbb{R}^3} (g \cdot h)dx = \int_\Gamma \frac{\partial u}{\partial n} \psi d\gamma, \qquad (3.3.30)$$

where ψ is given by $\psi = [v]$. From the Stokes formula (2.5.194) (Theorem 2.5.19), follows

$$\operatorname{curl}(\varphi \vec{n} \delta_\Gamma) = \left(\overrightarrow{\operatorname{curl}}_\Gamma \varphi\right) \delta_\Gamma, \qquad (3.3.31)$$

$$\operatorname{curl} h = \operatorname{curl}(\psi \vec{n} \delta_\Gamma) = \left(\overrightarrow{\operatorname{curl}}_\Gamma \psi\right) \delta_\Gamma. \qquad (3.3.32)$$

Combining all the above results, we get

$$
\begin{cases}
\displaystyle \int_\Gamma \varphi \frac{\partial v}{\partial n} d\gamma = \langle E * \operatorname{curl} \operatorname{curl} (\varphi \vec{n} \delta_\Gamma), h \rangle \\[3mm]
\qquad\quad = \langle E * \operatorname{curl} (\varphi \vec{n} \delta_\Gamma), \operatorname{curl} h \rangle \\[3mm]
\qquad\quad = \displaystyle \frac{1}{4\pi} \int_\Gamma \int_\Gamma \frac{\overrightarrow{\operatorname{curl}}_\Gamma \varphi(x) \cdot \overrightarrow{\operatorname{curl}}_\Gamma \psi(y)}{|x-y|} d\gamma(x) d\gamma(y),
\end{cases}
\tag{3.3.33}
$$

which is (3.3.16). The estimate (3.3.17) follows from the coercivity property (3.3.23) and the continuity of the trace operator.

Starting from the already established formula (3.3.16), it follows that

$$
\int_\Gamma \frac{\partial u}{\partial n} \psi d\gamma = \frac{1}{4\pi} \int_\Gamma \overrightarrow{\operatorname{curl}}_y \left(\int_\Gamma \frac{\overrightarrow{\operatorname{curl}}_\Gamma \varphi(x)}{|x-y|} d\gamma(x) \right) \psi(y) d\gamma(y).
\tag{3.3.34}
$$

We use the calculus relation

$$
\operatorname{curl}_\Gamma (\alpha \vec{v}) = - \left(\overrightarrow{\operatorname{curl}}_\Gamma \alpha \cdot \vec{v} \right) + \alpha \operatorname{curl}_\Gamma \vec{v}
\tag{3.3.35}
$$

to expand the above identity as

$$
\begin{cases}
\operatorname{curl}_{\Gamma_y} \left(\displaystyle \int_\Gamma \frac{\overrightarrow{\operatorname{curl}}_\Gamma \varphi(x)}{|x-y|} d\gamma(x) \right) \\[3mm]
\qquad\quad = - \displaystyle \int_\Gamma \left(\overrightarrow{\operatorname{curl}}_{\Gamma_y} \frac{1}{|x-y|} \cdot \overrightarrow{\operatorname{curl}}_\Gamma \varphi(x) \right) d\gamma(x),
\end{cases}
\tag{3.3.36}
$$

$$
\begin{cases}
\overrightarrow{\operatorname{curl}}_{\Gamma_y} \dfrac{1}{|x-y|} = \nabla_y \left(\dfrac{1}{|x-y|} \right) \wedge n_y \\[3mm]
\qquad = -\nabla_x \left(\dfrac{1}{|x-y|} \right) \wedge n_x + \nabla_x \left(\dfrac{1}{|x-y|} \right) \wedge (n_x - n_y) \\[3mm]
\qquad = -\overrightarrow{\operatorname{curl}}_{\Gamma_x} \dfrac{1}{|x-y|} + \nabla_x \left(\dfrac{1}{|x-y|} \right) \wedge (n_x - n_y),
\end{cases}
\tag{3.3.37}
$$

and thus,

$$
\begin{cases}
\displaystyle \int_\Gamma (\overrightarrow{\operatorname{curl}}_{\Gamma_y} \frac{1}{|x-y|} \cdot \overrightarrow{\operatorname{curl}}_\Gamma \varphi(x)) d\gamma(x) \\[3mm]
\qquad = - \displaystyle \int_\Gamma \overrightarrow{\operatorname{curl}}_\Gamma \frac{1}{|x-y|} \cdot \overrightarrow{\operatorname{curl}}_\Gamma \varphi(x) d\gamma(x) \\[3mm]
\qquad + \displaystyle \int_\Gamma \left(\overrightarrow{\operatorname{curl}}_\Gamma \varphi(x) \wedge \nabla_x \left(\frac{1}{|x-y|} \right) \cdot (n_x - n_y) \right) d\gamma(x).
\end{cases}
\tag{3.3.38}
$$

From the Stokes formula (2.5.194) and $\Delta_\Gamma = -\operatorname{curl}_\Gamma \overrightarrow{\operatorname{curl}}_\Gamma$, follows

$$
\left\{
\begin{aligned}
\frac{\partial u}{\partial n}(y) = {} & -\frac{1}{4\pi}\left[\int_\Gamma \frac{\Delta_\Gamma \varphi(x)}{|x-y|} d\gamma(x) \right. \\
& + \int_\Gamma \left(\nabla_\Gamma \varphi(x) \cdot \nabla_\Gamma \frac{1}{|x-y|} \right) (1 - (n_x \cdot n_y)) \, d\gamma(x) \quad (3.3.39) \\
& + \left. \int_\Gamma \left(\nabla_x \left(\frac{1}{|x-y|} \right) \cdot n_x \right) (\nabla_\Gamma \varphi(x) \cdot n_y) \, d\gamma(x) \right].
\end{aligned}
\right.
$$

∎

Comments

- It appears that the two symmetric operators S and $-N$ are moreover positive in the case of the Laplace equation.

- In the above expression (3.3.39), the principal part is the first integral. In the two other parts, the kernels are regularizing for exactly two degrees more than the first one (see below). This expression has also the form

$$
-S^{-1}N = -\Delta_\Gamma + K_1 \qquad (3.3.40)
$$

where K_1 is an order zero operator which "vanishes" for a plane surface. It can be compared to the formula (3.1.44). It holds that

$$
S = (-\Delta_\Gamma)^{-1/2} + K_2, \qquad (3.3.41)
$$

$$
-N = (-\Delta_\Gamma)^{1/2} + K_3. \qquad (3.3.42)
$$

Although they are quite formal, these expressions help to understand why $(Sq, q)^{1/2}$ is a norm on $H^{-1/2}(\Gamma)$, while $(-N\varphi, \varphi)^{1/2}$ is a norm on $H^{1/2}(\Gamma)$.

3.4 Variational Formulations for the Helmholtz Problems

3.4.1 The operator S

The integral equation (3.2.2) solves both the interior and exterior Dirichlet problems. It has the expression

$$
\frac{1}{4\pi} \int_\Gamma \frac{e^{ik|x-y|}}{|x-y|} q(x) d\gamma(x) = u_d(y), \quad y \in \Gamma. \qquad (3.4.1)
$$

It is not invertible when $-k^2$ is an eigenvalue of the interior Dirichlet problem for the Laplacian. When the parameter k vanishes, this integral equation admits a variational formulation in the space $H^{-1/2}(\Gamma)$. The similar variational formulation for equation (3.4.1) is expressed in

Theorem 3.4.1 *The equation* (3.4.1) *admits the variational formulation*

$$
\begin{cases}
\dfrac{1}{4\pi}\displaystyle\int_\Gamma\!\!\int_\Gamma \dfrac{e^{ik\,|x-y|}}{|x-y|}q(x)q^t(y)d\gamma(x)d\gamma(y) \\[4mm]
\qquad\qquad = \displaystyle\int_\Gamma u_d(y)q^t(y)d\gamma(y), \quad \forall\, q^t \in H^{-1/2}(\Gamma).
\end{cases}
\tag{3.4.2}
$$

The associated operator S is an isomorphism from $H^{-1/2}(\Gamma)$ onto $H^{1/2}(\Gamma)$, when $-k^2$ is not an eigenvalue of the interior Dirichlet problem for the Laplacian.

Proof

Let us denote by S_0 the corresponding operator when $k = 0$. We know from Theorem 3.3.1 that it is an isomorphism from $H^{-1/2}(\Gamma)$ onto $H^{1/2}(\Gamma)$. We split this operator in the form

$$
S = S_0 + R,
\tag{3.4.3}
$$

and then the kernel of the operator R is

$$
\phi = ik + \frac{e^{ikr} - 1 - ikr}{r}.
\tag{3.4.4}
$$

As will result from the next section, this kernel generates a continuous operator from $H^{-1/2}(\Gamma)$ to $H^{5/2}(\Gamma)$. Thus, the operator S has the form

$$
S = \left(I + RS_0^{-1}\right) S_0.
\tag{3.4.5}
$$

The Fredholm alternative holds and shows that the operator $I + RS_0^{-1}$ is invertible, when $-k^2$ is not an eigenvalue of the interior Dirichlet problem for the Laplacian. In fact, RS_0^{-1} is continuous from $H^{1/2}(\Gamma)$ to $H^{5/2}(\Gamma)$, which is compactly imbedded in $H^{1/2}(\Gamma)$.

3.4.2 Fredholm operators

The four integral equations (3.2.6), (3.2.7) and (3.2.11), (3.2.12) are of the same type. The operators D and D^* are continuous from $L^2(\Gamma)$ to $H^1(\Gamma)$, as will be seen in the next section. The Fredholm alternative applies to these four equations when they operate in the space $L^2(\Gamma)$. They are isomorphisms from $L^2(\Gamma)$ into $L^2(\Gamma)$, except when $-k^2$ is one of the critical values that we have already identified. They admit variational formulations in the space $L^2(\Gamma)$. For example, equation (3.2.6) takes the form

$$
\begin{cases}
\dfrac{1}{4\pi}\displaystyle\int_\Gamma\!\!\int_\Gamma \dfrac{\partial}{\partial n_y}\left(\dfrac{e^{ik\,|x-y|}}{|x-y|}\right)q(x)q^t(y)d\gamma(x)d\gamma(y) \\[4mm]
\quad + \dfrac{1}{2}\displaystyle\int_\Gamma q(y)q^t(y)d\gamma(y) = \int_\Gamma u_n(y)q^t(y)d\gamma(y), \quad \forall\, q^t \in L^2(\Gamma).
\end{cases}
\tag{3.4.6}
$$

3.4.3 The operator N

The formal expression for this operator is (3.2.16) and in the case of the Laplace equation, we have exhibited an integrable bilinear form and an associated variational coercive formulation for this operator. We give now a similar variational formulation for the Helmholtz equation.

Theorem 3.4.2 *The operator $-N$ given by (3.2.16), defines an isomorphism from $H^{1/2}(\Gamma)$ onto $H^{-1/2}(\Gamma)$, when $-k^2$ is not an eigenvalue for the interior Neumann problem for the Laplace equation. It admits the variational formulation*

$$
\begin{cases}
\dfrac{1}{4\pi} \displaystyle\int_\Gamma\!\!\int_\Gamma \frac{e^{ik\,|x-y|}}{|x-y|} \left(\overrightarrow{\mathrm{curl}}_\Gamma\varphi(x) \cdot \overrightarrow{\mathrm{curl}}_\Gamma\psi(y) \right) d\gamma(x)d\gamma(y) \\[2mm]
\quad - \dfrac{k^2}{4\pi} \displaystyle\int_\Gamma\!\!\int_\Gamma \frac{e^{ik\,|x-y|}}{|x-y|} \varphi(x)\psi(y)(n_x \cdot n_y)d\gamma(x)d\gamma(y) \qquad (3.4.7) \\[2mm]
\quad = \displaystyle\int_\Gamma u_n(y)\psi(y)d\gamma(y), \quad \forall \psi \in H^{1/2}(\Gamma).
\end{cases}
$$

An expression in terms of usual integrals of the operator $-N$ is

$$
\begin{cases}
-N\varphi = \dfrac{1}{4\pi}\left[-\displaystyle\int_\Gamma \frac{e^{ik\,|x-y|}}{|x-y|} \Delta_\Gamma\varphi(x)d\gamma(x) \right. \\[2mm]
\quad + \displaystyle\int_\Gamma \left(\nabla_\Gamma\varphi(x) \cdot \nabla_\Gamma \frac{e^{ik\,|x-y|}}{|x-y|} \right)(1 - (n_x \cdot n_y))\, d\gamma(x) \\[2mm]
\hspace{6cm} (3.4.8) \\[2mm]
\quad + \displaystyle\int_\Gamma \frac{\partial}{\partial n_x}\left(\frac{e^{ik\,|x-y|}}{|x-y|} \right)(\nabla_\Gamma\varphi(x) \cdot n_y)\, d\gamma(x) \\[2mm]
\quad \left. - k^2 \displaystyle\int_\Gamma \frac{e^{ik\,|x-y|}}{|x-y|} \varphi(x)(n_x \cdot n_y)d\gamma(x) \right].
\end{cases}
$$

Proof

The double layer potential solves the system of equations

$$
\begin{cases}
\Delta u + k^2 u = 0, & \text{in } \Omega_i \text{ and } \Omega_e, \\[1mm]
[u]_\Gamma = \varphi, \\[1mm]
\left[\dfrac{\partial u}{\partial n}\right]_\Gamma = 0.
\end{cases}
\qquad (3.4.9)
$$

The solution u satisfies the radiation condition at infinity. It belongs to the space $H^1(\Omega_i) \times H$, where uniqueness holds except, of course, for the critical values of k.

Let us consider the vector σ which coincides in each domain Ω_i and Ω_e, with the vector ∇u. The vector σ satisfies the radiation condition at infinity. From $[(\nabla u \cdot n)]_\Gamma = 0$ follows that $\operatorname{div} \sigma$ is a function in $L^2(R^3)$ and that in the sense of distributions in R^3,

$$\operatorname{div} \sigma + k^2 u = 0. \tag{3.4.10}$$

The vector ∇u, in the sense of distributions, has two parts. One, the distributed part, is the vector σ. The other one, the concentrated part, is a distribution whose support is the surface Γ and density is $-\varphi \vec{n}$. The couple (u, σ) satisfies the system of equations in R^3,

$$\begin{cases} \operatorname{div} \sigma + k^2 u = 0, \\ \nabla u - \sigma = -(\varphi \vec{n}) \delta_\Gamma. \end{cases} \tag{3.4.11}$$

We can express the solution of this system as a convolution of the right-hand side with a fundamental solution of the system, which satisfies the radiation conditions.

Lemma 3.4.1 *The fundamental solution of the system* (3.4.11), *which satisfies the radiation conditions and is associated with a non-zero right-hand side in the second part of the system is*

$$\begin{cases} U = -\nabla E, \\ \Sigma = -\operatorname{curl} \operatorname{curl}(EI) + k^2 EI, \end{cases} \tag{3.4.12}$$

where E is given by (3.1.1). *Here, the curl of a matrix is to be understood column by column.*

Proof
It is a simple computation. We have

$$\operatorname{div}(EI)) = \nabla E \tag{3.4.13}$$

which yields

$$\operatorname{div} \Sigma + k^2 U = 0. \tag{3.4.14}$$

Besides, it holds that

$$\nabla U = -\nabla \operatorname{div}(EI) \tag{3.4.15}$$

which, with the help of the vectorial calculus formula

$$\Delta = \nabla \operatorname{div} - \operatorname{curl} \operatorname{curl} \tag{3.4.16}$$

yields (δ is the Dirac mass)

$$\nabla U - \Sigma = -(\Delta E + k^2 E)I = \delta I. \tag{3.4.17}$$

Proof of Theorem 3.4.2 continued

The following integral representation holds:

$$u(y) = -U * (\varphi \vec{n} \delta_\Gamma) = \int_\Gamma \frac{\partial}{\partial n_x} (E(x-y)) \, \varphi(x) d\gamma(x) \qquad (3.4.18)$$

$$\begin{cases} \sigma(y) = -\Sigma * (\varphi \vec{n} \delta_\Gamma) \\[2mm] = (\operatorname{curl} \operatorname{curl}(EI) * (\varphi \vec{n} \delta_\Gamma)) - k^2 \int_\Gamma E(x-y)\varphi(x)\vec{n_x} d\gamma(x). \end{cases} \qquad (3.4.19)$$

In a convolution product, we can commute a differential operator, and doing so with curl, identity (3.4.19) becomes

$$\langle \operatorname{curl} \operatorname{curl}(EI) * \varphi \vec{n} \delta_\Gamma \rangle = \langle \operatorname{curl}(EI) * \operatorname{curl}(\varphi \vec{n} \delta_\Gamma) \rangle. \qquad (3.4.20)$$

We have

Lemma 3.4.2 *In the distribution sense in R^3, any regular function φ defined in a neighbourhood of Γ is such that*

$$\operatorname{curl}(\varphi \vec{n} \delta_\Gamma) = (\overrightarrow{\operatorname{curl}}_\Gamma \varphi)\delta_\Gamma. \qquad (3.4.21)$$

Proof

From the distribution definition of the differential operator curl, it follows that for any vector $\vec{a}$ in $(\mathcal{D}(R^3))^3$,

$$\langle \operatorname{curl}(\varphi \vec{n} \delta_\Gamma), \vec{a} \rangle = \langle \varphi \vec{n} \delta_\Gamma, \operatorname{curl} \vec{a} \rangle = \int_\Gamma \varphi(\operatorname{curl} \vec{a} \cdot \vec{n}) d\gamma. \qquad (3.4.22)$$

On the surface Γ, the operator $(\operatorname{curl} \vec{a} \cdot \vec{n})$ is exactly the **surfacic rotational** of the tangent part of the vector $\vec{a}$. Its adjoint on the surface Γ is $\overrightarrow{\operatorname{curl}}_\Gamma$ and from Theorem 2.5.19 follows

$$\langle \operatorname{curl}(\varphi \vec{n} \delta_\Gamma), \vec{a} \rangle = \int_\Gamma (\overrightarrow{\operatorname{curl}}_\Gamma \varphi \cdot \vec{a}) d\gamma, \qquad (3.4.23)$$

which is the formula (3.4.21).

End of the proof of Theorem 3.4.2

The value of the operator $-N\varphi$ is $(\sigma \cdot n)$. Thus, a variational formulation associated with this operator is

$$\int_\Gamma (\sigma \cdot n)\psi d\gamma = \int_\Gamma u_n \psi d\gamma, \quad \forall \psi \in H^{1/2}(\Gamma). \qquad (3.4.24)$$

We have proved that

$$\begin{cases} \int_\Gamma (\sigma \cdot n)\psi d\gamma = \left\langle \operatorname{curl}(EI) * \left(\overrightarrow{\operatorname{curl}}_\Gamma \varphi \delta_\Gamma\right), \psi \vec{n} \delta_\Gamma \right\rangle \\[3mm] - k^2 \int_\Gamma\!\!\int_\Gamma E(x-y)\varphi(x)\psi(y)(n_x \cdot n_y) d\gamma(x) d\gamma(y). \end{cases} \qquad (3.4.25)$$

We commute the operator curl in the above product of convolution, from which follows (notice that there are two changes of sign)

$$\begin{cases} \left\langle \mathrm{curl}(EI) * \left(\overrightarrow{\mathrm{curl}}_\Gamma \varphi \delta_\Gamma\right), \psi \vec{n} \delta_\Gamma \right\rangle \\ \\ = \left\langle EI * \overrightarrow{\mathrm{curl}}_\Gamma \varphi \delta_\Gamma, \overrightarrow{\mathrm{curl}}_\Gamma \psi \delta_\Gamma \right\rangle \qquad\qquad (3.4.26) \\ \\ = \displaystyle\int_\Gamma\!\!\int_\Gamma E(x-y)\left(\overrightarrow{\mathrm{curl}}_\Gamma \varphi(x) \cdot \overrightarrow{\mathrm{curl}}_\Gamma \psi(y)\right) d\gamma(x)d\gamma(y), \end{cases}$$

which is the formula (3.4.7). Then, formula (3.4.8) results from expression (3.4.7), the same way as formula (3.3.18) was obtained in Theorem 3.3.2. The first term is different, but has exactly the same structure.

We also need to prove that $-N$ is an isomorphism except for the known critical values. We apply the Fredholm alternative. The difference between the two bilinear forms (3.4.7) and (3.3.16) has the expression

$$\begin{cases} b(\varphi,\psi) - b_0(\varphi,\psi) \\ \\ = \displaystyle\int_\Gamma\!\!\int_\Gamma \left(\frac{e^{ik\,|x-y|}-1}{4\pi\,|x-y|}\right)\left(\overrightarrow{\mathrm{curl}}_\Gamma \varphi(x) \cdot \overrightarrow{\mathrm{curl}}_\Gamma \psi(y)\right) d\gamma(x)d\gamma(y) \\ \hspace{8cm} (3.4.27) \\ \\ \quad - \dfrac{k^2}{4\pi} \displaystyle\int_\Gamma\!\!\int_\Gamma \frac{e^{ik\,|x-y|}}{|x-y|} \varphi(x)\psi(y)(n_x \cdot n_y)d\gamma(x)d\gamma(y). \end{cases}$$

The associated kernel $(e^{ik|x|}-1)/|x|$ is regularizing of order 3, as will be seen in the next section. The above bilinear form is bicontinuous on $H^{-1/2}(\Gamma) \times H^{-1/2}(\Gamma)$, and so the Fredholm alternative holds.   ∎

Remark
Exactly as for the Laplace equation, it is possible to obtain identities (3.3.41) and (3.3.42) where K_2 and K_3 are compact operators which depend on k.

3.4.4 Formulation with the far field

Following an original idea developed by B. Desprès in his Thesis [70], we can modify the above integral equations in order to introduce the far field as an auxiliary unknown. This is possible for all integral equations of the above type. Our presentation is different from that of B. Desprès. It is based on a technical result on the imaginary part of the fundamental solution E.

Lemma 3.4.3 *The imaginary part of the fundamental solution E of the Helmholtz equation satisfies the identity*

$$\frac{\sin(k\,|x-y|)}{4\pi k\,|(x-y)|} = \frac{1}{(4\pi)^2}\int_S e^{ik(y-x\,\cdot\,z)}d\sigma(z). \qquad (3.4.28)$$

Proof

This identity is the expression of the Bessel function j_0, which we had obtained when studying the expansion of a plane wave in spherical harmonics. It is also a direct consequence of formula (2.6.107), Lemma 2.6.2. It can also be proved directly by integration on the sphere. ∎

Let us consider a **single layer potential** of the form (3.2.37). It can be split into

$$u(y) = \frac{1}{4\pi} \int_\Gamma \left[\frac{\cos(k\,|x-y|)}{|x-y|} + i\frac{\sin(k\,|x-y|)}{|x-y|} \right] q(x)d\gamma(x). \qquad (3.4.29)$$

The far field associated with this potential has the asymptotic

$$u(y) \sim \frac{e^{ik|y|}}{|y|}\alpha(\frac{y}{|y|}), \qquad (3.4.30)$$

where α is

$$\alpha(\frac{y}{|y|}) = \frac{1}{4\pi} \int_\Gamma e^{-ik(x\cdot y)/|y|} q(x)d\gamma(x). \qquad (3.4.31)$$

From Lemma 3.4.3, the above single layer potential takes the form

$$\begin{cases} u(y) = \dfrac{1}{4\pi} \displaystyle\int_\Gamma \dfrac{\cos(k\,|x-y|)}{|x-y|} q(x)d\gamma(x) + \beta(y), \\[3mm] \beta(y) = \dfrac{ik}{4\pi} \displaystyle\int_S e^{ik(y\cdot z)}\alpha(\dfrac{z}{|z|})d\sigma(z). \end{cases} \qquad (3.4.32)$$

Let us denote by A^* this new operator

$$A^*\alpha(y) = \frac{1}{4\pi} \int_S e^{ik(y\cdot z)}\alpha(\frac{z}{|z|})d\sigma(z). \qquad (3.4.33)$$

It is continuous from $L^2(S)$ to $L^2(\Gamma)$. Its adjoint is the operator A which gives the expression of the far field

$$Aq(z) = \frac{1}{4\pi} \int_\Gamma e^{-ik(x\cdot z)} q(x)d\gamma(x), \quad z \in S. \qquad (3.4.34)$$

Let us denote by S_{Re} the operator

$$S_{Re}q(y) = \frac{1}{4\pi} \int_\Gamma \frac{\cos(k\,|x-y|)}{|x-y|} q(x)d\gamma(x). \qquad (3.4.35)$$

We **replace equation** (3.4.1), which solves the **Dirichlet problem**, by the system

$$\begin{cases} S_{Re}q + ikA^*\alpha = u_d, \\[3mm] -ikAq + ik\alpha = 0. \end{cases} \qquad (3.4.36)$$

This system permits a simultaneous computation for the far field and the solution. Besides, we have proved that the imaginary part of the operator

S is semi-positive definite and has the expression

$$S_{Im}q = kA^*Aq. \tag{3.4.37}$$

Let us consider now a **double layer potential** of the form (3.2.10). It can be split into

$$u(y) = \int_\Gamma \frac{\partial}{\partial n_x} \left[\frac{cos(k|x-y|)}{4\pi|x-y|} + i\frac{sin(k|x-y|)}{4\pi|x-y|} \right] \varphi(x)d\gamma(x). \tag{3.4.38}$$

The far field associated with this potential has the asymptotic

$$u(y) \sim \frac{e^{ik|y|}}{|y|}\alpha(\frac{y}{|y|}), \tag{3.4.39}$$

where α is

$$\alpha(\frac{y}{|y|}) = \frac{-ik}{4\pi}\int_\Gamma e^{-ik(x\cdot y)/|y|}(n_x \cdot \frac{y}{|y|})\varphi(x)d\gamma(x). \tag{3.4.40}$$

We want to **replace equation** (3.2.17), which solves the **Neumann problem**, by a system where the far field α appears. Let us denote by N_{Re} the operator

$$N_{Re}q(y) = \frac{1}{4\pi}\int_\Gamma \frac{\partial^2}{\partial n_x\partial n_y} \left\{ \frac{cos(k|x-y|)}{|x-y|} \right\} \varphi(x)d\gamma(x). \tag{3.4.41}$$

Lemma 3.4.4 *The imaginary part of the double normal derivative of the fundamental solution E of the Helmholtz equation satisfies the identity*

$$\left\{ \begin{array}{l} \dfrac{\partial^2}{\partial n_x\partial n_y} \left\{ \dfrac{sin(k|x-y|)}{4\pi k|x-y|} \right\} \\[3mm] \quad = \dfrac{k^2}{(4\pi)^2}\displaystyle\int_S e^{ik(y-x\cdot z)}(n_x \cdot z)(n_y \cdot z)\,d\sigma(z). \end{array} \right. \tag{3.4.42}$$

Proof
We differentiate twice the above identity (3.4.28) with respect to the variables x and y. It can also be proved by a direct integration.   ∎

From the above lemma, we can replace equation (3.2.17) by the system

$$\left\{ \begin{array}{l} N_{Re}\varphi - ikA^*\alpha = -u_n, \\[3mm] ikA\varphi - ik\alpha = 0 \end{array} \right. \tag{3.4.43}$$

where A^* is the operator

$$A^*\alpha(y) = \frac{ik}{4\pi}\int_S e^{ik(y\cdot z)}(n_y \cdot z)\alpha(z)d\sigma(z). \tag{3.4.44}$$

This operator is continuous from $L^2(S)$ to $L^2(\Gamma)$. Its adjoint is the operator A which gives the far field

$$A\varphi(z) = \frac{-ik}{4\pi} \int_\Gamma e^{-ik(x \cdot z)} (n_x \cdot z) \varphi(x) d\gamma(x), \quad z \in S. \qquad (3.4.45)$$

Besides, we have proved that the imaginary part of the operator N is semi-positive definite and has the expression

$$N_{Im}\varphi = -kA^*A\varphi. \qquad (3.4.46)$$

4

Singular Integral Operators

We present in this chapter the definition and the main properties of the singular integral operator. These results are quite classical and were first studied by Giraud [78] in France and then Calderon and Zygmund in the United States and Michlin in Russia. The present exposition uses some ideas of the notes of V. Neri and the book of E.M. Stein [142]. We establish some continuity results in Sobolev spaces for a class of singular integral operators and then deduce similar properties for the integral operators that we have encountered in the previous chapter.

4.1 The Hilbert Transform

The Hilbert transform is a singular integral operator which acts on functions of one real variable x. It is defined as the principal value

$$H(f)(y) = \lim_{\varepsilon \to 0} \frac{1}{\pi} \int_{|x| > \varepsilon} \frac{f(y - x)}{x} dx, \qquad (4.1.1)$$

where f is a continuous function with compact support. It admits the equivalent definition

$$H(f)(y) = \lim_{\varepsilon \to 0} \frac{1}{\pi} \int_{x > \varepsilon} \frac{f(y - x) - f(y + x)}{x} dx, \qquad (4.1.2)$$

where the limit exists when f satisfies a Hölder condition of order $\alpha > 0$.

Another equivalent form is

$$H(f)(y) = \frac{1}{\pi} \int_0^\infty \frac{f(y-x) - f(y+x)}{x} dx. \tag{4.1.3}$$

Theorem 4.1.1 *The Hilbert transform is defined for any function f in the space $L^2(\mathbb{R})$ by*

$$H(f)(y) = \lim_{\substack{\varepsilon \to 0 \\ R \to \infty}} \frac{1}{\pi} \int_{\varepsilon < |x| < R} \frac{f(y-x)}{x} dx. \tag{4.1.4}$$

It is an isometry from $L^2(R)$ into $L^2(R)$, whose inverse is $-H$.

Proof

We use the Fourier transform in one space variable. Let $K_{\varepsilon,R}$ be the truncated kernel

$$K_{\varepsilon,R}(x) = \begin{cases} \frac{1}{\pi x}, & \varepsilon < |x| < R, \\ 0 & \text{else.} \end{cases} \tag{4.1.5}$$

Let us denote by $H_{\varepsilon,R}(f)$ the associated operator. The kernel $K_{\varepsilon,R}$ belongs to the space L^1 and it follows from Young's inequality that

$$\|H_{\varepsilon,R}(f)\|_{L^2(\mathbf{R})} \le \|K_{\varepsilon,R}\|_{L^1(\mathbf{R})} \|f\|_{L^2(\mathbf{R})}. \tag{4.1.6}$$

The convolution product is such that

$$\widehat{H_{\varepsilon,R}(f)} = \sqrt{2\pi} \widehat{K_{\varepsilon,R}} \hat{f}. \tag{4.1.7}$$

Let us compute the Fourier transform $\widehat{K_{\varepsilon,R}}$.

$$\begin{cases} \widehat{K_{\varepsilon,R}}(\xi) = \frac{1}{\sqrt{2} \pi^{3/2}} \int_{\varepsilon < |x| < R} \frac{e^{-i\xi x}}{x} dx \\ \quad = -\frac{2i}{\sqrt{2\pi} \pi} \int_\varepsilon^R \frac{\sin x\xi}{x} dx; \end{cases} \tag{4.1.8}$$

or equivalently using a change of variable,

$$\widehat{K_{\varepsilon,R}}(\xi) = -\frac{2i}{\sqrt{2\pi} \pi} \text{sign}(\xi) \int_{\varepsilon|\xi|}^{R|\xi|} \frac{\sin x}{x} dx. \tag{4.1.9}$$

The limit when ε tends to zero and R tends to infinity is

$$\hat{H}(\xi) = -\frac{2i}{\sqrt{2\pi} \pi} \text{sign}(\xi) \int_0^\infty \frac{\sin x}{x} dx = -\frac{i\text{sign}(\xi)}{\sqrt{2\pi}}. \tag{4.1.10}$$

From (4.1.10) and (4.1.7) follows

$$\widehat{Hf} = -i\text{sign}(\xi)\hat{f}, \tag{4.1.11}$$

which gives the expected result using Plancherel's identity and

$$(i\text{sign}(\xi))^2 = -1.$$

Definition

A linear operator T acting from $L^p(\mathbb{R}^n)$ to $L^q(\mathbb{R}^n)$ is called **of weak type (p, q)**, when there exists a constant A such that

$$
\begin{cases}
\operatorname{mes}\{x; |Tf(x)| \geq \alpha\} \leq \left(\dfrac{A\,\|f\|_p}{\alpha}\right)^q, \\[2mm]
q < \infty, \forall \alpha > 0, \quad \forall f \in L^p(\mathbb{R}^n).
\end{cases}
\tag{4.1.12}
$$

We have

Theorem 4.1.2 *The Hilbert transform is of weak type (1,1), i.e., for any $\alpha > 0$, it holds that*

$$
\operatorname{mes}\{x; |H(f)(x)| \geq \alpha\} \leq \frac{A\,\|f\|_{L^1(\mathbb{R})}}{\alpha}.
\tag{4.1.13}
$$

Proof

We will establish a number of preliminary lemmas.

Lemma 4.1.1 (Marcinkiewicz) *Let F be a closed set of $\mathbb{R}$ whose complement is bounded and let $\delta(y)$ be the distance of the point y to F. The function*

$$
I(x) = \int_{\mathbb{R}} \frac{\delta(y)}{|x-y|^2}\,dy
\tag{4.1.14}
$$

has a bounded integral on F and

$$
\int_F \int_{\mathbb{R}} \frac{\delta(y)}{|x-y|^2}\,dy\,dx \leq 2 \operatorname{mes}(F^c).
\tag{4.1.15}
$$

Proof

When y belongs to F, the distance $\delta(y)$ vanishes. From Fubini's theorem, the above integral takes the form

$$
\begin{cases}
\displaystyle \int_{F^c} \delta(y)\,dy \int_F \frac{1}{|x-y|^2}\,dx \\[3mm]
\displaystyle \leq \int_{F^c} \delta(y)\,dy \int_{|x|>\delta(y)} \frac{dx}{|x|^2} = 2 \operatorname{mes}(F^c).
\end{cases}
\tag{4.1.16}
$$

Lemma 4.1.2 *Let F be a closed set of $\mathbb{R}$ whose complement is Ω. There exists a covering of Ω by a set of closed balls B_k with disjoint interiors such that*

$$
\operatorname{mes}(B_k) \leq \operatorname{dist}(B_k, F) \leq 4 \operatorname{mes}(B_k).
\tag{4.1.17}
$$

Proof

We divide $\mathbb{R}$ in a collection of non-overlapping end-to-end open segments with length 2^{-k}. Let $\mathcal{M}_k$ be this covering of $\mathbb{R}$. Consider Ω_k the layers

$$
\Omega_k = \{x; 2^{-k+1} < \operatorname{dist}(x, F) \leq 2^{-k+2}\}.
$$

From the very definition of Ω_k, the segments in $\mathcal{M}_k$ which intersect with Ω_k satisfy the inequalities (4.1.17). We keep them. As the union $\bigcup_{k=-\infty}^{\infty} \Omega_k$ is Ω, the set of such segments covers Ω. These segments are not disjoint, but either their intersection is empty, or one contains the other. A maximal covering for inclusion gives the expected covering.

Definition
Let us associate with any locally integrable function f, the following function $M(f)$,

$$M(f)(x) = \sup_{r>0} \frac{1}{\mathrm{mes}(B(x,r))} \int_{B(x,r)} |f(y)|\, dy, \qquad (4.1.18)$$

where $B(x,r)$ is the segment with center x and length $2r$.

Lemma 4.1.3 *Let f be a function in $L^1(\mathbb{R})$ and α be a positive number. The set $F_\alpha = \{x \mid M(f)(x) \leq \alpha\}$ is closed. Its complement E_α can be covered by a collection of closed segments $B_1, \ldots, B_k, \ldots$, whose interiors are disjoint and which satisfy (4.1.17). It holds that*

$$\mathrm{mes}\,(E_\alpha) \leq \frac{3}{\alpha} \int_{\mathbb{R}} |f|\, dx, \qquad (4.1.19)$$

$$\frac{1}{\mathrm{mes}\,(B_k)} \int_{B_k} |f(x)|\, dx \leq 10\,\alpha. \qquad (4.1.20)$$

Proof
For any x in E_α, there exists a ball with center x and radius r such that

$$\mathrm{mes}(B(x,r)) < \frac{1}{\alpha} \int_{B(x,r)} |f(x)|\, dx,$$

from which it follows that E_α is open. From Lemma 4.1.2, we can cover E_α by a collection of segments B_k satisfying (4.1.17). Let p_k be a point in F_α such that

$$\mathrm{dist}\,(B_k, F_\alpha) = \mathrm{dist}\,(p_k, B_k)\,.$$

Let Q_k be the smallest ball with center p_k enclosing B_k. From (4.1.17), its diameter is bounded by

$$\mathrm{mes}\,(Q_k) \leq 10\ \mathrm{mes}\,(B_k)\,.$$

It holds that

$$\begin{cases} \alpha \geq M(f)\,(p_k) \geq \dfrac{1}{\mathrm{mes}\,(Q_k)} \displaystyle\int_{Q_k} |f|\, dx \\[3mm] \qquad\qquad \geq \dfrac{1}{10\,\mathrm{mes}\,(B_k)} \displaystyle\int_{B_k} |f|\, dx. \end{cases} \qquad (4.1.21)$$

Besides, there exists a covering of E_α by a collection of balls with center x and radius r such that

$$\operatorname{mes}(B(x,r)) < \frac{1}{\alpha} \int_{B(x,r)} |f(x)| \, dx.$$

We extract from this covering a countable number of disjoint balls which are such that their homothetic, by a factor 3, covers E_α and then (4.1.19) follows from inequality.

Proof of Theorem 4.1.2

Let $\alpha > 0$ and F_α be the set associated with the function f, introduced in Lemma 4.1.3. We can split f into the form:

$$f = g + h, \tag{4.1.22}$$

$$g(x) = \begin{cases} f(x), & x \in F_\alpha, \\ \dfrac{1}{\operatorname{mes}(B_k)} \displaystyle\int_{B_k} f(x)dx, & x \in B_k. \end{cases} \tag{4.1.23}$$

From the definition of g follows

$$\begin{cases} h(x) = 0, & x \in F_\alpha, \\ \displaystyle\int_{B_k} h(x)dx = 0. \end{cases} \tag{4.1.24}$$

The following inclusion holds,

$$\{x; |Hf| \geq \alpha\} \subset \left\{ \left\{ x; |Hg| \geq \frac{\alpha}{2} \right\} \cup \left\{ x; |Hh| \geq \frac{\alpha}{2} \right\} \right\}, \tag{4.1.25}$$

and we only have to show the estimate (4.1.12) separately for h and g. From (4.1.20) and $|f(x)| \leq M(f)(x)$, in each Lebesgue point of the function f, the function g satisfies

$$|g(x)| \leq 10\,\alpha, \tag{4.1.26}$$

and thus belongs to the space $L^2(\mathbb{R})$ and from Theorem 4.1.1, it holds that

$$\|Hg\|_{L^2(\mathbb{R})} = \|g\|_{L^2(\mathbb{R})}. \tag{4.1.27}$$

We have

$$\begin{cases} \operatorname{mes} \left\{ x; |Hg| \geq \frac{\alpha}{2} \right\} = \operatorname{mes} \left\{ x; |Hg|^2 \geq \frac{\alpha^2}{4} \right\} \\ \qquad\qquad \leq \frac{4}{\alpha^2} \|Hg\|^2_{L^2(\mathbb{R})} = \frac{4}{\alpha^2} \|g\|^2_{L^2(\mathbb{R})}. \end{cases} \tag{4.1.28}$$

Besides, from (4.1.26) follows

$$\|g\|^2_{L^2(\mathbb{R})} \leq 10\,\alpha \, \|g\|_{L^1(\mathbb{R})}. \tag{4.1.29}$$

Combining the estimates (4.1.28) and (4.1.29) leads to

$$\operatorname{mes} \left\{ x; |Hg| \geq \frac{\alpha}{2} \right\} \leq \frac{40}{\alpha} \|g\|_{L^1(\mathbb{R})} \leq \frac{40}{\alpha} \|f\|_{L^1(\mathbb{R})}. \tag{4.1.30}$$

Let us examine the function h. We denote by h_k its restriction to B_k. Let y_k be the nearest point in F_α belonging to the segment B_k. It holds that

$$
\begin{cases}
Hh_k(x) = \dfrac{1}{\pi} \displaystyle\int_{B_k} h(y) \left(\dfrac{1}{x-y} - \dfrac{1}{x-y_k} \right) dy \\[3mm]
\qquad\quad = \dfrac{1}{\pi} \displaystyle\int_{B_k} h(y) \dfrac{y - y_k}{(x-y)(x-y_k)} dy.
\end{cases}
\tag{4.1.31}
$$

From (4.1.17), we know that when x is in the domain F_α, we have

$$
\begin{cases}
|y_k - y| \le \mathrm{mes}\,(B_k) \le \mathrm{dist}\,(y_k, F_\alpha) \le \mathrm{dist}(y, F_\alpha), \\[3mm]
\dfrac{1}{|x-y|} \le \dfrac{1}{|x-y_k|} \le \dfrac{2}{|x-y|}.
\end{cases}
\tag{4.1.32}
$$

Thus, when x belongs to F_α, we obtain

$$
|Hh_k(x)| \le \frac{\mathrm{mes}\,(B_k)}{\pi} \int_{B_k} \frac{|h(y)|}{|x-y_k|^2} dy.
\tag{4.1.33}
$$

From the definition (4.1.22), and from (4.1.20) and (4.1.26) follow

$$
\begin{cases}
\displaystyle\int_{B_k} |h(y)|\,dy \le \int_{B_k} |f(y)|\,dy + \int_{B_k} |g(y)|\,dy \\[3mm]
\qquad\qquad\quad \le 20\,\alpha\,\mathrm{mes}\,(B_k),
\end{cases}
\tag{4.1.34}
$$

and thus

$$
|Hh_k(x)| \le \frac{20}{\pi} \frac{\alpha\,(\mathrm{mes}\,(B_k))^2}{|x-y_k|^2} \le \frac{20}{\pi}\alpha \int_{B_k} \frac{\delta(y)}{|x-y|^2} dy.
\tag{4.1.35}
$$

Adding all these inequalities for all the sets B_k yields

$$
|Hh(x)| \le \frac{20}{\pi}\alpha \int_{\mathbf{R}} \frac{\delta(y)}{|x-y|^2} dy.
\tag{4.1.36}
$$

From the Marcinkiewicz lemma, Lemma 4.1.1 and inequality (4.1.19) follows

$$
\int_{F_\alpha} |Hh(x)|\,dx \le \frac{20}{\pi}\alpha\,\mathrm{mes}\,(E_\alpha) \le \frac{120}{\pi} \|f\|_{L^1(\mathbf{R})}.
\tag{4.1.37}
$$

This estimate implies

$$
\mathrm{mes}\left\{ x \in F_\alpha; |Hh(x)| \ge \frac{\alpha}{2} \right\} \le \frac{240}{\pi}\frac{1}{\alpha} \|f\|_{L^1(\mathbf{R})},
\tag{4.1.38}
$$

which together with inequality (4.1.19) yields

$$
\mathrm{mes}\left\{ x \in \mathbf{R}; |Hh(x)| \ge \frac{\alpha}{2} \right\} \le \frac{80}{\alpha} \|f\|_{L^1(\mathbf{R})}.
\tag{4.1.39}
$$

∎

Theorem 4.1.3 (theorem of interpolation) *Let the real r be such that $1 < r \le \infty$. Any linear operator T from the space $L^1(\mathbb{R}^n) + L^r(\mathbb{R}^n)$ into*

the space of measurable functions, which is simultaneously of weak type
(1,1) and of weak type (r,r), i.e., satisfies

$$\text{mes}\,\{x;\,|Tf(x)| \geq \alpha\} \leq \left(\frac{A_r}{\alpha}\,\|f\|_r\right)^r,\quad f \in L^r(\mathbb{R}^n),\qquad (4.1.40)$$

is continuous from $L^p(\mathbb{R}^n)$ into $L^p(\mathbb{R}^n)$, i.e. satisfies

$$\begin{cases} \|Tf\|_{L^p(\mathbb{R}^n)} \leq A_p\,\|f\|_{L^p(\mathbb{R}^n)},\quad f \in L^p(\mathbb{R}^n),\\[2mm] \forall p\,\text{such that } 1 < p < r. \end{cases}\qquad (4.1.41)$$

Proof

Let α be a positive real. We split the function f in

$$\begin{cases} f_2(x) = \begin{cases} f(x), & \text{when } |f(x)| \leq \alpha,\\[1mm] 0, & \text{else}, \end{cases}\\[4mm] f = f_1 + f_2. \end{cases}\qquad (4.1.42)$$

We have the estimate

$$\begin{cases} \lambda(\alpha) = \text{mes}\,\{x;\,|Tf(x)| \geq \alpha\}\\[2mm] \quad \leq \text{mes}\,\left\{x;\,|Tf_1(x)| \geq \dfrac{\alpha}{2}\right\} + \text{mes}\,\left\{x;\,|Tf_2(x)| \geq \dfrac{\alpha}{2}\right\}. \end{cases}\qquad (4.1.43)$$

From (4.1.40) and (4.1.42) follow

$$\begin{cases} \lambda(\alpha) \leq \dfrac{2A_1}{\alpha}\displaystyle\int_{\mathbb{R}} |f_1(x)|\,dx + \dfrac{(2A_r)^r}{\alpha^r}\int_{\mathbb{R}} |f_2(x)|^r\,dx\\[4mm] \quad \leq \dfrac{2A_1}{\alpha}\displaystyle\int_{|f|>\alpha} |f(x)|\,dx + \dfrac{(2A_r)^r}{\alpha^r}\int_{|f|\leq\alpha} |f(x)|^r\,dx. \end{cases}\qquad (4.1.44)$$

It holds that

$$\int_{\mathbb{R}^n} |Tf|^p\,dx = -\int_0^\infty \alpha^p d\lambda(\alpha) = p\int_0^\infty \alpha^{p-1}\lambda(\alpha)d\alpha.\qquad (4.1.45)$$

We multiply inequality (4.1.44) by $p\alpha^{p-1}$ and integrate the two corresponding terms to get

$$\begin{cases} \displaystyle\int_0^\infty \alpha^{p-2}\int_{|f|>\alpha} |f(x)|\,dx\,d\alpha = \int_{\mathbb{R}^n} |f|\int_0^{|f|} \alpha^{p-2}d\alpha\,dx\\[4mm] \hspace{5.5cm} = \dfrac{1}{p-1}\displaystyle\int_{\mathbb{R}^n} |f||f|^{p-1}dx, \end{cases}\qquad (4.1.46)$$

$$\begin{cases} \displaystyle\int_0^\infty \alpha^{p-1-r}\int_{|f|\leq\alpha} |f(x)|^r\,dx = \int_{\mathbb{R}^n} |f|^r\int_{|f|}^\infty \alpha^{p-1-r}d\alpha\,dx\\[4mm] \hspace{5.5cm} = \dfrac{1}{r-p}\displaystyle\int_{\mathbb{R}^n} |f|^r|f|^{p-r}dx. \end{cases}\qquad (4.1.47)$$

Combining (4.1.44), (4.1.45), (4.1.46), and (4.1.47), we obtain (4.1.41) with

$$(A_p)^p = \frac{2pA_1}{p-1} + \frac{p(2A_r)^r}{r-p}. \tag{4.1.48}$$

Theorem 4.1.4 *The Hilbert transform is continuous from $L^p(\mathbb{R})$ onto $L^p(\mathbb{R})$, for any p such that $1 < p < \infty$, and the continuity constant explodes as $1/(p-1)$ when p tends to 1, and as p when p tends to infinity.*

Proof

From the interpolation theorem and from Theorems 4.1.1 and 4.1.2, we deduce that the continuity is satisfied when $1 < p \leq 2$. Then by duality, it follows for $2 \leq p < \infty$. The constants are given by (4.1.48) .

Definition

For any given positive integer m and for $1 < p < \infty$, we denote by $W^{m,p}(\mathbb{R})$ the space of functions in $L^p(\mathbb{R})$, all of whose weak derivatives up to order m are also in $L^p(\mathbb{R})$. For a given negative integer m and for $1 < p < \infty$, we denote by $W^{m,p}(\mathbb{R})$, the dual space of $W^{-m,q}(\mathbb{R})$ with

$$\frac{1}{p} + \frac{1}{q} = 1.$$

Theorem 4.1.5 *For any integer s and any p such that $1 < p < \infty$, the Hilbert transform is an isomorphism from $W^{s,p}(\mathbb{R})$ onto $W^{s,p}(\mathbb{R})$. When $p = 2$, the result is valid for any real s.*

Proof

For any given positive integer m, a derivation under the integral gives

$$g_\varepsilon(y) = \frac{1}{\pi} \int_{|x|>\varepsilon} \frac{f(y-x)}{x} dx, \tag{4.1.49}$$

$$\frac{d^m}{dy^m} g_\varepsilon(y) = \frac{1}{\pi} \int_{|x|>\varepsilon} \frac{\frac{d^m f}{dy^m}(y-x)}{x} dx. \tag{4.1.50}$$

From Theorem 4.1.4, this shows the existence, when $\varepsilon \to 0$, of a limit in $L^p(\mathbb{R})$, which is $d^m g/dy^m$. This result is extended to any integer m by duality. When $p = 2$, the spaces $H^s(\mathbb{R})$ for real s are defined through interpolation, and thus the results known for s integer can be extended to any real s.

4.2 Singular Integral Operators in $\mathbb{R}^n$

We examine in this section the continuity properties of a singular operator of the form

$$g(y) = \int_{\mathbb{R}^n} K(y, y-x) f(x) dx, \tag{4.2.1}$$

where the kernel $K(y, z)$ is homogeneous of degree $-n$ with respect to the variable z and regular with respect to the variable y. The above integral is not convergent in the usual sense, and will be defined as a finite part. This is not possible for general kernels. Thus, we consider kernels K, whose mean values are zero on the unit sphere S^{n-1} of $\mathbb{R}^n$. This property is expressed as

$$\int_{S^{n-1}} K(y, z')dz' = 0, \quad \text{for any } y. \tag{4.2.2}$$

4.2.1 Odd kernels

A kernel $K(y, z)$, homogeneous of degree $-n$ with respect to the variable z, has the form

$$K(y, z) = \frac{1}{|z|^n} K\left(y, \frac{z}{|z|}\right). \tag{4.2.3}$$

Let us denote by $z' = z/|z|$ the angular variable which corresponds to a point of the unit sphere. Being odd, the kernel K satisfies

$$K(y, -z') = -K(y, z'). \tag{4.2.4}$$

We introduce the following approximate integral

$$g_\varepsilon(y) = \int_{|x-y|\geq\varepsilon} K(y, y-x)f(x)dx. \tag{4.2.5}$$

In spherical coordinates with origin in y, the above integral takes the form (we denote by dz' the area element on the unit sphere)

$$\begin{cases} y - x = rz', \\ g_\varepsilon(y) = \int_{S^{n-1}} \int_{r\geq\varepsilon} K(y, z')\frac{1}{r}f(y - rz')drdz'. \end{cases} \tag{4.2.6}$$

From relation (4.2.4), this integral can be written as an integral on the unit sphere S^{n-1},

$$g_\varepsilon(y) = \frac{1}{2} \int_{S^{n-1}} \left[K(y, z') \int_\varepsilon^\infty \frac{1}{r} (f(y - rz') - f(y + rz')) \, dr \right] dz'. \tag{4.2.7}$$

Theorem 4.2.1 *Let $K(y, z)$ be a kernel, homogeneous of degree $-n$ and odd with respect to the variable z, which moreover satisfies*

$$\sup_{y\in\mathbb{R}^n} \int_{S^{n-1}} |K(y, z')| \, dz' \leq C. \tag{4.2.8}$$

The operator associated with this kernel K by the expression (4.2.1), can be defined as the limit of the expression (4.2.7) when $\varepsilon \to 0$. It is a continuous operator from $L^p(\mathbb{R}^n)$ into $L^p(\mathbb{R}^n)$, for any p such that $1 < p < \infty$; i.e.,

$$\|g\|_{L^p(\mathbb{R}^n)} \leq C_p \|f\|_{L^p(\mathbb{R}^n)}. \tag{4.2.9}$$

The continuity constant explodes as $1/(p-1)$ when p tends to 1 and as p when p tends to infinity.

Proof

From classical arguments of density, and a limit process when ε tends to zero, we need to consider only the case where f has a compact support and to prove inequality (4.2.9) in that case.

It holds that

$$\left\{ \begin{array}{l} 2 \, \|g_\varepsilon\|_{L^p(\mathbb{R}^n)} \\[2mm] \le \left[\displaystyle\sup_{y \in \mathbb{R}^n} \int_{S^{n-1}} |K(y,z')| \, dz' \right] \left[\displaystyle\sup_{z' \in S^{n-1}} \|F_\varepsilon(\cdot, z')\|_{L^p(\mathbb{R}^n)} \right], \end{array} \right. \tag{4.2.10}$$

$$F_\varepsilon(y, z') = \int_\varepsilon^\infty \frac{1}{r} \left(f(y - rz') - f(y + rz') \right) dr. \tag{4.2.11}$$

By a change of variables, F_ε can written as a Hilbert transform, for which we can use Theorem 4.1.4. Let us denote by H the orthogonal plane to the vector z'. We decompose y in the form

$$y = h + sz'.$$

A new expression for (4.2.11) follows, namely,

$$F_\varepsilon(h + sz', z') = \int_\varepsilon^\infty \left(f(h + (s-r)z') - f(h + (r+s)z') \right) \frac{1}{r} dr. \tag{4.2.12}$$

The $L^p(\mathbb{R}^n)$ norm of F_ε is bounded by

$$\left\{ \begin{array}{l} \|F_\varepsilon(\cdot, z')\|^p_{L^p(\mathbb{R}^n)} \le A_p \displaystyle\int_H dh \int_{-\infty}^\infty ds \\[3mm] \left[\displaystyle\int_\varepsilon^\infty \left(f(h + (s-r)z') - f(h + (r-s)z') \right) \frac{dr}{r} \right]^p . \end{array} \right. \tag{4.2.13}$$

From Theorem 4.1.4 follows

$$\left\{ \begin{array}{l} \|F_\varepsilon(\cdot, z')\|^p_{L^p(\mathbb{R}^n)} \\[3mm] \le A_p \displaystyle\int_H \int_{-\infty}^\infty |f(h + sz')|^p \, dh \, ds = A_p \|f\|^p_{L^p(\mathbb{R}^n)}, \end{array} \right. \tag{4.2.14}$$

and from (4.2.10) and (4.2.14) follows the expected result.

Remark

This kind of operator is not bounded on L^1 and L^∞, exactly as the Hilbert transform. We showed that the Hilbert transform was of weak type (1, 1). The above operators are also of weak type (1, 1), when the kernel K satisfies some extra properties (see E.M. Stein [142]).

4.2.2   The M. Riesz transforms

M. Riesz kernels are given by

$$R_i(z) = \frac{1}{\gamma_n} \frac{z_i}{|z|^{n+1}}, \quad 1 \le i \le n, \tag{4.2.15}$$

and can also be written as

$$R_i(z) = -\frac{1}{(n-1)\gamma_n} \frac{\partial}{\partial z_i} \frac{1}{|z|^{n-1}}. \tag{4.2.16}$$

Theorem 4.2.2 *For any integer m and any p such that $1 < p < \infty$, the M. Riesz transforms associated with the kernels R_i,*

$$(R_i f)(y) = \int_{\mathbf{R}^n} \frac{1}{\gamma_n} \frac{y_i - x_i}{|x - y|^{n+1}} f(x)dx, \tag{4.2.17}$$

with

$$\gamma_n = \frac{i}{n-1} \frac{\displaystyle\int_0^\infty e^{-r^2/2} dr}{\displaystyle\int_0^\infty r^{n-2} e^{-r^2/2} dr}, \tag{4.2.18}$$

are continuous from $W^{m,p}(\mathbf{R}^n)$ into $W^{m,p}(\mathbf{R}^n)$. Moreover, they satisfy the identity

$$\sum_{i=1}^n R_i \circ R_i = I. \tag{4.2.19}$$

Proof
The M. Riesz transforms are of the type considered in Section 4.2.1. From Theorem 4.2.1, they act on $L^p(\mathbf{R}^n)$. They are convolution operators and thus commute with any differentiation. It follows that they are continuous on $W^{m,p}(\mathbf{R}^n)$ for any positive integer m. By duality, they are also continuous on the dual spaces.

We prove formula (4.2.19) in the space $L^2(\mathbf{R}^n)$. Using the Fourier transform, it follows that

$$\widehat{R_i f} = (2\pi)^{n/2} \widehat{R_i}\, \hat{f}. \tag{4.2.20}$$

Thus, (4.2.19) is equivalent to

$$\sum_{i=1}^n \widehat{R_i}^2 = \frac{1}{(2\pi)^n}. \tag{4.2.21}$$

From (4.2.16) follows

$$\widehat{R_i} = -i\xi_i \frac{1}{(n-1)\gamma_n} \widehat{\frac{1}{|z|^{n-1}}}. \tag{4.2.22}$$

It is easy to check that the Fourier transform of a radial function is radial, and that the Fourier transform of a homogeneous function α is homogeneous with degree $-n - \alpha$. Thus,

$$\widehat{\frac{1}{|z|^{n-1}}} = C\frac{1}{|\xi|}, \tag{4.2.23}$$

and also

$$\widehat{\mathcal{R}_i}(\xi) = -i\frac{C}{(n-1)\gamma_n}\frac{\xi_i}{|\xi|}. \tag{4.2.24}$$

The only remaining part to check is that the constant C has the announced value. This comes from the well-known identity

$$\frac{1}{(2\pi)^{n/2}}\int_{\mathbb{R}^n} e^{-i(x\cdot\xi)}e^{-|x|^2/2}dx = e^{-|\xi|^2/2}, \tag{4.2.25}$$

which, combined with Plancherel's identity, yields

$$\int_{\mathbb{R}^n}\frac{1}{|x|^{n-1}}e^{-|x|^2/2}dx = C\int_{\mathbb{R}^n}\frac{1}{|\xi|}e^{-|\xi|^2/2}d\xi \tag{4.2.26}$$

and hence also the expected relation using the expression of γ_n.   ∎

Let us examine an integral operator of the form (4.2.1), with a kernel $K(y, z)$ homogeneous of degree $-n$ with respect to the variable z, which moreover satisfies (4.2.2). It can be split into its odd part and its even part in the form

$$\begin{cases} K = K_1 + K_2, \\ K_1(y, z) = (K(y, z) - K(y, -z))/2, \\ K_2(y, z) = (K(y, z) - K(y, -z))/2. \end{cases} \tag{4.2.27}$$

Theorem 4.2.1 has been proved for the odd part. It remains to prove it for the even part.

We use the M. Riesz transforms. From formula (4.2.19) follows

$$Kf = \sum_{i=1}^{n} K \circ \mathcal{R}_i \circ \mathcal{R}_i f. \tag{4.2.28}$$

The kernels $K \circ \mathcal{R}_i$ are odd and, using Theorem 4.2.2, it is sufficient to prove that they act in $L^p(\mathbb{R}^n)$.

Lemma 4.2.1 *Let K be an even kernel, homogeneous of degree $-n$ with respect to the variable z, which moreover satisfies for one value of $q > 1$,*

$$\sup_{y\in\mathbb{R}^n}\int_{S^{n-1}}|K(y, z')|^q dz' \leq C . \tag{4.2.29}$$

Then, the product operator $K \circ \mathcal{R}_i$ has an odd kernel and satisfies (4.2.8).

Proof

Let us define the kernels K_ε and $\mathcal{R}_{i\delta}$ by

$$K_\varepsilon(z, w) = \begin{cases} 0, & \text{when } |w| \leq \varepsilon, \\ K(z, w), & \text{else,} \end{cases} \qquad (4.2.30)$$

$$\mathcal{R}_{i\delta}(x) = \begin{cases} 0, & \text{when } |x| \leq \delta, \\ \mathcal{R}_i(x), & \text{else.} \end{cases} \qquad (4.2.31)$$

For any regular function f with compact support, we have

$$(K_\varepsilon \circ \mathcal{R}_{i\delta}) f(z) = \int_{\mathbf{R}^n} \int_{\mathbf{R}^n} K_\varepsilon(z, z - y) \mathcal{R}_{i\delta}(y - x) f(x) dx dy. \quad (4.2.32)$$

A change of variables gives the kernel associated with this operator,

$$L_{\varepsilon,\delta}(z, w) = \int_{\mathbf{R}^n} K_\varepsilon(z, w - t) \mathcal{R}_{i\delta}(t) dt. \qquad (4.2.33)$$

The change of variable $w \to -w$ and $t \to -t$ shows that $L_{\varepsilon,\delta}$ is odd with respect to the variable w. It follows from the hypotheses of Lemma 4.2.1 that the kernel $K_\varepsilon(z, w)$, for given z and ε, belongs to $L^q(\mathbf{R}^n)$.

From the properties of the kernel $\mathcal{R}_i$ and for given z and ε, the limit of $L_{\varepsilon,\delta}(z, w)$ when δ tends to zero belongs to $L^q(\mathbf{R}^n)$. Let us denote by L_ε the operator

$$L_\varepsilon(z, w) = \int_{\mathbf{R}^n} K_\varepsilon(z, w - t) \mathcal{R}_i(t) dt. \qquad (4.2.34)$$

For ε and η given, it holds that

$$L_\varepsilon(z, w) - L_\eta(z, w) = \int_{\varepsilon \leq |w-t| \leq \eta} K(z, w - t) \mathcal{R}_i(t) dt, \qquad (4.2.35)$$

which from identity (4.2.2) can be written

$$L_\varepsilon(z, w) - L_\eta(z, w) = \int_{\varepsilon \leq |w-t| \leq \eta} K(z, w - t) \left(\mathcal{R}_i(t) - \mathcal{R}_i(w) \right) dt. \quad (4.2.36)$$

For a value of θ such that $0 < \theta < 1$, Taylor's remainder expression gives

$$|\mathcal{R}_i(t) - \mathcal{R}_i(w)| \leq \frac{1}{\gamma_n} \frac{(n+2)|t - w|}{|w + \theta(t - w)|^{n+1}}. \qquad (4.2.37)$$

Let us choose w such that

$$\varepsilon \leq |w - t| \leq \eta \leq \frac{|w|}{2}, \qquad (4.2.38)$$

and thus

$$|w + \theta(t - w)| \geq |w| - |t - w| \geq \frac{|w|}{2}, \qquad (4.2.39)$$

and also

$$|\mathcal{R}_i(t) - \mathcal{R}_i(w)| \le C \frac{|t - w|}{|w|^{n+1}}. \tag{4.2.40}$$

Combining (4.2.40) and (4.2.36), for a non-vanishing w, yields

$$|L_\varepsilon(z,w) - L_\eta(z,w)| \le \frac{C}{|w|^{n+1}} |\eta - \varepsilon| \int_{S^{n-1}} |K(z,w')| dw'. \tag{4.2.41}$$

The sequence of kernels $L_\varepsilon(z,w)$, for a non-vanishing w, is Cauchy in L^∞. It converges almost everywhere to a limit $L(z,w)$. For any positive λ,

$$L_{\varepsilon,\delta}(z, \lambda w) = \lambda^{-n} L_{\varepsilon/\lambda, \delta/\lambda}(z, w), \tag{4.2.42}$$

and thus, taking the limit,

$$L(z, \lambda w) = \lambda^{-n} L(z, w). \tag{4.2.43}$$

It remains to show that the operator L satisfies the estimate (4.2.8). It holds that

$$\int_{S^{n-1}} |L(z,w')| dw' = \frac{1}{\text{Log } 2} \int_{1 \le |w| \le 2} |L(z,w)| dw. \tag{4.2.44}$$

Pick w such that $|w| \ge 1$. The limit when ε tends to zero in (4.2.41) yields

$$\left| (K_{1/2} \circ \mathcal{R}_i)(z,w) - L(z,w) \right| \le \frac{C}{|w|^{n+1}} \int_{S^{n-1}} |K(z,w')| dw', \tag{4.2.45}$$

or else

$$\begin{cases} \int_{1 \le |w| \le 2} \left| (K_{1/2} \circ \mathcal{R}_i)(z,w) - L(z,w) \right| dx \\ \qquad \le C \int_{S^{n-1}} |K(z,w')| \, dw'. \end{cases} \tag{4.2.46}$$

From (4.2.44), (4.2.46), and the triangle inequality follow

$$\begin{cases} \int_{S^{n-1}} |L(z,w')| \, dw' \le C \left(\int_{S^{n-1}} |K(z,w')| \, dw' \right. \\ \qquad \left. + \int_{1 \le |w| \le 2} \left| (K_{1/2} \circ \mathcal{R}_i)(z,w) \right| dw \right), \end{cases} \tag{4.2.47}$$

which, combined with the fact that $K_{1/2} \in L^q$, yields

$$\int_{S^{n-1}} |L(z,w')| \, dw' \le C \int_{S^{n-1}} |K(z,w')|^q \, dw'. \tag{4.2.48}$$

∎

The following theorem summarizes all the previous results:

Theorem 4.2.3 Let $K(y,z)$ be a homogeneous kernel of degree $-n$ with respect to the variable w, regular with respect to the variable y, which *satisfies*

$$\int_{S^{n-1}} K(z, w') dw' = 0, \quad \text{for any } z, \qquad (4.2.49)$$

$$\sup_{z \in \mathbf{R}^n} \int_{S^{n-1}} |K(z, w')|^q \, dz' \leq C, \quad \text{for a value of } q > 1. \qquad (4.2.50)$$

Then the associated operator K, defined as the limit

$$K_\varepsilon(z, w) = \begin{cases} 0, & \text{when } |w| \leq \varepsilon, \\ K(z, w), & \text{else;} \end{cases} \qquad (4.2.51)$$

$$(Kf)(z) = \lim_{\varepsilon \to 0} \int_{\mathbf{R}^n} K_\varepsilon(z, y - z) f(y) dy \qquad (4.2.52)$$

is continuous from $L^p(\mathbf{R}^n)$ into $L^p(\mathbf{R}^n)$ i.e.,

$$\|Kf\|_{L^p(\mathbf{R}^n)} \leq C_p \|f\|_{L^p(\mathbf{R}^n)}, \quad 1 < p < \infty. \qquad (4.2.53)$$

The continuity constants explode as $1/(p-1)$ when p tends to 1 and as p when p tends to infinity. For an odd kernel, condition (4.2.48) can be weakened by choosing $q = 1$.

Proof
Let us summarize the steps. The case of an odd kernel is the object of Theorem 4.2.1. To prove the general case, we split the kernel into its odd and its even part. Separately, they satisfy (4.2.2). For the even part, we write

$$K_\varepsilon f = \sum_{i=1}^n (K_\varepsilon \mathcal{R}_i) \mathcal{R}_i f. \qquad (4.2.54)$$

From Lemma 4.2.1, the product operator $K_\varepsilon \mathcal{R}_i$ is associated with an odd kernel $L_\varepsilon(z, w)$ whose limit L satisfies Theorem 4.2.1. We must show that the truncated operator $\widetilde{L_\varepsilon}$ associated with L,

$$\widetilde{L_\varepsilon}(z, w) = \begin{cases} 0, & \text{when } |w| \leq \varepsilon, \\ L(z, w), & \text{else,} \end{cases} \qquad (4.2.55)$$

is bounded from L^p to L^p for any ε. We will show that

$$\sup_{z \in \mathbf{R}^n} \int \left| L_\varepsilon(z, w) - \widetilde{L_\varepsilon}(z, w) \right| dw \leq C_q. \qquad (4.2.56)$$

From (4.2.42) follows

$$L_\varepsilon(z, w) = \varepsilon^{-n} L_1 \left(z, \frac{w}{\varepsilon} \right). \qquad (4.2.57)$$

The homogeneity of $L(z, w)$ yields the same equality for $\widetilde{L_\varepsilon}$,

$$\widetilde{L_\varepsilon}(z, w) = \varepsilon^{-n} \widetilde{L_1} \left(z, \frac{w}{\varepsilon} \right). \qquad (4.2.58)$$

Subtracting, it follows that

$$L_\varepsilon(z,w) - \widetilde{L_\varepsilon}(z,w) = \varepsilon^{-n}\left(L_1\left(z,\frac{w}{\varepsilon}\right) - \widetilde{L_1}\left(z,\frac{w}{\varepsilon}\right)\right), \qquad (4.2.59)$$

and thus,

$$\int_{\mathbb{R}^n}\left|L_\varepsilon(z,w) - \widetilde{L_\varepsilon}(z,w)\right| dw = \int_{\mathbb{R}^n}\left|L_1(z,w) - \widetilde{L_1}(z,w)\right| dw. \quad (4.2.60)$$

We estimate the right-hand side integral as

$$\begin{cases} \int_{\mathbb{R}^n}\left|L_1(z,w) - \widetilde{L_1}(z,w)\right| dw \leq \int_{|w|\leq 2} |(K_1\circ R_i)(z,w)|\, dw \\[2mm] + \int_{1\leq|w|\leq 2} |L(z,w)|\, dw + \int_{|w|\geq 2}\left|L_1(z,w) - \widetilde{L_1}(z,w)\right| dw. \end{cases} \qquad (4.2.61)$$

As K_1 belongs to $L^p(\mathbb{R}^n)$, the first term is bounded. From (4.2.43) and (4.2.48), we can estimate the second one. From (4.2.45) follows

$$\begin{cases} \int_{|w|\geq 2}\left|L_1(z,w) - \widetilde{L_1}(z,w)\right| dw \\[2mm] \leq C\left(\int_{|w|\geq 2}\frac{dw}{|w|^{n+1}}\right)\left(\int_{S^{n-1}} |K(z,w)|\, dw\right), \end{cases} \qquad (4.2.62)$$

and thus,

$$\int_{\mathbb{R}^n}\left|L_\varepsilon(z,w) - \widetilde{L_\varepsilon}(z,w)\right| dw \leq C_q\int_{S^{n-1}} |K(z,w)|^q\, dw'. \qquad (4.2.63)$$

From Young's inequality, the corresponding operator acts in $L^p(\mathbb{R}^n)$. To conclude, it remains to show that, when $\varepsilon \to 0$,

$$g_\varepsilon(z) = \int_{\mathbb{R}^n} K_\varepsilon(z, y-z)f(y)dy \qquad (4.2.64)$$

converges to a function g in L^p. We need to do so only when f is a differentiable function with compact support and then use a density argument.

Let h be a radial function with compact support and value 1 at the origin. From (4.2.2) follows

$$g_\varepsilon(z) = \int_{\mathbb{R}^n} K_\varepsilon(z, y-z)(f(y) - f(z)h(y-z))dy. \qquad (4.2.65)$$

For a small $y - z$, it holds that $|f(y) - f(z)h(y-z)| \leq c\,|y-z|$, and thus, for a fixed z,

$$|g_\varepsilon(z)| \leq C.$$

When z is big enough, $f(z) = 0$, and thus,

$$|g_\varepsilon(z)| \leq \frac{c}{|z|^n}\int_{f\neq 0} K\left(z, \frac{y-z}{|y-z|}\right)f(y)dy \leq \frac{c}{|z|^n}; \qquad (4.2.66)$$

then combining,

$$|g_\varepsilon(z)| \le \frac{c(f)}{1 + |z|^n}. \qquad (4.2.67)$$

As the right-hand side is in L^p, this estimate shows that the classical Lebesgue theorem can be used for any $p \ge 1$. The almost everywhere convergence follows from expression (4.2.64).   ∎

4.2.3   Adjoint operators

Let us examine now the family of operators which are the adjoints of the ones considered previously in Theorem 4.2.3. They take the form

$$g(z) = \int_{\mathbf{R}^n} K(y, y - z) f(y) dy, \qquad (4.2.68)$$

where the kernel $K(y, w)$ satisfies all the hypotheses of Theorem 4.2.3. We have

Theorem 4.2.4 *Let $K(y, w)$ be a homogeneous kernel of degree $-n$ with respect to the variable w satisfying (4.2.49) and (4.2.50). Then the associated operator, defined as the limit*

$$K_\varepsilon(y, w) = \begin{cases} 0, & \text{when } |w| \le \varepsilon, \\ K(y, w), & \text{else,} \end{cases} \qquad (4.2.69)$$

$$(Kf)(z) = \lim_{\varepsilon \to 0} \int_{\mathbf{R}^n} K_\varepsilon(y, y - z) f(y) dy, \qquad (4.2.70)$$

is continuous from $L^p(\mathbf{R}^n)$ into $L^p(\mathbf{R}^n)$, i.e.,

$$\|Kf\|_{L^p(\mathbf{R}^n)} \le C_p \|f\|_{L^p(\mathbf{R}^n)}, \quad 1 < p < \infty. \qquad (4.2.71)$$

The continuity constant explodes as $1/(p-1)$ when p tends to 1 and as p when p tends to infinity.

Proof
We easily check that the adjoint with respect to the scalar product in $L^2(\mathbf{R}^n)$ of the operator K_ε, given by (4.2.70), is the operator K_ε given by (4.2.52). Estimate (4.2.71) then follows from estimate (4.2.53).

Theorem 4.2.5 *Let $K(z, w)$ be a kernel which satisfies, besides the hypotheses of theorem (4.2.4), the hypothesis*

$$\sup_{z \in \mathbf{R}^n} \int_{S^{n-1}} \left| \frac{\partial^{|\alpha|}}{\partial z^\alpha} K(z, w') \right|^q dz' \le C; \quad q > 1; \quad \forall \alpha; \ |\alpha| \le m. \qquad (4.2.72)$$

Then the associated operators, either through (4.2.52), or through (4.2.70), are continuous from $W^{m,p}(\mathbf{R}^n)$ into $W^{m,p}(\mathbf{R}^n)$ and from $W^{-m,p}(\mathbf{R}^n)$

into $W^{-m,p}(\mathbb{R}^n)$, for any real p such that $1 < p < \infty$, and any positive integer m.

Proof

Consider the associated operator K_ε. In the first case (4.2.52),

$$g_\varepsilon(y) = \int_{|t| \ge \varepsilon} K(y,t) f(t+y) dt \qquad (4.2.73)$$

and, differentiating,

$$\frac{\partial g_\varepsilon}{\partial y_i}(y) = \int_{|t| \ge \varepsilon} \left(\frac{\partial K}{\partial y_i}(y,t) f(t+y) + K(y,t) \frac{\partial f}{\partial y_i}(t+y) \right) dt. \qquad (4.2.74)$$

From hypothesis (4.2.72), each of the two operators on the right-hand side satisfy the hypothesis of Theorem 4.2.4 and thus, taking the limit when ε tends to zero, we obtain the theorem for $m = 1$. A recursion with respect to m, gives the theorem for positive integer m.

In the second case (4.2.70),

$$g_\varepsilon(y) = \int_{|t| \ge \varepsilon} K(y+t,t) f(y+t) dt \qquad (4.2.75)$$

and, differentiating,

$$\frac{\partial g_\varepsilon}{\partial y_i} = \int_{|t| \ge \varepsilon} \left[\frac{\partial K}{\partial y_i}(y+t,t) f(y+t) + K(y+t,t) \frac{\partial f}{\partial y_i}(y+t) \right] dt. \qquad (4.2.76)$$

From hypothesis (4.2.72), each of the two operators in the right-hand side satisfy the hypotheses of Theorem 4.2.5.

A recursion with respect to m, gives the theorem for a positive integer m. We conclude for a negative integer m using the duality and the fact that the first family contains the adjoint of the second family and vice versa. ∎

4.3   Application to Integral Equations

4.3.1   Introduction

Let Γ be a bounded regular surface of dimension $n-1$ contained in $\mathbb{R}^n$. We consider integral operators of the form

$$g(y) = \int_\Gamma K(y, x-y) \varphi(x) d\gamma(x) ; \quad y \in \Gamma. \qquad (4.3.1)$$

We establish continuity properties of these operators from $W^{m,p}(\Gamma)$ to similar Sobolev spaces, for any real p such that $1 < p < \infty$, and any positive integer m. The associated kernels K are regular with respect to the variable y and quasi-homogeneous with respect to the variable $x - y$.

We establish similar continuity properties for operators of the form

$$g(y) = \int_\Gamma K(x, y - x)\varphi(x)d\gamma(x), \quad y \in \Gamma, \tag{4.3.2}$$

which is the family of adjoint operators with respect to the scalar product in $L^2(\Gamma)$.

4.3.2 Homogeneous kernels

We examine kernels K which are the restriction to the surface Γ of kernels defined in $\mathbb{R}^n$.

Definition
A **homogeneous kernel** $K(y, z)$ defined in $\mathbb{R}^n$ is of **class** $-m$, for an integer m such that $m \geq 0$, when

$$\sup_{y \in \mathbb{R}^n} \sup_{|z|=1} \left| \frac{\partial^{|\alpha|}}{\partial y^\alpha} \frac{\partial^{|\beta|}}{\partial z^\beta} K(y, z) \right| \leq C_{\alpha,\beta}, \quad \forall \alpha \quad \text{and} \quad \forall \beta. \tag{4.3.3}$$

$$\begin{cases} \dfrac{\partial^{|\beta|}}{\partial z^\beta} K(y, z) \text{ is homogeneous of degree } -n + 1 \\[2mm] \text{with respect to the variable } z, \text{ for } |\beta| = m. \end{cases} \tag{4.3.4}$$

For any hyperplane H whose equation is $(h, z) = 0$, and any m-uple of vectors $z_1, \ldots, z_m$ in H,

$$\int_{S^{n-2}} D_z^m K(y, z')(z_1, \ldots, z_m) \, dz' = 0, \tag{4.3.5}$$

where S^{n-2} is the intersection of the sphere S^{n-1} with the hyperplane H. ∎

We give some examples of these operators among those that we will encounter.

Example 4.1
Let m be a positive integer. We consider the kernel

$$K(x) = |x|^{2m-n}. \tag{4.3.6}$$

This operator is of class $1 - 2m$. Condition (4.3.4) is satisfied, since the partial derivatives of order $2m - 1$ are odd. When n is even and $2m - n > 0$, this operator is of infinite class, since then K is a polynomial.

Example 4.2
Let α be a multi-index and ℓ an integer. A more general example, which includes the previous one, is (notation: $(x)^\alpha = x_1^{\alpha_1} \cdots x_n^{\alpha_n}$)

$$K(x) = (x)^\alpha |x|^\ell. \tag{4.3.7}$$

It is of class $m = -(|\alpha| + \ell + n - 1)$, for a positive m. Condition (4.3.5) is satisfied when m **is odd and** $|\alpha|$ **even** or when m **is even and** $|\alpha|$ **odd**, since in both cases, the derivatives of order $-m$ are odd.

Example 4.3

Let α be a multi-index and ℓ a positive integer. The kernel

$$K(x) = (x)^\alpha |x|^{2\ell} \log |x| \tag{4.3.8}$$

is of class $m = -(|a| + 2\ell + n - 1)$. Condition (4.3.4) is satisfied when n *is even and $|\alpha|$ even* or when n *is odd and $|\alpha|$ odd*, since in both cases the derivatives of order $-m$ are odd.

In all the above examples, the derivatives of order m were odd, which yielded immediately condition (4.3.5). The following lemma will allow us to consider cases where the derivatives of order $-m$ are even.

Lemma 4.3.1 *Let $G(z)$ be a homogeneous function of degree $-(n - 2)$ defined in $\mathbb{R}^{n-1}$. Then the function*

$$K(z) = \frac{\partial}{\partial z_i} G(z) \tag{4.3.9}$$

satisfies

$$\int_{S^{n-2}} K(z) dz' = 0. \tag{4.3.10}$$

Proof

Let us denote by γ^{n-2} the area element on the sphere S^{n-2}. It holds that

$$\begin{cases} \displaystyle\int_{S^{n-2}} K(z) dz' = \lim_{\varepsilon \to 0} \frac{1}{\gamma^{n-2}} \frac{1}{\varepsilon} \int_{1 \leq |z| \leq 1+\varepsilon} \frac{\partial}{\partial z_i} G dz \\[2mm] \displaystyle= \lim_{\varepsilon \to 0} \frac{1}{\gamma^{n-2} \varepsilon} \left[\int_{|z'|=1+\varepsilon} G(z') \frac{z_i'}{|z'|} dz' - \int_{|z'|=1} G(z') \frac{z_i'}{|z'|} dz' \right]. \end{cases} \tag{4.3.11}$$

The homogeneity of the function G and the change of variable $z' = (1+\varepsilon)y'$ in the first integral show that the right-hand side vanishes. ∎

Lemma (4.3.1) allows us to consider the following examples:

Example 4.4

Let α be a multi-index and ℓ a positive integer. The kernel

$$K(x) = (x)^\alpha |x|^\ell (\text{Log} |x|)^\beta \tag{4.3.12}$$

is of class m with

$$m = -(|\alpha| + \ell + n - 1)$$

when $\beta = 0$ **and** $m > 0$, or when $\beta = 1$ and ℓ **is a positive even integer**. In this example, lemma (4.3.1) shows condition (4.3.5). When $\beta = 1$ and ℓ is an odd integer, we can check that the logarithmic term disappears after $-m$ derivations.

The following theorem gives the continuity properties for these examples:

Theorem 4.3.1 *The operator associated with the kernel K by either (4.3.1) or (4.3.2) is continuous from $H^r(\Gamma)$ into $H^{m+r}(\Gamma)$, for any real positive r, when the kernel K is of class $-m$ and the surface Γ regular enough. It also acts from $W^{r,p}(\Gamma)$ into $W^{r+m,p}(\Gamma)$, for any integer r and any p such that $1 < p < \infty$.*

Proof
We only need to establish the result for integal operators of the type (4.3.2) and then use the duality for integral operators of the type (4.3.1).

Let us consider an atlas covering the surface Γ and an associated partition of unity. Let ϕ_i denote the diffeomorphism that maps the set $\omega_i \cap \Gamma$ into $\mathbb{R}^{n-1}$, and ϕ_i^{-1} the inverse diffeomorphism that maps a bounded set of $\mathbb{R}^{n-1}$ onto $\omega_i \cap \Gamma$. Let α_i denote the functions of the partition of unity that satisfy

$$\sum_{i=1}^p \alpha_i(x) \equiv 1, \quad x \in \Gamma. \tag{4.3.13}$$

We decompose a function φ defined on Γ into

$$\begin{cases} \varphi = \displaystyle\sum_{i=1}^p \alpha_i \varphi, \\ \varphi_i = \alpha_i \varphi. \end{cases} \tag{4.3.14}$$

The spaces $W^{r,p}(\Gamma)$ are defined on the surface Γ in the same exact way as the spaces $H^m(\Gamma)$ were defined, using a transport through the mapping ϕ_i and a localisation through the functions α_i. From (4.3.14), the function g given by (4.3.1) is decomposed in

$$\begin{cases} g = \displaystyle\sum_{i=1}^p g^i, \\ g^i = \displaystyle\int_\Gamma K(x, y - x)\varphi_i(x)d\gamma(x). \end{cases} \tag{4.3.15}$$

We only need to prove regularity for all the g^i. Let β_i be a function with support in Γ_i whose value is 1 on the support of α_i. The difference between g^i and the function

$$g_i = \int_{\Gamma_i} \beta_i(y)K(x, y - x)\varphi_i(x)d\gamma(x) \tag{4.3.16}$$

has the expression

$$g_i - g^i = \int_\Gamma (\beta_i(y) - 1)K(x, y - x)\varphi_i(x)d\gamma(x). \tag{4.3.17}$$

Thus, g_i has the same regularity as g^i since the kernel in (4.3.17) has no singularity in z.

Using this localization, we only have to study the integral (4.3.16). The next step consists in writing them as integrals on $\mathbb{R}^{n-1}$ through the change of variables

$$\begin{cases} \xi = \phi_i(x), \\ \eta = \phi_i(y). \end{cases} \qquad (4.3.18)$$

To simplify notation, the corresponding points in the space will be denoted $x(\xi)$ and $y(\eta)$, and the jacobian of this transform will be denoted by J. It holds that

$$g_i(y(\eta)) = \int_{R^{n-1}} \beta_i(y(\eta)) K(x(\xi), y(\eta) - x(\xi)) \varphi_i(x(\xi)) J(\xi) d\xi. \qquad (4.3.19)$$

This function of the variable η has the same regularity as the function

$$h(\eta) = \int_{\mathbf{R}^{n-1}} K(x(\xi), y(\eta) - x(\xi)) f(\xi) d\xi, \qquad (4.3.20)$$

where f has a compact support and the same regularity as φ.

To show that the operator (4.3.20) acts from $W^{r,p}(\mathbb{R}^{n-1})$ into the space $W^{r+m,p}(\mathbb{R}^{n-1})$, it is sufficient to show that its m order differentials map $W^{r,p}(\mathbb{R}^{n-1})$ into the space $W^{r,p}(\mathbb{R}^{n-1})$. For any given m-tuple of vectors in $\mathbb{R}^{n-1}$, denoted $(\alpha_1, \ldots, \alpha_m)$, this differential takes the form

$$\begin{cases} \left\langle \dfrac{\partial^m h}{\partial \eta^m}, \alpha \right\rangle \\ = \displaystyle\int_{\mathbf{R}^{n-1}} \left\langle \dfrac{\partial^m K}{\partial z^m}(x(\xi), y(\eta) - x(\xi)), Dy(\eta) \cdot \alpha \right\rangle f(\xi) d\xi, \end{cases} \qquad (4.3.21)$$

where $Dy(\eta) \cdot \alpha$ is the m-tuple of tangent vectors to the surface Γ at the point $y(\eta)$ given by $(Dy(\eta)\alpha_1, \ldots, Dy(\eta)\alpha_m)$.

The operator given by (4.3.21) is a sum, with coefficients depending regularly on the variable η of operators of the form

$$\ell(\eta) = \int_{\mathbf{R}^{n-1}} L(x(\xi), y(\eta) - x(\xi)) f(\xi) d\xi, \qquad (4.3.22)$$

where the kernel L is homogeneous of degree $-(n-1)$ with respect to the second variable and satisfies definition (4.3.3).

This integral operator is not yet of the type considered in Theorems 4.2.3 and 4.2.4 since its kernel is not exactly homogeneous with respect to the variable $\eta - \xi$. Using a Taylor expansion with an integral remainder in the variable z, we can write the kernel L as

$$\begin{cases} L(x, z^*) = L(x, z) + \dfrac{\partial}{\partial z} L(x, z) (z^* - z) + \cdots + \\ + \dfrac{1}{(j-1)!} \displaystyle\int_0^1 \dfrac{\partial^j}{\partial z^j} L(x, (1-t)z + tz^*) (z^* - z)^j (1-t)^{j-1} dt. \end{cases} \qquad (4.3.23)$$

From the choice

$$\begin{cases} z = Dy(\xi) \cdot (\eta - \xi), \\ z^* = y(\eta) - x(\xi), \end{cases} \tag{4.3.24}$$

it follows that

$$L(x(\xi), y(\eta) - x(\xi)) = L(x(\xi), Dy(\xi) \cdot (\eta - \xi)) + R(\xi, \eta). \tag{4.3.25}$$

The right-hand side has a first kernel which is homogeneous of degree $-n+1$ with respect to the second variable, and regular with respect to the first variable. The second kernel is the remainder which is more regular.

We have already proved that the principal part in (4.3.25), satisfies all the hypotheses of Theorem 4.2.5, except (4.2.2). From (4.3.5) and a change of variable follows

$$\int_{|Dy(\xi)\theta'|=1} L(x(\xi), Dy(\xi)\theta')d\theta' = 0. \tag{4.3.26}$$

If we specify the charts in the atlas in a way such that the basis vectors in the tangent plane are orthogonal and have the same length, the surface $|Dy(\xi) \cdot \theta'| = 1$ is a sphere in $\mathbb{R}^{n-1}$, and then (4.2.2) results from (4.3.5). From Theorem 4.2.5, the operator associated with this principal part is continuous from $W^{r,p}(\mathbb{R}^{n-1})$ into $W^{r,p}(\mathbb{R}^{n-1})$. The kernel $R(\xi, \eta)$ is more regular.

To study the L^p regularity, we choose $j = 1$ in the Taylor expansion (4.3.23), and then

$$|R(\xi, \eta)| \leq \frac{c}{|\xi - \eta|^{n-2}}. \tag{4.3.27}$$

Now, Young's inequality shows the continuity of the associated operator from $L^p(\mathbb{R}^{n-1})$ into $L^p(\mathbb{R}^{n-1})$. To study the $W^{r,p}(\mathbb{R}^{n-1})$ regularity with a positive r, we expand the kernel $R(\xi, \eta)$. A Taylor expansion of the quantity $z^* - z$ yields

$$\begin{cases} z^* - z = \frac{1}{2!}D^2y(\xi)(\eta - \xi)^2 + \cdots \\ \quad + \frac{1}{(k-1)!}\int_0^1 (D^k y(t\eta + (1-t)\xi))(\eta - \xi)^k(1-t)^{k-1}dt. \end{cases} \tag{4.3.28}$$

Combining this expansion with (4.3.23), it follows that

$$L(x, z^*) = L_0(\xi, \eta) + L_1(\xi, \eta) + \cdots + L_\ell(\xi, \eta), \tag{4.3.29}$$

where the kernel L_0 is homogeneous of degree $-n + 1$ and is the principal part, and the successive kernels are homogeneous of increasing degree $-n + 2, \ldots, 0, \ldots$. The last kernel is not homogeneous. We can choose the integers j and k such that this last kernel is r times differentiable. Then, it maps $W^{r,p}(\mathbb{R}^{n-1})$ into $W^{r,p}(\mathbb{R}^{n-1})$.

This shows the theorem for integer r. Using duality and interpolation, the general case follows.   ∎

4.3.3 Pseudo-homogeneous kernels

Definition

For an integer m such that $m \geq 0$, a **kernel** $K(y, z)$ **is pseudo-homogeneous of class** $-m$ when it admits for any positive integer s an expansion of the form

$$K(y, z) = K_m(y, z) + \sum_{j=1}^{\ell-1} K_{m+j}(y, z) + K_{m+\ell}(y, z), \qquad (4.3.30)$$

where K_{m+j} is of class $-(m+j)$ for $j = 0, \ell - 1$, and l is chosen such that $K_{m+\ell}$ is s times differentiable.   ∎

Theorem 4.3.2 *Let K be a pseudo-homogeneous kernel chosen of class $-m$. The associated operators given by (4.3.1) or by (4.3.2), are continuous from $W^{r,p}(\Gamma)$ into $W^{r+m,p}(\Gamma)$ for any integer r and any p such that $1 < p < \infty$.*

This theorem is a direct consequence of Theorem 4.3.1 using the expression (4.3.30) for s big enough.   ∎

We describe now some examples of this situation:

Example 4.5

Let n_x be the unit normal to the surface Γ at the point x. The kernel (n is the space dimension)

$$K(x, z) = \frac{(n_x \cdot z)}{|z|^n} \qquad (4.3.31)$$

looks homogeneous of degree $-n + 1$, but is rather homogeneous of degree $-n + 2$. The vector z which appears in the integral operators (4.3.1) and (4.3.2) is either $x - y$ or $y - z$. When x and y are close, this vector is asymptotically tangent, and thus, the kernel is close to being homogeneous of degree $-n + 2$.

Let us be more precise. Let $\mathcal{P}$ be the projection operator on the surface Γ, which is defined in a tubular neighbourhood of this surface (see Section 2.5.6). Then, for a z of the form $y - x$ with y and x on the surface Γ, the kernel K takes the form

$$K^*(x, z) = \frac{(n_x \cdot (\mathcal{P}(x + z) - \mathcal{P}(x)))}{|z|^n}. \qquad (4.3.32)$$

A Taylor expansion of the numerator is

$$\mathcal{P}(x + z) = \mathcal{P}(x) + D\mathcal{P}(x) \cdot z + \frac{1}{2}D^2\mathcal{P}(x)(z, z) + \cdots. \qquad (4.3.33)$$

Using the moving frame on the surface, $(x = \mathcal{P}(x) + sn(\mathcal{P}(x)))$, the differential of $D\mathcal{P}(x)$ is $(I + s\mathcal{R})^{-1}$ which acts in the tangent plane to the surface at the point $\mathcal{P}(x)$, and thus,

$$(n_x \cdot (D\mathcal{P}(x) \cdot z)) = 0. \tag{4.3.34}$$

This shows that the operator associated with the kernel K^* is pseudo-homogeneous of class -1, the first non-zero term of the Taylor expansion being homogeneous of degree $-n + 2$.

Example 4.6
In space dimension 3, the single layer potential kernel associated with the Helmholtz equation is

$$K(z) = \frac{e^{ik|z|}}{|z|}. \tag{4.3.35}$$

A Taylor expansion of the exponential in a neighbourhood of zero is

$$e^{ik|z|} = 1 + ik|z| - k^2|z|^2 + \cdots, \tag{4.3.36}$$

from which it is clear that the kernel (4.3.35) is pseudo-homogeneous of class -1.

Example 4.7
In space dimension 3, the double layer potential kernel associated with the Helmholtz equation is

$$K(x, z) = \frac{\partial}{\partial n_x} \left(\frac{e^{ik|x - y|}}{|x - y|} \right). \tag{4.3.37}$$

Combining the two previous examples, it is also pseudo-homogeneous of class -1.

Example 4.8
In space dimension 2, the single layer potential kernel associated with the Helmholtz equation is

$$K(z) = H_0^{(1)}(k|z|), \tag{4.3.38}$$

where $H_0^{(1)}$ is the Hankel function of order zero. A Taylor expansion of this function in a neighbourhood of zero is

$$H_0^{(1)}(z) = \frac{2i}{\pi} \mathrm{Log}\,|z| + \gamma + \cdots, \tag{4.3.39}$$

which is an expansion in homogeneous functions. This kernel is pseudo-homogeneous of class -1.

Example 4.9
In space dimension 2, the double layer potential kernel associated with the
Helmholtz equation is

$$K(z) = \frac{\partial}{\partial n_x} \left(H_0^{(1)}(k|z|) \right). \tag{4.3.40}$$

This kernel is pseudo-homogeneous of class -1.

4.4   Application to Integral Equations

We summarize here all the continuity results for the integral operators
examined in Chapter 3.

Theorem 4.4.1 *The operators S, D, D^* defined in Chapter 3 by (3.1.33),
(3.1.34), and (3.1.35) are continuous from $H^s(\Gamma)$ into $H^{s+1}(\Gamma)$, for any
real s. They are also continuous from $W^{m,p}(\Gamma)$ into $W^{m+1,p}(\Gamma)$ for any p
such that $1 < p < \infty$, and any positive integer m.*

*The operator N formally defined by (3.1.36) in Chapter 3, and more
precisely in Theorem 3.4.2, is continuous from $H^{s+1}(\Gamma)$ into $H^s(\Gamma)$, for
any real s. It is also continuous from $W^{m+1,p}(\Gamma)$ into $W^{m,p}(\Gamma)$ for any p
such that $1 < p < \infty$, and any positive integer m.*

Proof
We established in the previous section that the operators S, D, D^* are
pseudo-homogeneous of class -1, from which comes the theorem.

Concerning the operator N, we can use the Calderon relations (Theorem
3.1.3) and in particular

$$-SN = \frac{I}{4} - D^2, \tag{4.4.1}$$

$$-NS = \frac{I}{4} - D^{*2}. \tag{4.4.2}$$

When k is such that the operator S is invertible, this gives the theorem,
using the previous results on S, D, D^*.

It would be better to use the expression (3.4.8) in Theorem 3.4.2 and
the operator Δ_Γ which maps $H^s(\Gamma)$ into $H^{s-2}(\Gamma)$ and $W^{m+2,p}(\Gamma)$ into
$W^{m,p}(\Gamma)$. This proof is also valid for the critical values of k. ∎

Comments
The above theorem shows in particular the regularity of the solutions of all
the problems that we have examined in Chapter 3. From Theorem 4.4.1,
we know the regularity of the solution of the associated integral equations
as well as those of the inverse operator. The regularity of the corresponding
density is linked to that of the Dirichlet or Neumann boundary data. The
representation formula then gives the solution which is analytic outside the

boundary. Its behavior in a neighbourhood of this boundary is also known. This technique gives a different proof of the regularity results obtained in Chapter 2, and in particular, it proves some results in non-Hilbertian spaces of type L^p.

5

Maxwell Equations and Electromagnetic Waves

5.1 Introduction

Electromagnetic waves are defined by the **electric field** E and the **magnetic field** H at each point in R^3. We start by describing their laws of propagation in an **isotropic dielectric medium**, which is characterized by **the electric permittivity** ε **and the magnetic permeability** μ.

The speed of waves in this dielectric medium is $1/\sqrt{\varepsilon\mu}$. We denote by ε_0, μ_0 respectively the permittivity and the permeability of the vacuum, and by c the speed of light in a vacuum, which is

$$c = \frac{1}{\sqrt{\varepsilon_0\mu_0}}. \tag{5.1.1}$$

The relative permittivity and permeability of the medium are defined by

$$\begin{cases} \varepsilon = \varepsilon_r\varepsilon_0, & \varepsilon_r \geq 1, \\[2mm] \mu = \mu_r\mu_0, & \mu_r \geq 1. \end{cases} \tag{5.1.2}$$

In the absence of electric and magnetic charges and currents, the electric and magnetic fields in a dielectric medium are governed by the **Maxwell system of equations**

$$\begin{cases} -\varepsilon\dfrac{\partial E}{\partial t} + \operatorname{curl} H = 0, \\[3mm] \mu\dfrac{\partial H}{\partial t} + \operatorname{curl} E = 0. \end{cases} \tag{5.1.3}$$

We must complete these equations by the transmission conditions at interfaces that separate different dielectric media. **The tangential components of the fields E and H** are continuous across a surface Γ of discontinuity of ε or μ. We denote by n the unit normal to Γ and then these jump conditions take the form

$$\begin{cases} [E \wedge n]_\Gamma = 0, \\ \\ [H \wedge n]_\Gamma = 0. \end{cases} \tag{5.1.4}$$

Isotropic conducting media are also quite common. They are characterized by ε and μ and by **the conductivity** σ, which is a real positive number. Maxwell equations in such a medium are

$$\begin{cases} -\varepsilon \frac{\partial E}{\partial t} + \operatorname{curl} H - \sigma E = 0, \quad & \text{(Ampère–Maxwell law)}, \\ \\ \mu \frac{\partial H}{\partial t} + \operatorname{curl} E = 0, \quad & \text{(Faraday law)}. \end{cases} \tag{5.1.5}$$

The interface conditions (5.1.4) on dielectric media are unchanged for conducting media.

We introduce the **electric induction** D and the **magnetic induction** B:

$$D = \varepsilon E, \tag{5.1.6}$$

$$B = \mu H. \tag{5.1.7}$$

From the Maxwell equations, it follows that

$$\operatorname{div} B = 0. \tag{5.1.8}$$

The **harmonic solutions** of Maxwell equations are complex-valued fields E and H such that the following fields (ω **is the pulsation**)

$$\begin{cases} E(t,x) = \Re\left(E(x)e^{-i\omega t}\right), \\ \\ H(t,x) = \Re\left(H(x)e^{-i\omega t}\right), \end{cases} \tag{5.1.9}$$

satisfy the Maxwell system. They satisfy the system of harmonic Maxwell equations

$$\begin{cases} i\omega\varepsilon E + \operatorname{curl} H - \sigma E = 0, \\ \\ -i\omega\mu H + \operatorname{curl} E = 0. \end{cases} \tag{5.1.10}$$

The **perfectly conducting medium** is a commonly used model. It can be obtained by a limit process when the conductivity σ tends to infinity in the conducting medium. For a fixed pulsation ω, the two fields E and H

tend to zero in this medium. The integral representation shows that at the interface Γ between the conductor and the dielectric, the interior limits are

$$\begin{cases} E \wedge n|_\Gamma = 0, \\[2mm] H \wedge n|_\Gamma = 0, \end{cases} \tag{5.1.11}$$

and so is the exterior limit $E \wedge n$, while the exterior limits $H \wedge n$ and $E \cdot n$ are non-zero. We introduce in Chapter 1 the plane waves solutions of Maxwell's equations.

A **classical problem** is that of a perfectly conducting object surrounded by a dielectric medium, lit by an incident plane wave. Similarly to the case of the Helmholtz equation, we need to describe precisely the behaviour of the solutions at infinity, which is equivalent to defining the radiation conditions. They appear in the expression of the fundamental solution, which we will examine in the next section.

5.2   Fundamental Solution and Radiation Conditions

We compute the fundamental harmonic solution of the Maxwell system in a homogeneous medium. The constants ε and μ are given and the harmonic system takes the form

$$\begin{cases} i\omega\varepsilon E + \operatorname{curl} H = 0, \\[2mm] -i\omega\mu H + \operatorname{curl} E = 0. \end{cases} \tag{5.2.1}$$

The fundamental solution for a system of equations with n unknowns and n equations of the form

$$au = 0, \tag{5.2.2}$$

where a is a differential operator with constant coefficients, is defined as a matrix $n \times n$, denoted by A, such that

$$aA = \delta I, \tag{5.2.3}$$

where δ is the Dirac mass at the origin and I is the $n \times n$ identity matrix.

The terms of A are in general distributions, and this will be the case here, in contrast with the fundamental solution of the harmonic Helmholtz equation which is an integrable function. Thus, the fundamental solution of the Maxwell system is a 6×6 matrix.

Due to the symmetry of the operator, we only need to compute half of this matrix or equivalently two 3×3 matrices, solutions of the system

$$\begin{cases} i\omega\varepsilon E + \operatorname{curl} H = \delta I, \quad I \text{ is the } 3 \times 3 \text{ identity matrix}, \\[2mm] -i\omega\mu H + \operatorname{curl} E = 0. \end{cases} \tag{5.2.4}$$

The other half of the fundamental solution is obtained by exchanging the role of the fields E and H, and of the parameters ε and μ. The parameter k is called the **wave number**,

$$k = \omega\sqrt{\varepsilon\mu}. \tag{5.2.5}$$

We introduce the **scalar potential** V and the **vector potential** A which satisfy

$$\begin{cases} E = \nabla V + A, \\[2mm] \operatorname{div} A - k^2 V = 0 \quad (\text{Lorentz gauge condition}). \end{cases} \tag{5.2.6}$$

In the expression of the fundamental solution, given by equation (5.2.4), the corresponding quantities are respectively the scalar potential V, which is a vector in $\mathbb{R}^3$, and the vector potential $\mathbf{A}$ which is a 3×3 matrix. To avoid the conflict of notation with the electric field, we denote by G (instead of E in the previous chapters), the fundamental solution of the scalar Helmholtz equation which satisfies the outgoing radiation condition:

$$G(x) = \frac{1}{4\pi} \frac{e^{ik\,|x|}}{|x|}. \tag{5.2.7}$$

Theorem 5.2.1 *The first half of the fundamental solution of the Maxwell system, solution of (5.2.4), is given by*

$$\begin{cases} E(x) = i\omega\mu G(x)I + \frac{i}{\omega\varepsilon} D^2\,G(x), \\[2mm] H(x) = \operatorname{curl}(G(x)I). \end{cases} \tag{5.2.8}$$

The second half of the fundamental solution of the Maxwell system is given by

$$\begin{cases} E(x) = \operatorname{curl}(G(x)I), \\[2mm] H(x) = -i\omega\varepsilon G(x)I - \frac{i}{\omega\mu} D^2\,G(x). \end{cases} \tag{5.2.9}$$

The second derivative Γ'' should be understood in the sense of distributions in $\mathcal{D}'(R^3)$.

Proof
Taking the divergence in the decomposition (5.2.6), it follows that

$$\operatorname{div} E = \Delta V + \operatorname{div} A, \tag{5.2.10}$$

which is obtained from the divergence in the first part of equation (5.2.4),

$$\operatorname{div} E = \frac{1}{i\omega\varepsilon} \operatorname{div}(\delta I) = -\frac{i}{\omega\varepsilon}\nabla\delta. \tag{5.2.11}$$

From the gauge condition (5.2.6), we can eliminate A and we obtain an equation for V:

$$\Delta V + k^2 V = -\frac{i}{\omega\varepsilon}\nabla\delta, \qquad (5.2.12)$$

which outgoing fundamental solution is

$$V(x) = \frac{i}{\omega\varepsilon}\nabla G(x). \qquad (5.2.13)$$

Taking the curl in the decomposition (5.2.6), it follows that

$$\operatorname{curl} E = \operatorname{curl} A. \qquad (5.2.14)$$

Eliminating H in (5.2.4) yields

$$\operatorname{curl}\operatorname{curl} E - k^2 E = i\omega\mu\delta I. \qquad (5.2.15)$$

From (5.2.6), we have

$$\nabla\operatorname{div} A - k^2(E - A) = 0, \qquad (5.2.16)$$

which in connection with (5.2.15) and (5.2.14) gives

$$\nabla\operatorname{div} A - \operatorname{curl}\operatorname{curl} A + k^2 A = -i\omega\mu\delta I, \qquad (5.2.17)$$

or equivalently

$$\Delta A + k^2 A = -i\omega\mu\delta I. \qquad (5.2.18)$$

The outgoing fundamental solution of (5.2.18) is

$$A(x) = i\omega\mu\, G(x)I. \qquad (5.2.19)$$

Thus, we have obtained (5.2.8). The expression (5.2.9) follows by exchanging the quantities E and H, and ε and μ, without forgetting to change i into $-i$. ∎

It is important to notice that the coefficients of the fundamental matrix are not integrable functions, nor usual finite parts. In particular, the trace of $D^2 G$ is ΔG, which has a Dirac mass singularity at the origin. We twice used the Sommerfeld condition to select a unique fundamental solution for the scalar potential and the vector potential. It is possible to choose in both cases the ingoing fundamental solution, which yields the **ingoing fundamental solution** of the Maxwell system.

Remark

The previous expressions are still valid for the Maxwell system (5.1.10) in a conducting medium. We have only to notice that the value of ε is then

$$\widetilde{\varepsilon} = \varepsilon + i\frac{\sigma}{\omega} \qquad (5.2.20)$$

and thus the corresponding value of k is

$$\widetilde{k} = \omega\sqrt{\mu\widetilde{\varepsilon}}. \qquad (5.2.21)$$

Its imaginary part is positive and tends to infinity as $\sqrt{\mu/2}\sqrt{\omega\sigma}$, when σ tends to infinity. Then, all the terms of the fundamental solution decrease as $e^{-\sqrt{\mu\omega\sigma/2}|x|}$.

Using this expression and the above integral representation, it follows that the fields in a conducting medium tend to zero, when $\omega\sigma$ tends to infinity. This behavior of the fundamental solution gives also a precise description of the boundary layer, whose thickness is proportional to $1/\sqrt{\omega\sigma}$.

Theorem 5.2.2 *The fundamental solution* (5.2.8) *and* (5.2.9) *(Theorem 5.2.1) satisfies the usual radiation conditions* ($r = |x|$)

$$\left|\frac{\partial E}{\partial r} - ikE\right| \leq \frac{c}{r^2}, \quad \text{for large } r, \tag{5.2.22}$$

$$\left|\frac{\partial H}{\partial r} - ikH\right| \leq \frac{c}{r^2}, \quad \text{for large } r, \tag{5.2.23}$$

and the following **radiation conditions of Silver-Müller***:* ($n = \vec{r}/r$)

$$\left|\sqrt{\varepsilon}\, E - \sqrt{\mu}\, H \wedge n\right| \leq \frac{c}{r^2}, \quad \text{for large } r, \tag{5.2.24}$$

$$\left|\sqrt{\varepsilon}\, E \wedge n + \sqrt{\mu}\, H\right| \leq \frac{c}{r^2}, \quad \text{for large } r. \tag{5.2.25}$$

Any of these four radiation conditions selects a unique fundamental solution. Moreover, each component of E and H behaves as $1/r$ at infinity, and the following components behave as

$$|(E \cdot x)| \leq \frac{c}{r}, \tag{5.2.26}$$

$$|(H \cdot x)| \leq \frac{c}{r}, \tag{5.2.27}$$

$$|(H \cdot E)| \leq \frac{c}{r^3}, \tag{5.2.28}$$

where c denotes a generic constant.

Proof
Using the expressions (5.2.8) and (5.2.9), conditions (5.2.22) and (5.2.23) follow from the fact that the fundamental solution G of the scalar Helmholtz equation satisfies the Sommerfeld condition, as well as all its partial derivatives. Besides, the fields E and H satisfy at any point, except the origin

$$\Delta E + k^2 E = 0, \tag{5.2.29}$$

$$\Delta H + k^2 H = 0. \tag{5.2.30}$$

From the condition (5.2.22) follows the uniqueness of the fundamental solution for E, and thus for H. The same is true with condition (5.2.23), just exchanging E and H. We will check the Silver-Müller condition (5.2.24) for the first half of the fundamental solution given by (5.2.8). The proof for the second half (5.2.25) given by (5.2.9), is similar.

The main parts at infinity of ∇G and $D^2 G$ behave as

$$
\begin{cases}
\dfrac{\partial}{\partial x_i} G \sim \dfrac{ik}{4\pi} \dfrac{e^{ikr}}{r} \dfrac{x_i}{r}, \\[3mm]
\dfrac{\partial^2}{\partial x_i \partial x_j} G \sim -\dfrac{k^2}{4\pi} \dfrac{e^{ikr}}{r} \dfrac{x_i x_j}{r^2}.
\end{cases}
\tag{5.2.31}
$$

From (5.2.6), the main parts of E behave as

$$
E_{ij} \sim \frac{i\omega\mu}{4\pi} \frac{e^{ikr}}{r} \left(\delta_i^j - \frac{x_i x_j}{r^2} \right).
\tag{5.2.32}
$$

It holds that

$$
H(x) = \nabla G(x) \wedge I,
\tag{5.2.33}
$$

and thus,

$$
H \sim \frac{ik}{4\pi} \frac{e^{ikr}}{r} \frac{\vec{r}}{r} \wedge I.
\tag{5.2.34}
$$

Using the double exterior product, it follows that

$$
\left(H \wedge \frac{\vec{r}}{r} \right)_{ij} \sim \frac{ik}{4\pi} \frac{e^{ikr}}{r} \left(\delta_i^j - \frac{x_i x_j}{r^2} \right),
\tag{5.2.35}
$$

from which follows condition (5.2.24). Relations (5.2.26) (5.2.27) and (5.2.28) are easy consequences of (5.2.24) and (5.2.25).

It remains to prove that the Silver-Müller condition (5.2.24) selects a unique fundamental solution. We introduce the scalar and vector potential. The Silver-Müller condition (5.2.24) takes the form

$$
\sqrt{\varepsilon}\,(\nabla V + A) - \frac{1}{i\omega\sqrt{\mu}} \operatorname{curl} A \wedge \frac{\vec{r}}{r} = o\left(\frac{1}{r^2}\right).
\tag{5.2.36}
$$

It follows that

$$
\frac{\partial V}{\partial r} + \left(A \cdot \frac{\vec{r}}{r} \right) = o\left(\frac{1}{r^2}\right)
\tag{5.2.37}
$$

and from the relation (2.5.211), Theorem 2.5.20, (the curvature terms are equivalent to $1/r$ and the index T stands for the tangent part to the sphere of radius r), we obtain

$$
\begin{cases}
ik\,(\nabla_T V + A_T) \\[2mm]
\quad - \left[\nabla_T \left(A \cdot \frac{\vec{r}}{r} \right) - \left(\frac{\partial}{\partial r} \left(A \wedge \frac{\vec{r}}{r} \right) \right) \wedge \frac{\vec{r}}{r} \right] = o\left(\frac{1}{r^2}\right).
\end{cases}
\tag{5.2.38}
$$

We already know that V and A behave as $1/r$, and that the terms $\nabla_T V$ and $\nabla_T (A \cdot (\vec{r}/r))$ behave as $1/r^2$, and thus,

$$ikA_T - \frac{\partial}{\partial r} A_T = o\left(\frac{1}{r^2}\right). \tag{5.2.39}$$

The gauge condition (5.2.6) yields

$$\nabla V = \frac{1}{k^2} \nabla \operatorname{div} A, \tag{5.2.40}$$

while the relation (2.5.210), Theorem 2.5.20, implies

$$\operatorname{div} A - \frac{\partial}{\partial r}\left(A \cdot \frac{\vec{r}}{r}\right) = o\left(\frac{1}{r^2}\right). \tag{5.2.41}$$

From (5.2.41) (5.2.40) and (5.2.37), it follows that

$$\frac{1}{k^2} \frac{\partial^2}{\partial r^2}\left(A \cdot \frac{\vec{r}}{r}\right) + \left(A \cdot \frac{\vec{r}}{r}\right) = o\left(\frac{1}{r^2}\right). \tag{5.2.42}$$

This last relation and (5.2.39) shows that the vector potential satisfies the usual Sommerfeld condition. From the gauge condition (5.2.6), it results that the scalar potential also satisfies the usual Sommerfeld condition, and this proves uniqueness.   ∎

Remarks
When the parameter ε has a positive imaginary part, the fundamental solutions are exponentially decreasing at infinity. It is also the case when the parameter μ has a positive imaginary part.

An interpretation of the radiation conditions is the following: r times the terms of order $1/r$ in the fundamental solutions behave as plane waves going outwards (although the plane waves do not satisfy the radiation condition). In particular, the fields E and H are asymptotically orthogonal, and the ratio of their lenghts is the vacuum impedence $\sqrt{\varepsilon/\mu}$.   ∎

We are able to give a precise expression of the **radiation conditions** for the Maxwell system. It has the form (c denotes a generic constant which we will avoid to mix up with the speed of the light)

$$\begin{cases} |E(x)| \leq \frac{c}{r}, & \text{for large } r, \\[2mm] |H(x)| \leq \frac{c}{r}, & \text{for large } r, \\[2mm] \left| \sqrt{\varepsilon} E - \sqrt{\mu}\, H \wedge \frac{\vec{r}}{r} \right| \leq \frac{c}{r^2}, & \text{for large } r. \end{cases} \tag{5.2.43}$$

A formulation of the **exterior harmonic Maxwell problem**, for a perfectly conducting medium, is:

Find E and H such that

$$\begin{cases} i\omega\varepsilon E + \operatorname{curl} H = 0, & x \in \Omega e, \\[2mm] -i\omega\mu H + \operatorname{curl} E = 0, & x \in \Omega e, \\[2mm] E \wedge n|_{\Gamma} = g, & x \in \Gamma, \quad g \text{ given, \quad tangent to } \Gamma, \\[2mm] E \text{ and } H \text{ satisfy (5.2.43).} \end{cases} \qquad (5.2.44)$$

In the case of an object lit by an incident plane wave, the total field takes the form

$$E = E^{inc} + E_d, \qquad (5.2.45)$$

and thus,

$$g = -E^{inc} \wedge n. \qquad (5.2.46)$$

The **interior harmonic Maxwell problem** is

$$\begin{cases} i\omega\varepsilon E + \operatorname{curl} H = 0, & x \in \Omega i, \\[2mm] -i\omega\mu E + \operatorname{curl} E = 0, & x \in \Omega i, \\[2mm] E \wedge n|_{\Gamma} = g. \end{cases} \qquad (5.2.47)$$

We will hereafter consider boundary conditions of the form

$$((E \wedge n) + \beta n \wedge (H \wedge n))|_{\Gamma} = g, \quad \text{where } \beta, \ g \text{ are given}, \qquad (5.2.48)$$

which are called impedance conditions or **Leontovitch conditions**.

5.3 Multipole Solutions

5.3.1 Multipoles

Similarly to the Helmholtz equation, in the case of an interior or an exterior domain delimited by a unit sphere S, the Maxwell system admits solutions which are expressed as a sum of separated variable solutions in the variable r on one side, and (θ, φ) on the other side. They are called **multipole solutions.** They are built with the help of the **vectorial spherical harmonics** which were introduced in Section 2.4.4.

Theorem 5.3.1 *In an exterior homogeneous medium, characterized by the constants ε and μ, the following harmonic Maxwell system in the exterior B_e of the unit ball*

$$\begin{cases} i\omega\varepsilon E + \operatorname{curl} H = 0, \\[2mm] -i\omega\mu H + \operatorname{curl} E = 0, \end{cases} \qquad (5.3.1)$$

*admits two families of specific solutions that satisfy the outgoing radia-
tion condition given in theorem (5.2.2), and in particular the Silver-Müller
condition.*

These solutions are on one side, **the transverse electric multipoles**
given by

$$
\begin{cases}
E(x) = h_\ell^{(1)}(kr) T_\ell^m(\theta, \varphi), \\[2mm]
H(x) = -\dfrac{i}{\omega\mu} \operatorname{curl}\left(h_\ell^{(1)}(kr) T_\ell^m(\theta, \varphi) \right) \\[4mm]
\qquad = -\dfrac{i\sqrt{\varepsilon/\mu}}{(2\ell+1)} \left[(\ell+1) h_{\ell-1}^{(1)}(kr) I_{\ell-1}^m(\theta, \varphi) + \ell h_{\ell+1}^{(1)}(kr) N_{\ell+1}^m(\theta, \varphi) \right],
\end{cases}
\tag{5.3.2}
$$

which moreover satisfy

$$
(E \cdot x) = 0,
\tag{5.3.3}
$$

and on the other side, **the transverse magnetic multipoles** *given by*

$$
\begin{cases}
E(x) = \dfrac{i}{\omega\varepsilon} \operatorname{curl}\left(h_\ell^{(1)}(kr) T_\ell^m(\theta, \varphi) \right) \\[4mm]
\qquad = \dfrac{i\sqrt{\mu/\varepsilon}}{(2\ell+1)} \left[(\ell+1) h_{\ell-1}^{(1)}(kr) I_{\ell-1}^m(\theta, \varphi) + \ell h_{\ell+1}^{(1)}(kr) N_{\ell+1}^m(\theta, \varphi) \right], \\[4mm]
H(x) = h_\ell^{(1)}(kr) T_\ell^m(\theta, \varphi),
\end{cases}
\tag{5.3.4}
$$

which moreover satisfy

$$
(H \cdot x) = 0.
\tag{5.3.5}
$$

*These solutions are, up to a multiplicative constant, the only solutions of the
Maxwell system that satisfy the Silver-Müller condition and have normal
components in the form of separated variables.*

*Similarly, the following harmonic Maxwell system in the interior of the
unit ball admits two families of specific solutions which are* **the transverse
electric multipoles** *given by*

$$
\begin{cases}
E(x) = j_\ell(kr) T_\ell^m(\theta, \varphi), \\[2mm]
H(x) \\[2mm]
\qquad = -\dfrac{i\sqrt{\varepsilon/\mu}}{(2\ell+1)} \left[(\ell+1) j_{\ell-1}(kr)\, I_{\ell-1}^m(\theta, \varphi) + \ell j_{\ell+1}(kr)\, N_{\ell+1}^m(\theta, \varphi) \right],
\end{cases}
\tag{5.3.6}
$$

and **the transverse magnetic multipoles** *given by*

$$
\begin{cases}
E(x) \\[2mm]
\qquad = \dfrac{i\sqrt{\mu/\varepsilon}}{(2\ell+1)} \left[(\ell+1) j_{\ell-1}(kr)\, I_{\ell-1}^m(\theta, \varphi) + \ell j_{\ell+1}(kr)\, N_{\ell+1}^m(\theta, \varphi) \right], \\[4mm]
H(x) = j_\ell(kr) T_\ell^m(\theta, \varphi).
\end{cases}
\tag{5.3.7}
$$

These solutions are, up to a multiplicative constant, the only regular solutions of the Maxwell system that have normal components in the form of separated variables.

Proof

We give a constructive proof of these solutions, using the following property: When E and H are solutions of the Maxwell system (5.3.1), $(E \cdot x)$ and $(H \cdot x)$ satisfy

$$\Delta(E \cdot x) + k^2(E \cdot x) = 0, \tag{5.3.8}$$

$$\Delta(H \cdot x) + k^2(H \cdot x) = 0. \tag{5.3.9}$$

(5.3.8) and (5.3.9) are consequences of the identity

$$\Delta(u \cdot x) = (\Delta u \cdot x) + 2 \operatorname{div} u, \tag{5.3.10}$$

in addition to the Maxwell system of equations and the identities $\operatorname{div} E = \operatorname{div} H = 0$.

When we seek a transverse electric multipole, $(H \cdot x)$ is a separated variables solution of the scalar Helmholtz equation. Thus, for an exterior problem, it takes the form

$$(H \cdot x) = \frac{\ell(\ell+1)}{i\omega\mu} h_\ell^{(1)}(kr) Y_\ell^m(\theta, \varphi), \tag{5.3.11}$$

whereas for an interior problem it has the expression

$$(H \cdot x) = \frac{\ell(\ell+1)}{i\omega\mu} j_\ell(kr) Y_\ell^m(\theta, \varphi). \tag{5.3.12}$$

We recall the two fundamental formulas in Theorem 2.5.20, in the case of a sphere S_r of radius r, which are

$$\begin{cases} \operatorname{curl} u = (\operatorname{curl}_{S_r} u_{S_r}) \frac{x}{r} + \overrightarrow{\operatorname{curl}}_{S_r} \left(u \cdot \frac{x}{r} \right) \\[2mm] \qquad\quad - \frac{\partial}{\partial r} \left(u \wedge \frac{x}{r} \right) - \frac{1}{r} \left(u \wedge \frac{x}{r} \right), \end{cases} \tag{5.3.13}$$

$$\operatorname{div} u = \operatorname{div}_{S_r} u_{S_r} + \frac{1}{r}(u \cdot x) + \frac{\partial}{\partial r} \left(u \cdot \frac{x}{r} \right). \tag{5.3.14}$$

From (5.3.13) and (5.3.14), we infer that the field E which is tangent to the sphere S and has a zero divergence satisfies

$$\operatorname{div}_S E_S = 0, \tag{5.3.15}$$

$$\operatorname{curl} E = (\operatorname{curl}_S E_S) \frac{x}{r} - \frac{\partial}{\partial r} \left(E \wedge \frac{x}{r} \right) - \frac{1}{r} \left(E \wedge \frac{x}{r} \right). \tag{5.3.16}$$

Thus, the normal component of $\operatorname{curl} E$, which is also that of $i\omega\mu H$, is

$$\operatorname{curl}_S E_S = \ell(\ell+1) h_\ell^{(1)}(k) Y_\ell^m(\theta, \varphi). \tag{5.3.17}$$

From (5.3.15) and (5.3.17), (see (2.4.189)), it follows that

$$\Delta_S E_S = -\ell(\ell+1)\overrightarrow{curl}_S\left(h_\ell^{(1)}(k)Y_\ell^m(\theta,\varphi)\right), \tag{5.3.18}$$

and using Theorem 2.4.8, it holds that

$$E_S = \overrightarrow{curl}_S\left(h_\ell^{(1)}(k)Y_\ell^m(\theta,\varphi)\right), \tag{5.3.19}$$

which shows that for any point x not on S, we have

$$E(x) = h_\ell^{(1)}(kr)T_\ell^m(\theta,\varphi). \tag{5.3.20}$$

The Maxwell system shows that the field H is curl $E/(iw\mu)$, which is given by (5.3.13). We have

$$\text{curl}_{S_r}\ E_{S_r} = \frac{1}{r}h_\ell^{(1)}(kr)\ \text{curl}_{S_r}\ T_\ell^m = \frac{\ell(\ell+1)}{r}h_\ell^{(1)}(kr)Y_\ell^m(\theta,\varphi), \tag{5.3.21}$$

$$E \wedge \frac{x}{r} = -h_\ell^{(1)}(kr)\nabla_S Y_\ell^m(\theta,\varphi),$$

$$\tag{5.3.22}$$

from which follows

$$\begin{cases} H(x) = \frac{1}{iw\mu}\left[\left[k\frac{d}{dr}h_\ell^{(1)}(kr)+\frac{1}{r}h_\ell^{(1)}(kr)\right]\nabla_S Y_\ell^m(\theta,\varphi)\right. \\ \\ \qquad\qquad \left. +\frac{\ell(\ell+1)}{r}h_\ell^{(1)}(kr)Y_\ell^m(\theta,\varphi)\frac{x}{r}\right]. \end{cases} \tag{5.3.23}$$

From the following recursion relations linking $I_{\ell-1}^m$, $N_{\ell+1}^m$ and Y_ℓ^m,

$$\begin{cases} N_{\ell+1}^m + I_{\ell-1}^m = (2\ell+1)Y_\ell^m\frac{x}{r}, \\ \\ (\ell+1)I_{\ell-1}^m - \ell N_{\ell+1}^m = (2\ell+1)\nabla_S Y_\ell^m, \end{cases} \tag{5.3.24}$$

and from the recursion relations in Theorem 2.6.1, the expression (5.3.23) takes the form

$$\begin{cases} H(x) = \frac{1}{iw\mu}\left[\frac{(\ell+1)}{(2\ell+1)}\left[k\frac{d}{dr}h_\ell^{(1)}(kr)+\frac{\ell+1}{r}h_\ell^{(1)}(kr)\right]I_{\ell-1}^m(\theta,\varphi)\right. \\ \\ \qquad\quad \left. -\frac{\ell}{(2\ell+1)}\left[k\frac{d}{dr}h_\ell^{(1)}(kr)-\frac{\ell}{r}h_\ell^{(1)}(kr)\right]N_{\ell+1}^m(\theta,\varphi)\right], \end{cases} \tag{5.3.25}$$

or equivalently

$$H(x) = \frac{-i\sqrt{\varepsilon/\mu}}{(2\ell+1)}\left[(\ell+1)h_{\ell-1}^{(1)}(kr)I_{\ell-1}^m(\theta,\varphi)+\ell h_{\ell+1}^{(1)}(kr)N_{\ell+1}^m(\theta,\varphi)\right]. \tag{5.3.26}$$

The same process leads to the expression of the transverse magnetic multipoles for the exterior domain. It also leads to the expression of the transverse multipoles for the interior problem.   ∎

We can exhibit the solutions of the Maxwell system in the interior and the exterior of the unit ball, in the form of a sum of multipoles.

Theorem 5.3.2 *The solution of the* **exterior problem**

$$
\begin{cases}
i\omega\varepsilon\, E + \operatorname{curl} H = 0, & x \in \Omega_e, \\[2mm]
-i\omega\mu\, H + \operatorname{curl} E = 0, & x \in \Omega_e, \\[2mm]
E \wedge n|_\Gamma = g,
\end{cases}
\tag{5.3.27}
$$

which moreover satisfies the radiation condition (5.2.43), is a sum of transverse electric multipoles and transverse magnetic multipoles, which is given by

$$
\begin{cases}
E(x) = \displaystyle\sum_{\ell=1}^{\infty}\sum_{m=-\ell}^{\ell}\left[u_\ell^m \frac{h_\ell^{(1)}(kr)}{h_\ell^{(1)}(k)} T_\ell^m(\theta,\varphi) + \frac{i\sqrt{\mu}}{\sqrt{\varepsilon}} v_\ell^m \right. \\[4mm]
\qquad\qquad \left. \times \left[\frac{\ell+1}{2\ell+1}\frac{h_{\ell-1}^{(1)}(kr)}{h_\ell^{(1)}(k)} I_{\ell-1}^m(\theta,\varphi) + \frac{\ell}{2\ell+1}\frac{h_{\ell+1}^{(1)}(kr)}{h_\ell^{(1)}(k)} N_{\ell+1}^m(\theta,\varphi) \right] \right],
\end{cases}
\tag{5.3.28}
$$

$$
\begin{cases}
H(x) = \displaystyle\sum_{\ell=1}^{\infty}\sum_{m=-\ell}^{\ell}\left[v_\ell^m \frac{h_\ell^{(1)}(kr)}{h_\ell^{(1)}(k)} T_\ell^m(\theta,\varphi) - \frac{i\sqrt{\varepsilon}}{\sqrt{\mu}} u_\ell^m \right. \\[4mm]
\qquad\qquad \left. \times \left[\frac{\ell+1}{2\ell+1}\frac{h_{\ell-1}^{(1)}(kr)}{h_\ell^{(1)}(k)} I_{\ell-1}^m(\theta,\varphi) + \frac{\ell}{2\ell+1}\frac{h_{\ell+1}^{(1)}(kr)}{h_\ell^{(1)}(k)} N_{\ell+1}^m(\theta,\varphi) \right] \right].
\end{cases}
\tag{5.3.29}
$$

The coefficients u and v admit the following expressions (where the function z_ℓ is given by $z_\ell(k) = k(dh_\ell^{(1)}(k)/dk)/(h_\ell^{(1)}(k))$):

$$
u_\ell^m = -\frac{1}{\ell(\ell+1)}\int_S (g\cdot\nabla_S Y_\ell^m)\,d\sigma = \frac{1}{\ell(\ell+1)}\int_S \operatorname{div}_S g\, Y_\ell^m\,d\sigma, \tag{5.3.30}
$$

$$
v_\ell^m = -\frac{1}{\ell(\ell+1)}\frac{i\omega\varepsilon}{z_\ell(k)+1}\int_S \left(g\cdot\overrightarrow{\operatorname{curl}}_S Y_\ell^m\right)d\sigma
$$

$$
= -\frac{1}{\ell(\ell+1)}\frac{i\omega\varepsilon}{z_\ell(k)+1}\int_S \operatorname{curl}_S g\; Y_\ell^m\,d\sigma.
\tag{5.3.31}
$$

The solution of the Maxwell **interior problem** *is a sum of transverse electric multipoles and transverse magnetic multipoles, which is given by*

$$
\begin{cases}
E(x) = \displaystyle\sum_{\ell=1}^{\infty}\sum_{m=-\ell}^{\ell}\left[u_\ell^m \frac{j_\ell(kr)}{j_\ell(k)} T_\ell^m(\theta,\varphi) + \frac{i\sqrt{\mu}}{\sqrt{\varepsilon}} v_\ell^m \right. \\[4mm]
\qquad\qquad \left. \times \left[\frac{\ell+1}{2\ell+1}\frac{j_{\ell-1}(kr)}{j_\ell(k)} I_{\ell-1}^m(\theta,\varphi) + \frac{\ell}{2\ell+1}\frac{j_{\ell+1}(kr)}{j_\ell(k)} N_{\ell+1}^m(\theta,\varphi) \right] \right],
\end{cases}
\tag{5.3.32}
$$

$$
\left\{
\begin{aligned}
H(x) &= \sum_{\ell=1}^{\infty} \sum_{m=-\ell}^{\ell} \left[v_\ell^m \frac{j_\ell(kr)}{j_\ell(k)} T_\ell^m(\theta, \varphi) - \frac{i\sqrt{\varepsilon}}{\sqrt{\mu}} u_\ell^m \right. \\
&\quad \times \left. \left[\frac{\ell+1}{2\ell+1} \frac{j_{\ell-1}(kr)}{j_\ell(k)} I_{\ell-1}^m(\theta, \varphi) + \frac{\ell}{2\ell+1} \frac{j_{\ell+1}(kr)}{j_\ell(k)} N_{\ell+1}^m(\theta, \varphi) \right] \right].
\end{aligned}
\right. \tag{5.3.33}
$$

The coefficients u and v admit the expressions

$$
u_\ell^m = -\frac{1}{\ell(\ell+1)} \int_S (g \cdot \nabla_S Y_\ell^m) \, d\sigma, \tag{5.3.34}
$$

$$
v_\ell^m = -\frac{1}{\ell(\ell+1)} \frac{i\omega\varepsilon \, j_\ell(k)}{k \frac{d}{dr} j_\ell(k) + j_\ell(k)} \int_S \left(g \cdot \overrightarrow{\mathrm{curl}}_S Y_\ell^m \right) d\sigma. \tag{5.3.35}
$$

Proof
As these solutions satisfy the Maxwell equations and the radiation conditions, we have only to check that the boundary condition is satisfied on the sphere S. Its expression is computed using the recursion relation satisfied by the Bessel functions $h_\ell^{(1)}$ and the property of the vectorial spherical harmonics. It holds that

$$
\left\{
\begin{aligned}
&E \wedge n \\
&= \sum_{\ell=1}^{\infty} \sum_{m=-\ell}^{\ell} \left[-u_\ell^m \nabla_S Y_\ell^m(\theta, \varphi) - v_\ell^m \frac{z_\ell(k)+1}{i\omega\varepsilon} \overrightarrow{\mathrm{curl}}_S Y_\ell^m(\theta, \varphi) \right].
\end{aligned}
\right. \tag{5.3.36}
$$

We conclude using the expression of the coefficients u and v and the Stokes formulas on Γ. ∎

Theorem 2.6.1 shows that the function $z_\ell(k) + 1$ does not vanish. Thus, the above formula is valid for any value of the frequency k, in the case of the exterior problem. In the case of the interior problem, the function z_ℓ is replaced by the logarithm derivative of j_ℓ, which vanishes for some values of the frequency k. They are the critical values of the interior problem.

Remarks
   − From (5.3.34) and the Stokes formula, it is clear that the coefficients u_ℓ^m vanish when we have

$$
\mathrm{div}_S \, g = 0. \tag{5.3.37}
$$

In that case the solution is transverse magnetic, i.e., satisfies

$$
(H \cdot x) = 0. \tag{5.3.38}
$$

   − From (5.3.35) and the Stokes formula, it is clear that the coefficients v_ℓ^m vanish when we have

$$
\mathrm{curl}_S \, g = 0, \tag{5.3.39}
$$

and in that case the solution is transverse electric, i.e., it satisfies

$$(E \cdot x) = 0. \tag{5.3.40}$$

– The interior problem does not have a unique solution when k is a zero of the Bessel function j_ℓ. It is also the case when k is a zero of $k(dj_\ell(k)/dk) + j_\ell(k)$.

These zeros are all the eigenvalues of the interior Maxwell problem. The corresponding eigenvectors are respectively the associated transverse electric and transverse magnetic multipoles.   ∎

The multipole solutions will be used later on to build variational formulations of the Maxwell system. These variational formulations are based on Hilbert spaces. Thus, we examine some of the many possible choices of Hilbert spaces that give a functional setting to the multipole solutions. The most natural choices depend on the regularity of E and H separately. We will use the spaces $H^s(S)$ introduced in Subsection 2.5.1.

We introduce a number of spaces. First denote by $TH^s(S)$ **the space of vectors tangent to** S and such that the two components in a local basis in the tangent plane are in the space $H^s(S)$. This definition can be extended to define $TH^s(\Gamma)$ for any smooth surface Γ.

Now we introduce the two Hilbert spaces

$$H_{\text{curl}}^{-1/2}(S) = \left\{ g \in TH^{-1/2}(S); \text{curl}_S\, g \in H^{-1/2}(S) \right\},$$

$$H_{\text{div}}^{-1/2}(S) = \left\{ g \in TH^{-1/2}(S); \text{div}_S\, g \in H^{-1/2}(S) \right\}.$$

These two spaces satisfy the following fundamental property.

Lemma 5.3.1 *The two spaces $H_{\text{curl}}^{-1/2}(S)$ and $H_{\text{div}}^{-1/2}(S)$ are mutually adjoint with respect to the scalar product in $TL^2(S)$.*

Proof

We expand g and h on the eigenbasis of the vectorial Laplace-Beltrami operator, constituted by $T_\ell^m = \overrightarrow{\text{curl}}_S Y_\ell^m$ and $\nabla_S Y_\ell^m$ (cf. Theorem 2.4.8);

$$g(x) = \sum_{\ell=1}^{\infty} \sum_{m=-\ell}^{\ell} g_{1\ell}^m\, \overrightarrow{\text{curl}}_\Gamma Y_\ell^m(\theta, \varphi) + g_{2\ell}^m\, \nabla_\Gamma Y_\ell^m(\theta, \varphi), \tag{5.3.41}$$

$$h(x) = \sum_{\ell=1}^{\infty} \sum_{m=-\ell}^{\ell} h_{1\ell}^m \overrightarrow{\text{curl}}_\Gamma Y_\ell^m(\theta, \varphi) + h_{2\ell}^m \nabla_\Gamma Y_\ell^m(\theta, \varphi). \tag{5.3.42}$$

We compute the hermitian product of these two vectors in $TL^2(S)$. The properties of the vectorial spherical harmonics show that it is given by

$$(g, h) = \sum_{\ell=1}^{\infty} \sum_{m=-\ell}^{\ell} \ell(\ell+1) \left(g_{1\ell}^m \overline{h}_{1\ell}^m + g_{2\ell}^m \overline{h}_{2\ell}^m \right). \tag{5.3.43}$$

Let us compute the norms and quantities

$$\|g\|_{TL^2(S)}^2 = \sum_{\ell=1}^{\infty} \sum_{m=-\ell}^{\ell} \ell(\ell+1)\left(|g_{1\ell}^m|^2 + |g_{2\ell}^m|^2\right), \tag{5.3.44}$$

$$\|g\|_{TH^{-1/2}(S)}^2 \leq c \sum_{\ell=1}^{\infty} \sum_{m=-\ell}^{\ell} (\ell(\ell+1))^{1/2}\left(|g_{1\ell}^m|^2 + |g_{2\ell}^m|^2\right), \tag{5.3.45}$$

$$\text{curl}_S\, g = \sum_{\ell=1}^{\infty} \sum_{m=-\ell}^{\ell} \ell(\ell+1)g_{1\ell}^m Y_\ell^m(\theta,\varphi), \tag{5.3.46}$$

$$\text{div}_S\, g = -\sum_{\ell=1}^{\infty} \ell(\ell+1)g_{2\ell}^m Y_\ell^m(\theta,\varphi), \tag{5.3.47}$$

$$\|\text{curl}_S\, g\|_{H^{-1/2}(S)}^2 \leq c \sum_{\ell=1}^{\infty} \sum_{m=-\ell}^{\ell} (\ell(\ell+1))^{3/2} |g_{1\ell}^m|^2, \tag{5.3.48}$$

$$\|\text{div}_S\, h\|_{H^{-1/2}(S)}^2 \leq c \sum_{\ell=1}^{\infty} \sum_{m=-\ell}^{\ell} (\ell(\ell+1))^{3/2} |h_{2\ell}^m|^2. \tag{5.3.49}$$

In view of the expansions of curl_S and div_S, the above hermitian product (5.3.43) can be rewritten as

$$\begin{cases} (g,h)_{TL^2(S)} = \left((-\Delta_S)^{-1}\text{curl}_S\, g, \text{curl}_S\, \overline{h}\right)_{L^2(S)} \\[2mm] \qquad\qquad + \left((-\Delta_S)^{-1}\text{div}_S\, g, \text{div}_S\, \overline{h}\right)_{L^2(S)}, \end{cases} \tag{5.3.50}$$

which expresses that the following operator

$$\overrightarrow{\text{curl}}_S(-\Delta_S)^{-1}\text{curl}_S - \nabla_S(-\Delta_S)^{-1}\text{div}_S \tag{5.3.51}$$

is the identity in $TL^2(S)$.

The hermitian product in $H_{\text{div}}^{-1/2}(S)$ is given by

$$\begin{cases} (g,h)_{H_{\text{div}}^{-1/2}(S)} \\[2mm] \quad = \sum_{\ell=1}^{\infty} \sum_{m=-\ell}^{\ell} \left[(\ell(\ell+1))^{3/2} g_{2\ell}^m \overline{h}_{2\ell}^m + (\ell(\ell+1))^{1/2} g_{1\ell}^m \overline{h}_{1\ell}^m\right], \end{cases} \tag{5.3.52}$$

and similarly the hermitian product in $H_{\text{curl}}^{-1/2}(S)$ is given by

$$
\left\{
\begin{aligned}
&(g,h)_{H_{\text{curl}}^{-1/2}(S)} \\
&= \sum_{\ell=1}^{\infty} \sum_{m=-\ell}^{\ell} (\ell(\ell+1))^{3/2}\, g_{1\ell}^m\, \overline{h}_{1\ell}^m + (\ell(\ell+1))^{1/2}\, g_{2\ell}^m\, \overline{h}_{2\ell}^m.
\end{aligned}
\right.
\tag{5.3.53}
$$

In view of the expansions of curl_S and div_S and the identities (5.3.51) (5.3.52) and (5.3.53), it appears that the duality operator between $H_{\text{div}}^{-1/2}(S)$ and $H_{\text{curl}}^{-1/2}(S)$ is given by

$$
\overrightarrow{\text{curl}}_S(-\Delta_S)^{-1/2}\,\text{curl}_S - \nabla_S(-\Delta_S)^{-3/2}\,\text{div}_S,
\tag{5.3.54}
$$

while the duality operator between $H_{\text{curl}}^{-1/2}(S)$ and $H_{\text{div}}^{-1/2}(S)$ is given by

$$
\overrightarrow{\text{curl}}_S(-\Delta_S)^{-3/2}\,\text{curl}_S - \nabla_S(-\Delta_S)^{-1/2}\,\text{div}_S.
\tag{5.3.55}
$$

The continuity of these operators is a consequence of the Cauchy-Schwartz inequality. It is quite easy to check that their product is the identity in $TL^2(S)$, using the properties of the differential operators $\overrightarrow{\text{curl}}_S$, curl_S, and div_S. ∎

We introduce the Hilbert space

$$
X = \left\{ E, H : E/r \in \left(L^2(\Omega_e)\right)^3, H/r \in \left(L^2(\Omega_e)\right)^3, \right.
$$

$$
r\left(\sqrt{\varepsilon}\,E_T - \sqrt{\mu}H \wedge \tfrac{x}{r}\right) \in \left(L^2(\Omega_e)\right)^3,
$$

$$
\left. \left(E \cdot \tfrac{x}{r}\right) \in L^2(\Omega_e), \left(H \cdot \tfrac{x}{r}\right) \in L^2(\Omega_e) \right\},
$$

which is in some sense the poorest Hilbert space where the exterior Maxwell equation (5.2.44) has a unique solution. It corresponds to the **space of the bounded energy**.

Theorem 5.3.3 *The solution of the exterior problem* (5.3.27), *given by* (5.3.28) *and* (5.3.29) *is the unique solution of this problem in the* **space** X, *when* $g \in H_{\text{div}}^{-1/2}(S)$. *It satisfies*

$$
\left\{
\begin{aligned}
&\int_{B_e} \frac{|E(x)|^2}{r^2}\,dx \le c\Bigg[\left(\|\text{div}_S\, g\|_{H^{-3/2}(S)}^2 + k^2 \|\text{div}_S\, g\|_{H^{-2}(S)}^2 \right) \\
&\qquad + \Big[\|\text{curl}_S\, g\|_{H^{-3/2}(S)}^2 + k^2 \|\text{curl}_S\, g\|_{H^{-2}(S)}^2 \\
&\qquad + k^4 \|\text{curl}_S\, g\|_{H^{-5/2}(S)}^2 + k^6 \|\text{curl}_S\, g\|_{H^{-3}(S)}^2 \Big] \Bigg].
\end{aligned}
\right.
\tag{5.3.56}
$$

$$\begin{cases} \displaystyle\int_{B_r} \frac{|H(x)|^2}{r^2}dx \le \frac{c}{k^2}\frac{\epsilon}{\mu}\left[\|\mathrm{div}_S\, g\|^2_{H^{-1/2}(S)}+k^2\|\mathrm{div}_S\, g\|^2_{H^{-1}(S)}\right. \\[2mm] \qquad\qquad +k^2\left[\|\mathrm{curl}_S\, g\|^2_{H^{-5/2}(S)}+k^4\|\mathrm{curl}_S\, g\|^2_{H^{-3}(S)}\right. \\[2mm] \qquad\qquad\qquad \left.\left.+k^4\|\mathrm{curl}_S\, g\|^2_{H^{-7/2}(S)}+k^6\|\mathrm{curl}_S\, g\|^2_{H^{-4}(S)}\right]\right]. \end{cases} \tag{5.3.57}$$

$$\begin{cases} \displaystyle\int_{B_r} r^2\left|\sqrt{\varepsilon}E_T-\sqrt{\mu}\,(H\wedge\frac{x}{r})\right|^2 dx \\[2mm] \le c\,\epsilon\left[\frac{1}{k^2}\left(\|\mathrm{div}_S\, g\|^2_{H^{-1/2}(S)}+k^2\|\mathrm{div}_S\, g\|^2_{H^{-1}(S)}\right)\right. \\[2mm] \qquad +\left[\|\mathrm{curl}_S\, g\|^2_{H^{-3/2}(S)}+k^2\|\mathrm{curl}_S\, g\|^2_{H^{-2}(S)}\right. \\[2mm] \qquad\qquad \left.\left.+k^4\|\mathrm{curl}_S\, g\|^2_{H^{-5/2}(S)}+k^6\|\mathrm{curl}_S\, g\|^2_{H^{-3}(S)}\right]\right]. \end{cases} \tag{5.3.58}$$

$$\begin{cases} \displaystyle\int_{B_r}\frac{1}{r^2}(E(x)\cdot x)^2 dx \\[2mm] \le c\,k^2\left[\|\mathrm{curl}_S\, g\|^2_{H^{-3/2}(S)}+k^2\|\mathrm{curl}_S\, g\|^2_{H^{-2}(S)}\right. \\[2mm] \qquad\qquad \left.+k^4\|\mathrm{curl}_S\, g\|^2_{H^{-5/2}(S)}+k^6\|\mathrm{curl}_S\, g\|^2_{H^{-3}(S)}\right]. \end{cases} \tag{5.3.59}$$

$$\begin{cases} \displaystyle\int_{B_r}\frac{1}{r^2}\,(H(x)\cdot x)^2 dx \\[2mm] \le c\,\frac{\epsilon}{\mu}\left[\|\mathrm{div}_S\, g\|^2_{H^{-1/2}(S)}+k^2\|\mathrm{div}_S\, g\|^2_{H^{-1}(S)}\right]. \end{cases} \tag{5.3.60}$$

Proof

The vectorial spherical harmonics are an orthonormal basis in $TL^2(S)$, which norms were computed in Theorem 2.4.7. Using these properties, it follows that

$$\begin{cases} \displaystyle\int_{B_r}\frac{|E(x)|^2}{r^2}dx = \sum_{\ell=1}^{\infty}\sum_{m=-\ell}^{\ell}\left[|u_\ell^m|^2\,\ell(\ell+1)\int_1^\infty\frac{\left|h_\ell^{(1)}(kr)\right|^2}{\left|h_\ell^{(1)}(k)\right|^2}dr\right. \\[2mm] \qquad +\frac{k^2}{\omega^2\varepsilon^2}|v_\ell^m|^2\,\ell(\ell+1)\int_1^\infty\frac{\left|h_{\ell-1}^{(1)}(kr)\right|^2+\left|h_{\ell+1}^{(1)}(kr)\right|^2}{\left|h_\ell^{(1)}(k)\right|^2}dr\Bigg], \end{cases} \tag{5.3.61}$$

$$
\left\{
\begin{aligned}
&\int_{B_r} \frac{|H(x)|^2}{r^2}\, dr = \sum_{\ell=1}^{\infty} \sum_{m=-\ell}^{\ell} \left[|v_\ell^m|^2\, \ell(\ell+1) \int_1^\infty \frac{\left|h_\ell^{(1)}(kr)\right|^2}{\left|h_\ell^{(1)}(k)\right|^2}\, dr \right. \\
&\left. + \frac{k^2}{\omega^2 \mu^2} |u_\ell^m|^2\, \ell(\ell+1) \int_1^\infty \frac{\left|h_{\ell-1}^{(1)}(kr)\right|^2 + \left|h_{\ell+1}^{(1)}(kr)\right|^2}{\left|h_\ell^{(1)}(k)\right|^2}\, dr \right].
\end{aligned}
\right.
\tag{5.3.62}
$$

Several integrals appear, some of which were already computed in Theorem 2.6.2. It is on one side $\gamma(k,\ell)$ given by (2.6.58) and on the other side ((q_ℓ given by (2.6.20))

$$
\left\{
\begin{aligned}
&\int_1^\infty \frac{\left|h_{\ell-1}^{(1)}(kr)\right|^2 + \left|h_{\ell+1}^{(1)}(kr)\right|^2}{\left|h_\ell^{(1)}(k)\right|^2}\, dr \\
&= \frac{q_{\ell-1}(k)\gamma(k,\ell-1) + q_{\ell+1}(k)\gamma(k,\ell+1)}{q_\ell(k)}.
\end{aligned}
\right.
\tag{5.3.63}
$$

The recursion relation (2.6.37) and the bounds (2.6.23) yield

$$
\left\{
\begin{aligned}
\frac{q_{\ell+1}}{q_\ell} &= \frac{1}{q_\ell^2} + \frac{1}{k^2}\left(\frac{p_\ell}{q_\ell} + \ell^2\right)^2 \le 1 + \frac{(2\ell+1)^2}{k^2}, \\
\frac{q_{\ell-1}}{q_\ell} &= \frac{1}{q_\ell^2} + \frac{1}{k^2}\left(\frac{p_\ell - (\ell+1)q_\ell}{q_\ell}\right)^2 \le 1 + \frac{\ell^2}{k^2}.
\end{aligned}
\right.
\tag{5.3.64}
$$

Thus, we obtain

$$
\left\{
\begin{aligned}
&\int_{B_r} \frac{|E(x)|^2}{r^2}\, dx \le \sum_{\ell=1}^{\infty} \sum_{m=-\ell}^{\ell} \ell(\ell+1)\left[\gamma((k,\ell)\, |u_\ell^m|^2 \right. \\
&\left. + \frac{k^2}{\omega^2 \varepsilon^2} |v_\ell^m|^2 \left[\frac{q_{\ell-1}(k)}{q_\ell(k)}\gamma(k,\ell-1) + \frac{q_{\ell+1}(k)}{q_\ell(k)}\gamma(k,\ell+1) \right] \right].
\end{aligned}
\right.
\tag{5.3.65}
$$

We need an upper bound for the quantity $1/(z_\ell(k)+1)$, or equivalently a lower bound for $(p_\ell - q_\ell)/q_\ell$. We seek a bound of the form $(\beta\ell)/(k^2+\beta)$, and then β is the lower bound of $(m+1/2)(m+\ell+1)$ from m satisfying $0 \le m \le \ell$. Thus, $(\beta = (\ell+1)/2)$

$$
\frac{p_\ell - q_\ell}{q_\ell} \ge \frac{\ell(\ell+1)}{2k^2 + (\ell+1)},
\tag{5.3.66}
$$

or equivalently

$$
\frac{1}{|z_\ell(k)+1|} \le \frac{1}{\ell} + \frac{2k^2}{\ell(\ell+1)}.
\tag{5.3.67}
$$

The bounds (2.6.59) and (2.6.68) on $\gamma(\ell, k)$, imply that

$$
\begin{cases}
\displaystyle \int_{B_e} \frac{|E(x)|^2}{r^2} dx \leq \sum_{\ell=1}^{\infty} \sum_{m=-\ell}^{\ell} \ell(\ell+1) \left[|u_\ell^m|^2 \gamma(k, \ell) \right. \\
\qquad\qquad\qquad\qquad \left. + \frac{2k^2}{\omega^2 |\varepsilon|^2} |v_\ell^m|^2 \left(1 + \frac{\ell + 1/2}{6k^2} \right) \right] \\[2mm]
\qquad \leq c \left[\|\mathrm{div}_S\, g\|^2_{H^{-3/2}(S)} + k^2 \|\mathrm{div}_S\, g\|^2_{H^{-2}(S)} \right. \\[2mm]
\qquad\qquad + \|\mathrm{curl}_S\, g\|^2_{H^{-3/2}(S)} + k^2 \|\mathrm{curl}_S\, g\|^2_{H^{-2}(S)} \\[2mm]
\qquad\qquad \left. + k^4 \|\mathrm{curl}_S\, g\|^2_{H^{-5/2}(S)} + k^6 \|\mathrm{curl}_S\, g\|^2_{H^{-3}(S)} \right],
\end{cases}
\tag{5.3.68}
$$

from which we obtain (5.3.56). Similarly, we obtain

$$
\begin{cases}
\displaystyle \int_{B_e} \frac{|H(x)|^2}{r^2} dx \leq \sum_{\ell=1}^{\infty} \sum_{m=-\ell}^{\ell} \ell(\ell+1) \left[\gamma(k, \ell) |v_\ell^m|^2 \right. \\[2mm]
\qquad \left. + \frac{k^2}{\omega^2 \mu^2} |u_\ell^m|^2 \left[\frac{q_{\ell-1}(k)}{q_\ell(k)} \gamma(k, \ell - 1) + \frac{q_{\ell+1}(k)}{q_\ell(k)} \gamma(k, \ell + 1) \right] \right],
\end{cases}
\tag{5.3.69}
$$

or equivalently

$$
\begin{cases}
\displaystyle \int_{B_e} \frac{|H(x)|^2}{r^2} dx \leq c \left[\frac{k^4 \varepsilon}{\mu} \left[\|\mathrm{curl}_S\, g\|^2_{H^{-5/2}(S)} + k^2 \|\mathrm{curl}_S\, g\|^2_{H^{-3}(S)} \right. \right. \\[2mm]
\qquad\qquad \left. + k^4 \|\mathrm{curl}_S\, g\|^2_{H^{-7/2}(S)} + k^6 \|\mathrm{curl}_S\, g\|^2_{H^{-4}(S)} \right] \\[2mm]
\qquad \left. + \frac{1}{\omega^2 \mu^2} \left[\|\mathrm{div}_S\, g\|^2_{H^{-1/2}(S)} + k^2 \|\mathrm{div}_S\, g\|^2_{H^{-1}(S)} \right] \right].
\end{cases}
\tag{5.3.70}
$$

Using moreover the expression $k = \omega \sqrt{\varepsilon \mu}$, (5.3.57) follows .

We can expand the quantity $\sqrt{\varepsilon} E_T - \sqrt{\mu} H \wedge (x/r)$, which appears in the radiation condition, on the vectorial spherical harmonics. We use the identities

$$
\begin{cases}
\frac{1}{r} T_\ell^m(\theta, \varphi) \wedge x = \frac{\ell}{2\ell+1} N_{\ell+1}^m(\theta, \varphi) - \frac{\ell+1}{2\ell+1} I_{\ell-1}^m(\theta, \varphi) \\[2mm]
\qquad\qquad\qquad = -\nabla_S Y_\ell^m(\theta, \varphi), \\[2mm]
\frac{1}{r} I_{\ell-1}^m(\theta, \varphi) \wedge x = T_\ell^m(\theta, \varphi), \\[2mm]
\frac{1}{r} N_{\ell+1}^m(\theta, \varphi) \wedge x = -T_\ell^m(t, \varphi),
\end{cases}
\tag{5.3.71}
$$

which lead to the expression (see also (5.3.23))

$$
\left\{
\begin{aligned}
&\sqrt{\varepsilon}E_T(x) - \sqrt{\mu}\tfrac{1}{r}(H(x)\wedge x) = \sum_{\ell=1}^{\infty}\sum_{m=-\ell}^{\ell}\Bigg[i\sqrt{\varepsilon}\frac{u_\ell^m}{h_\ell^{(1)}(k)}\\
&\quad\times\left[\left(\frac{d}{dr}h_\ell^{(1)}\right)(kr) - ih_\ell^{(1)}(kr) + \frac{1}{kr}h_\ell^{(1)}(kr)\right]T_\ell^m(\theta,\varphi)\\
&\quad+ i\sqrt{\mu}\frac{v_\ell^m}{h_\ell^{(1)}(k)}\\
&\quad\times\left[\left(\frac{d}{dr}h_\ell^{(1)}\right)(kr) - ih_\ell^{(1)}(kr) + \frac{1}{kr}h_\ell^{(1)}(kr)\right]\nabla_S Y_\ell^m(\theta,\varphi)\Bigg].
\end{aligned}
\right.
\tag{5.3.72}
$$

Thus, using the orthogonality of the spherical harmonics, it follows that

$$
\left\{
\begin{aligned}
&\int_{B_c}\left|\sqrt{\varepsilon}\,E_T(x) - \sqrt{\mu}\left(H(x)\wedge\frac{x}{r}\right)\right|^2 r^2 dx\\
&= \sum_{\ell=1}^{\infty}\sum_{m=-1}^{\ell}\Bigg[\ell(\ell+1)\left(\varepsilon|u_\ell^m|^2 + \mu|v_\ell^m|^2\right)\\
&\quad\times\int_1^{\infty}\left|\left(\frac{d}{dr}h_\ell^{(1)}\right)(kr) + i\,h_\ell^{(1)}(kr) + \frac{1}{kr}h_\ell^{(1)}(kr)\right|^2\frac{r^4 dr}{\left|h_\ell^{(1)}(k)\right|^2}\Bigg].
\end{aligned}
\right.
\tag{5.3.73}
$$

The function $z_\ell(kr) + 1 - ikr$, appearing above, has the value

$$
z_\ell(r) + 1 - ir = \frac{q_\ell - p_\ell}{q_\ell} + ir\frac{1 - q_\ell}{q_\ell}.
\tag{5.3.74}
$$

We seek an upper bound for $(p_\ell - q_\ell)/q_\ell$, of the form $\beta\ell/(k^2 + \beta)$, and as above, β is the upper bound of $(m + 1/2)(m + \ell + 1/2)$ and thus $\beta = 2(\ell + 1/2)^2$, and so

$$
\frac{p_\ell - q_\ell}{q_\ell} \le \frac{2\ell\left(\ell+\frac{1}{2}\right)^2}{r^2 + 2\left(\ell+\frac{1}{2}\right)^2}.
\tag{5.3.75}
$$

We also have

$$
\frac{q_\ell - 1}{q_\ell} \le \frac{4\ell(\ell+1)}{r^2 + 2\left(\ell+\frac{1}{2}\right)^2},
\tag{5.3.76}
$$

and thus,

$$
\left\{
\begin{aligned}
|z_\ell(r) + 1 - ir|^2 &\le \frac{(\ell(\ell+1))^2\left(16r^2 + 4\left(\ell+\frac{1}{2}\right)^2\right)}{\left(r^2 + 2\left(\ell+\frac{1}{2}\right)^2\right)^2}\\
&\le 16\frac{(\ell(\ell+1))^2}{r^2 + 2\left(\ell+\frac{1}{2}\right)^2},
\end{aligned}
\right.
\tag{5.3.77}
$$

or equivalently

$$
\left\{
\begin{aligned}
&\frac{\left|\left(\dfrac{d}{dr}h_\ell^{(1)}\right)(kr) - i\,h_\ell^{(1)}(kr) + \dfrac{1}{kr}h_\ell^{(1)}(kr)\right|^2}{\left|h_\ell^{(1)}(k)\right|^2} \\[2em]
&\qquad\qquad \leq 16\,\frac{\ell^2}{k^2r^2}\frac{(\ell+1)^2}{k^2r^2 + 2\,(\ell+1/2)^2}\frac{\left|h_\ell^{(1)}(kr)\right|^2}{\left|h_\ell^{(1)}(k)\right|^2}.
\end{aligned}
\right.
\tag{5.3.78}
$$

The last bound needed is

$$
\left\{
\begin{aligned}
&\int_{B_r}\left|\sqrt{\varepsilon}\,E_T(x) - \sqrt{\mu}\left(H(x)\wedge\frac{x}{r}\right)\right|^2 r^2 dx \\[1em]
&\qquad \leq \sum_{\ell=1}^{\infty}\sum_{m=-\ell}^{\ell}\Bigg[\frac{(\ell(\ell+1))^3}{k^2}\left(\varepsilon\,|u_\ell^m|^2 + \mu\,|v_\ell^m|^2\right) \\[1em]
&\qquad\qquad \times \int_1^{\infty}\frac{r^2}{k^2r^2 + 2\,(\ell+1/2)^2}\frac{\left|h_\ell^{(1)}(kr)\right|^2}{\left|h_\ell^{(1)}(r)\right|^2}dr\Bigg].
\end{aligned}
\right.
\tag{5.3.79}
$$

We then estimate the inside rational function by

$$
\frac{r^2}{k^2r^2 + 2\,(\ell+1)^2} \leq
\begin{cases}
\dfrac{1}{k^2}, & \text{when}\quad r^2 \geq \dfrac{2(\ell+1)^2}{k^2}, \\[1.5em]
\dfrac{r^2}{2(\ell+1)^2}, & \text{when}\quad r^2 \leq \dfrac{2(\ell+1)^2}{k^2},
\end{cases}
\tag{5.3.80}
$$

from which results (5.3.58).

The normal components of the fields are given by the expression (5.3.11) and the corresponding expression for $(E(x)\cdot n)$. It holds that

$$
(E(x)\cdot x) = \sum_{\ell=1}^{\infty}\sum_{\ell=-m}^{m}\frac{ik}{\omega\varepsilon}\ell(\ell+1)\frac{h_\ell^{(1)}(kr)}{h_\ell^{(1)}(r)}v_\ell^m Y_\ell^m(\theta,\varphi),
\tag{5.3.81}
$$

$$
(H(x)\cdot x) = \sum_{\ell=1}^{\infty}\sum_{\ell=-m}^{m}\frac{k}{i\omega\mu}\ell(\ell+1)\frac{h_\ell^{(1)}(kr)}{h_\ell^{(1)}(r)}u_\ell^m Y_\ell^m(\theta,\varphi),
\tag{5.3.82}
$$

from which (5.3.59) and (5.3.60) results.

From the expressions (5.3.81) and (5.3.82), it follows that $(E\cdot x)$ and $(H\cdot x)$ satisfy the scalar Helmholtz equation and the Sommerfeld radiation condition. Their traces on the sphere S are the quantities

$$(E(x) \cdot x)|_S = \sum_{\ell=1}^{\infty} \sum_{\ell=-m}^{m} \frac{ik}{\omega\varepsilon}\ell(\ell+1)v_\ell^m Y_\ell^m(\theta,\varphi), \qquad (5.3.83)$$

$$(H(x) \cdot x)|_S = \sum_{\ell=1}^{\infty} \sum_{\ell=-m}^{m} \frac{k}{i\omega\mu}\ell(\ell+1)u_\ell^m Y_\ell^m(\theta,\varphi), \qquad (5.3.84)$$

which are both in $H^{-1/2}(S)$. Therefore, uniqueness of the solution of the Helmholtz equation implies uniqueness of $(E(x) \cdot x)$ and $(H(x) \cdot x)$, which then implies uniqueness of the solution of the Maxwell equation. ∎

The following theorem is a **regularity result** on the solution of Maxwell's system:

Theorem 5.3.4 *The solution of the exterior problem* (5.3.27), *given by the expressions* (5.3.28) *and* (5.3.29), *satisfies the following properties:*
When $g \in TH^{1/2}(S)$ *(i.e.,* $g \in H_{\mathrm{div}}^{-1/2}(S)$ *and* $\mathrm{curl}_S\, g \in H^{-1/2}(S)$*),*

$$\left\{ \begin{aligned}
\int_{B_r} |\nabla_T E|^2 \, dx &\le c\bigg[\|\mathrm{div}_S\, g\|_{H^{-1/2}(S)}^2 + k^2 \|\mathrm{div}_S\, g\|_{H^{-1}(S)}^2 \\
&\quad + \|\mathrm{curl}_S\, g\|_{H^{-1/2}(S)}^2 + k^2 \|\mathrm{curl}_S\, g\|_{H^{-1}(S)}^2 \qquad (5.3.85) \\
&\quad + k^4 \|\mathrm{curl}_S\, g\|_{H^{-3/2}(S)}^2 + k^6 \|\mathrm{curl}_S\, g\|_{H^{-2}(S)}^2\bigg],
\end{aligned} \right.$$

$$\left\{ \begin{aligned}
\int_{B_r} \frac{1}{r^2}\left|\frac{\partial E}{\partial r}\right|^2 dx &\le c\bigg[\|\mathrm{div}_S\, g\|_{H^{-1/2}(S)}^2 + k^2 \|\mathrm{div}_S\, g\|_{H^{-1}(S)}^2 \\
&\quad + \|\mathrm{curl}_S\, g\|_{H^{-1/2}(S)}^2 + k^2 \|\mathrm{curl}_S\, g\|_{H^{-1}(S)}^2 \qquad (5.3.86) \\
&\quad + k^4 \|\mathrm{curl}_S\, g\|_{H^{-3/2}(S)}^2 + k^6 \|\mathrm{curl}_S\, g\|_{H^{-2}(S)}^2\bigg].
\end{aligned} \right.$$

Proof
We have already proved that, when $\mathrm{curl}_S\, g \in H^{-1/2}(S)$ and $\mathrm{div}_S\, g \in H^{-1/2}(S)$, $\Delta_S g \in TH^{-3/2}(S)$ and thus $g \in TH^{1/2}(S)$. Besides, the normal component of E on the sphere S, given by expression (5.3.83) also belongs to $H^{1/2}(S)$. The field E satisfies the scalar Helmholtz equation and the Sommerfeld radiation condition. Applying Theorem 2.6.2 to each of its components leads to (5.3.85) and (5.3.86).

The traces of the field H on the sphere S are in $H^{-1/2}(S)$. Thus, this field has no extra regularity under the above hypothesis. Conversely, when $g \in H_{\mathrm{div}}^{1/2}(S)$, the traces of the field H are respectively in $TH^{1/2}(S)$ and $H^{1/2}(S)$. It then satisfies estimates similar to the field E and in particular, it belongs to H_{loc}^1. ∎

5.3.2   The capacity operator

Definition

We denote by T the operator that associates to the value of E_T on the sphere S, the value of $H \wedge n$ on the sphere S, where E and H are the solutions of the exterior problem (5.3.27). We expand E_T and $H \wedge n$ on the basis of tangent vectors of the form T_ℓ^m and $\nabla_S Y_\ell^m$,

$$E_T(\theta, \varphi) = \sum_{\ell=0}^{\infty} \sum_{m=-\ell}^{\ell} \left[u_\ell^m T_\ell^m(\theta, \varphi) + w_\ell^m \nabla_S Y_\ell^m(\theta, \varphi) \right]. \tag{5.3.87}$$

The expansion of the capacity operator T is given by

$$\begin{cases} T E_T = H \wedge n = \sum_{\ell=0}^{\infty} \sum_{m=-\ell}^{\ell} \left[-i \sqrt{\frac{\varepsilon}{\mu}} \frac{1}{k} \left(z_\ell(k) + 1 \right) u_\ell^m T_\ell^m(\theta, \varphi) \right. \\ \\ \left. + i \sqrt{\frac{\varepsilon}{\mu}} \frac{k}{z_\ell(k) + 1} w_\ell^m \nabla_S Y_\ell^m(\theta, \varphi) \right]. \end{cases} \tag{5.3.88}$$

Theorem 5.3.5 *The capacity operator T is an isomorphism from the space $H_{\mathrm{curl}}^{-1/2}(S)$ onto the space $H_{\mathrm{div}}^{-1/2}(S)$. Moreover, it satisfies*

$$\Re \int_S \left(T E_T \cdot \overline{E}_T \right) d\sigma > 0, \tag{5.3.89}$$

and this quantity vanishes if and only if E_T vanishes.
It also holds that

$$\begin{cases} \int_S \Im \left(T E_T \cdot \overline{E}_T \right) d\sigma \geq \sqrt{\frac{\varepsilon}{\mu}} \frac{1}{k} \left[\| \mathrm{curl}_S\, E_T \|_{H^{-1/2}(S)}^2 \right. \\ \\ \left. - 2k^2 \| \mathrm{curl}_S\, E_T \|_{H^{-1}(S)}^2 - ck^2 \left[\| \mathrm{div}_S\, E_T \|_{H^{-3/2}(S)}^2 \right. \right. \\ \\ \left. \left. + k^2 \| \mathrm{div}_S\, E_T \|_{H^{-2}(S)}^2 + k^4 \| \mathrm{div}_S\, E_T \|_{H^{-5/2}(S)}^2 \right] \right]. \end{cases} \tag{5.3.90}$$

Proof

It is easy to check that

$$\mathrm{div}_S\, E_T = -\sum_{\ell=0}^{\infty} \sum_{m=-\ell}^{\ell} \ell(\ell+1) w_\ell^m Y_\ell^m(\theta, \varphi), \tag{5.3.91}$$

$$\mathrm{curl}_S\, E_T = \sum_{\ell=0}^{\infty} \sum_{m=-\ell}^{\ell} \ell(\ell+1) u_\ell^m Y_\ell^m(\theta, \varphi), \tag{5.3.92}$$

$$\mathrm{div}_S\, T E_T = -\sum_{\ell=0}^{\infty} \sum_{m=-\ell}^{\ell} i\ell(\ell+1) \sqrt{\frac{\varepsilon}{\mu}} \frac{k}{z_\ell(k)+1} w_\ell^m Y_\ell^m(\theta, \varphi), \tag{5.3.93}$$

$$\text{curl}_S \, T \, E_T = -\sum_{\ell=0}^{\infty} \sum_{m=-\ell}^{\ell} \frac{i\ell(\ell+1)}{k} \sqrt{\frac{\varepsilon}{\mu}} (z_\ell(k) + 1) \, u_\ell^m Y_\ell^m(\theta, \varphi). \quad (5.3.94)$$

The quantities $z_\ell(k) + 1$ and $1/(z_\ell(k) + 1)$ appear. From the bounds

$$\frac{1}{k^2} |z_\ell(k) + 1|^2 \le 1 + \frac{\ell^2}{k^2}, \quad (5.3.95)$$

$$\frac{k}{|z_\ell(k) + 1|} \le \frac{k}{\ell}\left(1 + \frac{2k^2}{\ell+1}\right), \quad (5.3.96)$$

we obtain

$$\begin{cases} \|\text{div}_S \, T \, E_T\|^2_{H^{-1/2}(S)} \le \frac{|\varepsilon|}{|\mu|} k^2 \Big[\|\text{div}_S \, E_T\|^2_{H^{-3/2}(S)} \\[2mm] \quad + c \Big[k^2 \|\text{div}_S \, E_T\|^2_{H^{-2}(S)} + k^4 \|\text{div}_S \, E_T\|^2_{H^{-5/2}(S)} \Big] \Big], \end{cases} \quad (5.3.97)$$

$$\begin{cases} \|\text{curl}_S \, T \, E_T\|^2_{H^{-3/2}(S)} \\[2mm] \quad \le \frac{|\varepsilon|}{|\mu|} \frac{1}{k^2} \Big[\|\text{curl}_S \, E_T\|^2_{H^{-1/2}(S)} + k^2 \|\text{curl}_S \, E_T\|^2_{H^{-3/2}(S)} \Big]. \end{cases} \quad (5.3.98)$$

The inverse operator is obtained by exchanging div_S and curl_S, and thus, it satisfies the same type of estimates. We compute

$$\begin{cases} \int_S (T \, E_T \cdot \overline{E}_T) \, d\sigma = \sum_{\ell=0}^{\infty} \sum_{m=-\ell}^{\ell} i\sqrt{\frac{\varepsilon}{\mu}} \ell(\ell+1) \\[2mm] \quad \times \Big[-\frac{1}{k} (z_\ell(k) + 1) |u_\ell^m|^2 + \frac{k}{z_\ell(k) + 1} |w_\ell^m|^2 \Big], \end{cases} \quad (5.3.99)$$

and thus,

$$\begin{cases} \Re \int_S (T \, E_T \cdot \overline{E}_T) \, d\sigma = \sum_{\ell=0}^{\infty} \sum_{m=-\ell}^{\ell} \sqrt{\frac{\varepsilon}{\mu}} \ell(\ell+1) \frac{1}{k} \Im (z_\ell(k)) \\[2mm] \quad \times \Big[|u_\ell^m|^2 + \frac{k^2}{|z_\ell(k) + 1|^2} |w_\ell^m|^2 \Big], \end{cases} \quad (5.3.100)$$

$$\begin{cases} \Im \int_S (T \, E_T \cdot \overline{E}_T) \, d\sigma = \sum_{\ell=0}^{\infty} \sum_{m=-\ell}^{\ell} \sqrt{\frac{\varepsilon}{\mu}} \ell(\ell+1) \frac{(\Re z_\ell(k) + 1)}{k} \\[2mm] \quad \times \Big[-|u_\ell^m|^2 + \frac{k^2}{|z_\ell(k) + 1|^2} |w_\ell^m|^2 \Big]. \end{cases} \quad (5.3.101)$$

We gave in (5.3.66) a lower bound for $-(\Re z_\ell + 1)$, which is

$$-(\Re z_\ell(k) + 1) \ge \frac{\ell(\ell+1)}{2k^2 + \ell + 1}, \quad (5.3.102)$$

which implies

$$-(\Re z_\ell(k) + 1) - \ell \geq -\frac{2\ell k^2}{2k^2 + \ell + 1} \geq -2k^2. \tag{5.3.103}$$

This lower bound leads to

$$\begin{cases} -\sum_{\ell=0}^{\infty} \sum_{m=-\ell}^{\ell} \sqrt{\frac{\varepsilon}{\mu}} \ell(\ell+1) \frac{\Re z_\ell(k) + 1}{k} |u_\ell^m|^2 \\ \geq \sqrt{\frac{\varepsilon}{\mu}} \left[\frac{1}{k} \|\mathrm{curl}_S E_T\|_{H^{-1/2}(S)}^2 - 2k \|\mathrm{curl}_S E_T\|_{H^{-1}(S)}^2 \right]. \end{cases} \tag{5.3.104}$$

Besides, from (5.3.96), it follows that

$$-\frac{\Re z_\ell(k) + 1}{(\Re z_\ell(k) + 1)^2 + |\Im z_\ell(k)|^2} \leq \frac{1}{\ell} \left(1 + \frac{2k^2}{\ell+1}\right)^2, \tag{5.3.105}$$

and thus,

$$\begin{cases} -\sum_{\ell=1}^{\infty} \sum_{m=-\ell}^{\ell} \sqrt{\frac{\varepsilon}{\mu}} \ell(\ell+1) \frac{k\,(\Re z_\ell(k) + 1)}{(\Re z_\ell(k) + 1)^2 + |\Im z_\ell(k)|^2} |w_\ell^m|^2 \\ \leq c\sqrt{\frac{\varepsilon}{\mu}} k \left[\|\mathrm{div}_S E_T\|_{H^{-3/2}(S)}^2 + k^2 \|\mathrm{div}_S E_T\|_{H^{-2}(S)}^2 \right. \\ \left. + k^4 \|\mathrm{div}_S E\|_{H^{-5/2}(S)} \right], \end{cases} \tag{5.3.106}$$

from which we obtain (5.3.90).   ∎

We now give some extra properties of the capacity operator that will be useful later.

Theorem 5.3.6 *The capacity operator T satisfies the following coercivity properties: For any $u_T \in H_{\mathrm{curl}}^{-1/2}(S)$ which moreover satisfies $\mathrm{div}_S\, u_T = 0$, it holds that*

$$\begin{cases} \Im \int_S (T\, u_T \cdot \bar{u}_T)\, d\gamma \geq 0 \\ \geq \sqrt{\frac{\varepsilon}{\mu}} \frac{1}{k} \left[\|\mathrm{curl}_S u_T\|_{H^{-1/2}(S)}^2 - 2k^2 \|\mathrm{curl}_S u_T\|_{H^{-1}(S)}^2 \right]. \end{cases} \tag{5.3.107}$$

For any $u_T \in H_{\mathrm{curl}}^{-1/2}(S)$ such that $u_T = \nabla_S\, p$, it holds that

$$\begin{cases} -\Im \int_S (T\, \nabla_S p \cdot \nabla_S \bar{p})\, d\gamma \geq 0 \\ \geq \sqrt{\frac{\varepsilon}{\mu}} k \left[\|p\|_{H^{1/2}(S)}^2 - k^2 \|p\|_{H^{-1/2}(S)}^2 \right], \end{cases} \tag{5.3.108}$$

$$\int_S (T\, u_T \cdot \nabla_S p)\, d\gamma = 0, \quad \text{when } \mathrm{div}_S\, u_T = 0, \tag{5.3.109}$$

and conversely, when this identity is true for any $p \in H^{1/2}(S)$, we have $\mathrm{div}_S\, u_T = 0$. Moreover, the inverse operator is given by

$$\left(T^{-1}(u \wedge n)\right) \wedge n = -\frac{\mu}{\varepsilon} T\, u_T. \tag{5.3.110}$$

Proof

We have already proved (5.3.107), which is nothing but a special case of (5.3.90). Similarly, (5.3.109) is a consequence of (5.3.91) (5.3.92) (5.3.93) and (5.3.94). Conversely, we use the Stokes formula and the expression (5.3.110) of the inverse operator.

Concerning (5.3.108), we use the expansion

$$p = \sum_{\ell=0}^{\infty} \sum_{m=-\ell}^{\ell} p_\ell^m Y_\ell^m(\theta, \varphi) \tag{5.3.111}$$

to obtain

$$\begin{cases} \Im \int_S \left(T\, \nabla_S\, p \cdot \nabla_S\, \bar{p}\right) d\gamma \\[2mm] = \sum_{\ell=1}^{\infty} \sum_{m=-\ell}^{\ell} \sqrt{\frac{\varepsilon}{\mu}} \ell(\ell+1) k\, \frac{\Re z_\ell(k)+1}{|z_\ell(k)+1|^2}\, |p_\ell^m|^2 \, . \end{cases} \tag{5.3.112}$$

This quantity is negative. It holds

$$\begin{cases} -\dfrac{\Re z_\ell + 1}{(\Re z_\ell + 1)^2 + |\Im z_\ell|^2} - \dfrac{1}{\ell} \\[3mm] = -\dfrac{(\Re z_\ell + 1)(\Re z_\ell + \ell + 1)}{\ell\left((\Re z_\ell + 1)^2 + (\Im z_\ell)^2\right)} - \dfrac{(\Im z_\ell)^2}{\ell\left((\Re z_\ell + 1)^2 + (\Im z_\ell)^2\right)} \, . \end{cases} \tag{5.3.113}$$

The first term of the right-hand side is positive. We estimate the second term using

$$-\frac{\Re z_\ell + 1}{\Im z_\ell} = \frac{p_\ell - q_\ell}{k} \geq \frac{\ell}{k}, \tag{5.3.114}$$

which yields

$$\frac{(\Im z_\ell)^2}{\ell\left((\Re z_\ell + 1)^2 + (\Im z_\ell)^2\right)} \leq \frac{k^2}{\ell\,(\ell^2 + k^2)} \leq \frac{k^2}{\ell^3}, \tag{5.3.115}$$

from which we obtain (5.3.108). Equality (5.3.110) follows from a simple computation.   ∎

Let us define $H_{\mathrm{curl}}^s(S)$ and $H_{\mathrm{div}}^s(S)$ by

$$H_{\mathrm{curl}}^s(S) = \{u \in TH^s(S);\, \mathrm{curl}_S\, u \in H^s(S)\},$$

$$H_{\mathrm{div}}^s(S) = \{u \in TH^s(S);\, \mathrm{div}_S\, u \in H^s(S)\}.$$

It follows from Theorem 5.3.5 that

Theorem 5.3.7 *For any real s, the capacity operator T given by (5.3.88) is an isomorphism from $H_{\text{curl}}^s(S)$ onto $H_{\text{div}}^s(S)$.*

Proof
We mimic the previous proof using the norms in $H^s(S)$ instead of the norms in $H^{-1/2}(S)$.

5.4   Exterior Problems

We describe and give the expression of the exterior harmonic Maxwell problem in a domain Ω_e whose boundary is Γ. We seek the electric field E and the magnetic field H. We exhibited in Section 5.2 the radiation condition that must be satisfied by the solution. Thus, a formulation of this problem is the

$$
\begin{cases}
\operatorname{curl} E - i\omega\mu H = 0, & x \in \Omega_e, \\
\operatorname{curl} H + i\omega\varepsilon E = 0, & x \in \Omega_e, \\
(E \wedge n - zH_T)|_\Gamma = g.
\end{cases}
\tag{5.4.1}
$$

The complex-valued functions $\varepsilon(x)$ and $\mu(x)$ satisfy the hypothesis

$$
\begin{cases}
\Re\varepsilon(x) > 0, \\
\Re\mu(x) > 0.
\end{cases}
\tag{5.4.2}
$$

Moreover, they are supposed to be piecewise analytic and such that the discontinuity surfaces are also analytic surfaces. Thus, they belong to $L^\infty(\Omega_e)$). Outside a ball of radius R enclosing Γ, these functions are constant with the values

$$
\begin{cases}
\varepsilon(x) = \varepsilon_0, & |x| \geq R, \\
\mu(x) = \mu_0, & |x| \geq R,
\end{cases}
\tag{5.4.3}
$$

which are such that

$$
\begin{cases}
\Re\varepsilon_0 > 0, & \Im\varepsilon_0 \geq 0, \\
\Re\mu_0 > 0, & \Im\mu_0 \geq 0.
\end{cases}
\tag{5.4.4}
$$

We look for fields E, H which satisfy the radiation conditions exhibited in Theorem 5.2.2, and especially the Silver-Müller condition which for $\Im\varepsilon_0 = \Im\mu_0 = 0$ takes the form

$$
\left| \sqrt{\varepsilon_0}\, E_T(x) - \sqrt{\mu_0}\, H \wedge \frac{x}{r} \right| \leq \frac{c}{r^3}, \quad \text{for large } r.
\tag{5.4.5}
$$

Thus, the problem consists in finding E and H which satisfy (5.4.1) and (5.4.5).

We have to choose adequate Hilbert spaces that contain E and H. Two different choices were examined in the case of a sphere and, correspondingly, two choices appear to be quite natural here also. We examine the first choice, which corresponds to the energy norm. We seek the solution in the space

$$X = \left\{ E/r \in \left(L^2(\Omega_e) \right)^3, H/r \in \left(L^2(\Omega_e) \right)^3, \left(E \cdot \frac{x}{r} \right) \in L^2(\Omega_e), \right.$$

$$\left. \left(H \cdot \frac{x}{r} \right) \in L^2(\Omega_e), \left(\sqrt{\varepsilon} \left(rE - (E \cdot x) \frac{x}{r} \right) - \sqrt{\mu} \left(H \wedge x \right) \right) \in \left(L^2(\Omega_e) \right)^3 \right\}.$$

The given right-hand side g belongs to the space $H_{\text{div}}^{-1/2}(\Gamma)$. We need to specify the spaces associated with this trace space. This is the object of the following trace and lifting theorems.

5.4.1 Trace and lifting associated with the space $H(\text{curl})$

We introduced in Section 2.5.2, the Hilbert spaces $H^s(\Gamma)$, and in Section 2.5.6 the surfacic operators ∇_Γ, div_Γ, curl_Γ and $\overrightarrow{\text{curl}}_\Gamma$ and then the scalar Laplace-Beltrami operator

$$\Delta_\Gamma u = \text{div}_\Gamma \nabla_\Gamma u = - \text{curl}_\Gamma \overrightarrow{\text{curl}}_\Gamma u, \qquad (5.4.6)$$

and the vectorial Laplace-Beltrami operator

$$\Delta_\Gamma v = \nabla_\Gamma \text{div}_\Gamma v - \overrightarrow{\text{curl}}_\Gamma \text{curl}_\Gamma v. \qquad (5.4.7)$$

The scalar operator Δ_Γ is self-adjoint with respect to the scalar product in $L^2(\Gamma)$ and is positive definite. It is also coercive in the space $H^1(\Gamma)$ and admits an inverse. As the injection of $H^1(\Gamma)$ into $L^2(\Gamma)$ is compact, this inverse operator is defined and compact in $L^2(\Gamma)$.

Thus, this operator admits a countable sequence of eigenfunctions in $L^2(\Gamma)$, denoted by Y_i such that

$$-\Delta_\Gamma Y_i = \lambda_i Y_i. \qquad (5.4.8)$$

The associated sequence of eigenvalues λ_i are real positive and tend to infinity. We normalize in $L^2(\Gamma)$ **the eigenfunctions Y_i which then constitute an orthonormal basis of $L^2(\Gamma)$.**

It follows easily from the expression of the vectorial Laplace-Beltrami operator given by (5.4.7) that

$$\begin{cases} -\Delta_\Gamma \overrightarrow{\text{curl}}_\Gamma Y_i = \lambda_i \overrightarrow{\text{curl}}_\Gamma Y_i, \\[2mm] -\Delta_\Gamma \nabla_\Gamma Y_i = \lambda_i \nabla_\Gamma Y_i. \end{cases} \qquad (5.4.9)$$

The vectors $\nabla_\Gamma Y_i$ and $\overrightarrow{\text{curl}}_\Gamma Y_i$ are eigenvectors of the vectorial Laplace-Beltrami operator associated with the eigenvalue λ_i.

From the Stokes formulas, it follows that any vector v such that

$$\begin{cases} \int_\Gamma (\nabla_\Gamma Y_i \cdot v)\, d\gamma = 0, \\[2mm] \int_\Gamma \left(\overrightarrow{\mathrm{curl}}_\Gamma Y_i \cdot v\right) d\gamma = 0, \end{cases} \tag{5.4.10}$$

satisfies

$$\mathrm{div}_\Gamma\, v = \mathrm{curl}_\Gamma\, v = 0. \tag{5.4.11}$$

All solutions of (5.4.10) and (5.4.11) are equal to zero when the **surface Γ is simply connected**, and then the sequence of vectors $\nabla_\Gamma Y_i$ and $\overrightarrow{\mathrm{curl}}_\Gamma Y_i$ constitute a basis in $TL^2(\Gamma)$. When the **surface Γ is not simply connected**, we denote by $\mathcal{N}$ the space of vectors which satisfy (5.4.11) and is thus the kernel of the vectorial Laplace-Beltrami operator

$$\Delta_\Gamma v = 0. \tag{5.4.12}$$

It has finite dimension. We choose an orthonormal basis of $\mathcal{N}$, which we denote by $\vec{u}_i$, $i = 1, \ldots, N$. The set of vectors constituted by $\vec{u}_i$, $i = 1, \ldots, N$, $\nabla_\Gamma Y_i$ and $\overrightarrow{\mathrm{curl}}_\Gamma Y_i$, span the space $TL^2(\Gamma)$ and is an orthogonal basis in $TL^2(\Gamma)$. It is also an eigenvector basis of the vectorial Laplace-Beltrami operator.

We can now redefine the spaces $H^s(\Gamma)$ and $TH^s(\Gamma)$, using this new basis. Any distribution u, defined on the surface Γ, can be formally expanded on the basis Y_i in the form

$$u(x) = \sum_{i=0}^\infty u_i Y_i(x), \quad u_i = \langle u, Y_i \rangle. \tag{5.4.13}$$

The space $H^s(\Gamma)$ is the space of distributions such that

$$H^s(\Gamma) = \left\{ u \in \mathcal{D}'(S),\ \sum_{i=0}^\infty (1 + \lambda_i)^s\, |u_i|^2 < \infty \right\}$$

equipped with the norm

$$\|u\|_{H^s(\Gamma)}^2 = \sum_{i=0}^\infty (1 + \lambda_i)^s\, |u_i|^2. \tag{5.4.14}$$

Remark

For $s = 0$ and $s = 1$, (5.4.14) is exactly the usual norm, but for other values of s, the new norms are only equivalent norms. We admit this last property which is directly linked to the regularity properties of the operator $I - \Delta_\Gamma$. For any s, this operator is an isomorphism from $H^{s+2}(\Gamma)$ onto $H^s(\Gamma)$, when the surface Γ is smooth (see for example I. Terrasse [144]). This property is not true for irregular surfaces. We already used this regularity, when writing the expansion (5.4.13), which supposes that the functions Y_i are in the space $\mathcal{D}(\Gamma)$. ∎

Any tangent vector field defined on the surface Γ can be expanded in the above eigenvector basis of $-\Delta_\Gamma$. Thus, it can be written

$$v(x) = \sum_{j=1}^{N} \gamma_j \, \vec{u}_j(x) + \sum_{i=0}^{\infty} \left[\alpha_i \, \nabla_\Gamma \, Y_i + \beta_i \, \overrightarrow{\text{curl}}_\Gamma \, Y_i(x) \right]. \tag{5.4.15}$$

The space $TL^2(\Gamma)$ is defined by

$$TL^2(\Gamma) = \left\{ v; \sum_{j=1}^{N} |\gamma_j|^2 + \sum_{i=0}^{\infty} \lambda_i \left(|\alpha_i|^2 + |\beta_i|^2 \right) < \infty \right\}$$

equipped with the norm

$$\|v\|^2_{TL^2(\Gamma)} = \sum_{j=1}^{N} |\gamma_j|^2 + \sum_{i=0}^{\infty} \lambda_i \left(|\alpha_i|^2 + |\beta_i|^2 \right). \tag{5.4.16}$$

We define **the space** $TH^s(\Gamma)$ by

$$TH^s(\Gamma) = \left\{ v; \sum_{j=1}^{N} |\gamma_j|^2 + \sum_{i=0}^{\infty} (\lambda_i)^{s+1} \left(|\alpha_i|^2 + |\beta_i|^2 \right) < \infty \right\}$$

equipped with the norm

$$\|v\|^2_{TH^s(\Gamma)} = \sum_{j=1}^{N} |\gamma_j|^2 + \sum_{i=0}^{\infty} (\lambda_i)^{s+1} \left(|\alpha_i|^2 + |\beta_i|^2 \right). \tag{5.4.17}$$

The expansion (5.4.15) yields

$$\text{div}_\Gamma \, v(x) = \sum_{i=0}^{\infty} \alpha_i \, \lambda_i \, Y_i(x), \tag{5.4.18}$$

$$\text{curl}_\Gamma \, v(x) = \sum_{i=0}^{\infty} \beta_i \, \lambda_i \, Y_i(x). \tag{5.4.19}$$

Thus, we **define**

$$H^{-1/2}_{\text{div}}(\Gamma) = \left\{ v \in TH^{-1/2}(\Gamma), \ \text{div}_\Gamma \, v \in H^{-1/2}(\Gamma) \right\}$$

equipped with the norm

$$\|v\|^2_{H^{-1/2}_{\text{div}}(\Gamma)} = \sum_{j=1}^{N} |\gamma_j|^2 + \sum_{i=0}^{\infty} \lambda_i^{1/2} \left(\lambda_i \, |\alpha_i|^2 + |\beta_i|^2 \right) \tag{5.4.20}$$

and **the space**

$$H^{-1/2}_{\text{curl}}(\Gamma) = \left\{ v \in TH^{-1/2}(\Gamma), \ \text{curl}_\Gamma \, v \in H^{-1/2}(\Gamma) \right\}$$

equipped with the norm

$$\|v\|^2_{H_{\mathrm{curl}}^{-1/2}(\Gamma)} = \sum_{j=1}^{N} |\gamma_j|^2 + \sum_{i=0}^{\infty} (\lambda_i)^{1/2} \left(|\alpha_i|^2 + \lambda_i |\beta_i|^2 \right). \tag{5.4.21}$$

Lemma 5.4.1 *The spaces $H_{\mathrm{div}}^{-1/2}(\Gamma)$ and $H_{\mathrm{curl}}^{-1/2}(\Gamma)$ are mutually adjoint with respect to the scalar product in $TL^2(\Gamma)$. The duality operator from $H_{\mathrm{div}}^{-1/2}(\Gamma)$ onto $H_{\mathrm{curl}}^{-1/2}(\Gamma)$ is*

$$\mathcal{R} = I_{\mathcal{N}} + \overrightarrow{\mathrm{curl}}_\Gamma(-\Delta_\Gamma)^{-1/2}\,\mathrm{curl}_\Gamma - \nabla_\Gamma(-\Delta_\Gamma)^{-3/2}\,\mathrm{div}_\Gamma \tag{5.4.22}$$

where $I_{\mathcal{N}}$ stands for the identity on the linear space $\mathcal{N}$.

Proof
Similarly to what was used in the case of the sphere S, it results from (5.4.18) and (5.4.19) that the scalar products admit the expression

$$\begin{cases} (v, w)_{TL^2(\Gamma)} = (v_{\mathcal{N}} \cdot w_{\mathcal{N}})_{TL^2(\Gamma)} \\ \qquad + \left((-\Delta_\Gamma)^{-1}\,\mathrm{curl}_\Gamma\,v, \mathrm{curl}_\Gamma\,w \right)_{L^2(\Gamma)} \\ \qquad + \left((-\Delta_\Gamma)^{-1}\,\mathrm{div}_\Gamma\,v, \mathrm{div}_\Gamma\,w \right)_{L^2(\Gamma)}, \end{cases} \tag{5.4.23}$$

$$\begin{cases} (v, w)_{H_{\mathrm{div}}^{-1/2}(\Gamma)} = (v_{\mathcal{N}} \cdot w_{\mathcal{N}})_{TL^2(\Gamma)} \\ \qquad + \left((-\Delta_\Gamma)^{-1/2}\,\mathrm{div}_\Gamma\,v, \mathrm{div}_\Gamma\,w \right)_{L^2(\Gamma)} \\ \qquad + \left((-\Delta_\Gamma)^{3/2}\,\mathrm{curl}_\Gamma\,v, \mathrm{curl}_\Gamma\,w \right)_{L^2(\Gamma)}, \end{cases} \tag{5.4.24}$$

$$\begin{cases} (v, w)_{H_{\mathrm{curl}}^{-1/2}(\Gamma)} = (v_{\mathcal{N}} \cdot w_{\mathcal{N}})_{TL^2(\Gamma)} \\ \qquad + \left((-\Delta_\Gamma)^{-1/2}\,\mathrm{curl}_\Gamma\,v, \mathrm{curl}_\Gamma\,w \right)_{L^2(\Gamma)} \\ \qquad + \left((-\Delta_\Gamma)^{3/2}\,\mathrm{div}_\Gamma\,v, \mathrm{div}_\Gamma\,w \right)_{L^2(\Gamma)}. \end{cases} \tag{5.4.25}$$

Thus, the lemma follows from

$$\mathrm{div}_\Gamma\,\mathcal{R}v = (-\Delta_\Gamma)^{-1/2}\,\mathrm{div}_\Gamma\,v, \tag{5.4.26}$$

$$\mathrm{curl}_\Gamma\,\mathcal{R}v = (-\Delta_\Gamma)^{1/2}\,\mathrm{curl}_\Gamma\,v. \tag{5.4.27}$$

■

We now have introduced all the necessary tools to prove the trace and lifting theorems. Let Ω be a regular bounded domain whose boundary is Γ

(Ω stands for Ω_i or $\Omega_e \cap B_R$ and the unit normal n is oriented toward the exterior of Ω). We define the **spaces**

$$H(\mathrm{div}) = \left\{ v \in \left(L^2(\Omega) \right)^3, \mathrm{div}\, v \in L^2(\Omega) \right\},$$

$$H(\mathrm{curl}) = \left\{ v \in \left(L^2(\Omega) \right)^3, \mathrm{curl}\, v \in L^2(\Omega)^3 \right\}.$$

Theorem 5.4.1 *The trace mapping which assigns to any $v \in H(\mathrm{div})$, its normal component on Γ denoted $(v \cdot n)$ is continuous and surjective from $H(\mathrm{div})$ onto $H^{-1/2}(\Gamma)$. There exists a lifting operator $\mathcal{R}$, continuous from $H^{-1/2}(\Gamma)$ into $H(\mathrm{div})$, which satisfies*

$$(\mathcal{R}g \cdot n)|_\Gamma = g. \tag{5.4.28}$$

Moreover, for any $u \in H^1(\Omega)$ and any $v \in H(\mathrm{div})$, we have

$$\int_\Omega [(\mathrm{grad}\, u \cdot v) + u\, \mathrm{div}\, v]\, dx = {}_{T}H^{-1/2} \langle (v \cdot n), u \rangle_{T}H^{1/2}. \tag{5.4.29}$$

Proof

The continuity of the trace mapping is a consequence of the Green formula (5.4.29). From the Cauchy-Scharwz inequality, it follows that

$$|_{T}H^{1/2} \langle (v \cdot n), u \rangle_{T}H^{1/2}| \leq \|u\|_{H^1(\Omega)} \|v\|_{H(\mathrm{div})}. \tag{5.4.30}$$

Then, choosing $u = \mathcal{R}\, u|_\Gamma$ given by the trace Theorem 2.5.3 in (5.4.29), we obtain

$$\|v \cdot n\|_{H^{-1/2}(\Gamma)} \leq \sup_{u \in H^{1/2}(\Gamma)} \frac{\langle (v \cdot n), u \rangle}{\|u\|_{H^{1/2}(\Gamma)}} \leq c \|v\|_{H(\mathrm{div})}. \tag{5.4.31}$$

We solve the Neumann problem

$$\begin{cases} -\Delta u + u = 0, & \text{in } \Omega, \\ \dfrac{\partial u}{\partial n} = g, \end{cases} \tag{5.4.32}$$

and then the vector $\mathcal{R}g = \nabla u$ is a lifting of g in $H(\mathrm{div})$. The continuity of $\mathcal{R}$ is easily checked.

Theorem 5.4.2 *The trace mapping which assigns to any $v \in H(\mathrm{curl})$ its tangential component on Γ, denoted v_T, is continuous and surjective from $H(\mathrm{curl})$ onto $H_{\mathrm{curl}}^{-1/2}(\Gamma)$, while the mapping which takes $v \in H(\mathrm{curl})$ to its tangential component $v \wedge n$ is continuous and surjective from $H(\mathrm{curl})$ onto $H_{\mathrm{div}}^{-1/2}(\Gamma)$.*

There exists, in both cases, a continuous lifting for these trace operators in $H(\mathrm{curl})$. Moreover, for any u and v in $H(\mathrm{curl})$, the following Stokes formula holds,

$$\int_\Omega [(u \cdot \mathrm{curl}\, v) - (v \cdot \mathrm{curl}\, u)]\, dx = {}_{H_{\mathrm{curl}}^{-1/2}(\Gamma)} \langle v_T \cdot (u \wedge n) \rangle_{H_{\mathrm{div}}^{-1/2}(\Gamma)}. \tag{5.4.33}$$

In both cases, there exists a lifting with zero divergence.

Proof

From the definition of $TH^{-1/2}(\Gamma)$, we have

$$\|v_T\|_{TH^{-1/2}(\Gamma)} = \sup_u \frac{\langle v_T \cdot (u \wedge n)\rangle}{\|u \wedge n\|_{TH^{1/2}(\Gamma)}}. \qquad (5.4.34)$$

We express the numerator of (5.4.34) using the Stokes formula (5.4.33), with u chosen as a lifting of $u|_\Gamma$ in the space $\left(H^1(\Omega)\right)^3$, which yields

$$\|v_T\|_{TH^{-1/2}(\Gamma)} \leq c\|v\|_{H(\mathrm{curl})}. \qquad (5.4.35)$$

Consider the vector $w = \mathrm{curl}\, v$. Its divergence is zero, and thus, it belongs to $H(\mathrm{div})$. From Theorem 5.3.6, its trace $(w \cdot n) = \mathrm{curl}_\Gamma v$ belongs to $H^{-1/2}(\Gamma)$. It follows that

$$\|v_T\|_{H^{-1/2}_{\mathrm{curl}}(\Gamma)} \leq c\|v\|_{H(\mathrm{curl})}.$$

For $v \wedge n$, the proof is similar, using the identity (cf. Theorem 2.5.19)

$$\mathrm{div}_\Gamma\, v_T = -\,\mathrm{curl}_\Gamma(v \wedge n). \qquad (5.4.36)$$

We have proved the continuity of the trace. We will use several steps to build the lifting.

Let $\varphi \in H^{-1/2}_{\mathrm{div}}(\Gamma)$. Consider the following equation on the surface Γ:

$$\theta + \overrightarrow{\mathrm{curl}}_\Gamma\, \mathrm{curl}_\Gamma\, \theta = \varphi. \qquad (5.4.37)$$

This problem is coercive and thus it has a unique solution. Taking the curl_Γ, we obtain

$$\mathrm{curl}_\Gamma\, \theta - \Delta_\Gamma\, \mathrm{curl}_\Gamma\, \theta = \mathrm{curl}_\Gamma\, \varphi, \qquad (5.4.38)$$

which shows that $\mathrm{curl}_\Gamma\, \theta \in H^{1/2}(\Gamma)$.

Taking the div_Γ, we obtain

$$\mathrm{div}_\Gamma\, \theta = \mathrm{div}_\Gamma\, \varphi \qquad (5.4.39)$$

and thus

$$\Delta_\Gamma \theta = \nabla_\Gamma\, \mathrm{div}_\Gamma\, \theta - \overrightarrow{\mathrm{curl}}_\Gamma\, \mathrm{curl}_\Gamma\, \theta \in H^{-3/2}(\Gamma), \qquad (5.4.40)$$

and moreover, $\theta \in TH^{1/2}(\Gamma)$ and $\mathrm{curl}_\Gamma\, \theta \in H^{1/2}(\Gamma)$.

Let u_1 denote a lifting in $\left(H^1(\Omega)\right)^3$ of $n \wedge \theta$. Let v denote a lifting in $H^1(\Omega)$ of $\mathrm{curl}_\Gamma\, \theta$. The vector

$$u = u_1 + \nabla v \qquad (5.4.41)$$

satisfies

$$\mathrm{curl}\, u = \mathrm{curl}\, u_1, \qquad (5.4.42)$$

$$u \wedge n = u_1 \wedge n + \nabla v \wedge n = \theta + \overrightarrow{\mathrm{curl}}_\Gamma\, \mathrm{curl}_\Gamma\, \theta = \varphi, \qquad (5.4.43)$$

and thus, $u \in H(\text{curl})$ and is a lifting of φ. The continuous dependence of this lifting upon φ is easily checked at each step of the contruction. To obtain a lifting with zero divergence, we add to the vector v, the gradient of a function w which vanishes on Γ and satisfies $\Delta w = -\,\text{div}\,u$.

The proof in the case of v_T is similar, multiplying first v_T by n. Lemma 5.4.1 shows that the Stokes formula still has a meaning when u and v are in $H(\text{curl})$, if the right-hand side is interpreted as a duality between $H_{\text{curl}}^{-1/2}(\Gamma)$ and $H_{\text{div}}^{-1/2}(\Gamma)$. ∎

Remark

The above proof is due to T. Abboud and I. Terrasse (see I. Terrasse [144]). The original result is due to L. Paquet [129]. The variational formulations for the Maxwell system in the space $H(\text{curl})$, are based on the Stokes formula (5.4.33), and also on the fundamental formula

Lemma 5.4.2 *For any vectors u and v in* $\left(H^1(\Omega)\right)^3$,

$$
\begin{cases}
\displaystyle\int_{\Omega}\Big[(\nabla u \cdot \nabla v) - (\text{curl}\,u \cdot \text{curl}\,v) - \text{div}\,u\,\text{div}\,v\Big]dx \\[2mm]
\qquad = -\displaystyle\int_{\Gamma}\Big[\text{div}_{\Gamma}\,u_T(v \cdot n) + \text{div}_{\Gamma}\,v_T(u \cdot n) \qquad\qquad (5.4.44) \\[2mm]
\qquad\qquad + 2H(u \cdot n)(v \cdot n) + (R\,u_T \cdot v_T)\Big]d\gamma.
\end{cases}
$$

Proof

We successively use the Green formulas

$$
\begin{cases}
\displaystyle\int_{\Omega}\text{div}(v\nabla u)dx \\[3mm]
\quad = \displaystyle\int_{\Omega}\Big[(\nabla u \cdot \nabla v) + (\Delta u \cdot v)\Big]dx = \displaystyle\int_{\Gamma}\left(\frac{\partial u}{\partial n} \cdot v\right)d\gamma,
\end{cases}
\qquad (5.4.45)
$$

$$
\begin{cases}
\displaystyle\int_{\Omega}\text{div}(v\,\text{div}\,u)dx = \displaystyle\int_{\Omega}\Big[\text{div}\,u \cdot \text{div}\,v + (\nabla\,\text{div}\,u \cdot v)\Big]dx \\[3mm]
\quad = \displaystyle\int_{\Gamma}(v \cdot n)\,\text{div}\,u d\gamma,
\end{cases}
\qquad (5.4.46)
$$

$$
\begin{cases}
\displaystyle\int_{\Omega}\text{div}(\text{curl}\,u \wedge v)dx \\[3mm]
\quad = \displaystyle\int_{\Omega}\Big[(\text{curl}\,\text{curl}\,u \cdot v) - (\text{curl}\,u \cdot \text{curl}\,v)\Big]dx \qquad (5.4.47) \\[3mm]
\quad = -\displaystyle\int_{\Gamma}((\text{curl}\,u \wedge n) \cdot v)d\gamma.
\end{cases}
$$

From the vectorial calculus formula $\Delta = \nabla\,\text{div} - \text{curl}\,\text{curl}$ we obtain, by adding the previous expressions,

$$\begin{cases} \displaystyle\int_{\Omega} \Big[(\nabla u \cdot \nabla v) - (\operatorname{curl} u \cdot \operatorname{curl} v) - \operatorname{div} u \operatorname{div} v \Big] dx \\[2mm] \displaystyle = \int_{\Gamma} \Big[\Big(\frac{\partial u}{\partial n} \cdot v \Big) - \operatorname{div} u (v \cdot n) - ((\operatorname{curl} u \wedge n) \cdot v) \Big] d\gamma. \end{cases} \tag{5.4.48}$$

We use the following identities on the surface Γ,

$$\frac{\partial}{\partial n}(u \cdot n) - \operatorname{div} u = -\operatorname{div}_\Gamma u_T - 2H(u \cdot n), \tag{5.4.49}$$

$$\frac{\partial}{\partial n} u_T - \operatorname{curl} u \wedge n = \nabla_\Gamma (u \cdot n) - \mathcal{R} u_T, \tag{5.4.50}$$

to modify the surface terms in (5.4.48). Then, a Stokes formula on the surface Γ leads to (5.4.44). ∎

As a consequence, we obtain the inclusion result:

Theorem 5.4.3 *The Hilbert space*

$$X = \Big\{ E \in H(\operatorname{curl}); E \in H(\operatorname{div}); E_T \in H_{\operatorname{div}}^{-1/2}(\Gamma) \Big\}$$

is included in $\left(H^1(\Omega) \right)^3$. *Similarly, the Hilbert space*

$$\widetilde{X} = \Big\{ E \in H(\operatorname{curl}); E \in H(\operatorname{div}); (E \cdot n) \in H^{1/2}(\Gamma) \Big\}$$

is included in $\left(H^1(\Omega) \right)^3$.

Proof
We use identity (5.4.44) with $u = E$ and $v = \overline{E}$. It follows that

$$\begin{cases} \displaystyle\int_{\Omega} |\nabla E|^2 \leq \int_{\Omega} \Big[|\operatorname{curl} E|^2 + |\operatorname{div} E|^2 \Big] dx \\[2mm] \qquad + c \, \|\operatorname{div}_\Gamma E_T\|_{H^{-1/2}(\Gamma)} \, \|E \cdot n\|_{H^{1/2}(\Gamma)} \\[2mm] \qquad + c \, \|E \cdot n\|_{H^{-1/2}(\Gamma)} \, \|E \cdot n\|_{H^{1/2}(\Gamma)} \\[2mm] \qquad + c \, \|E_T\|_{TH^{1/2}(\Gamma)} \, \|E_T\|_{TH^{-1/2}(\Gamma)}. \end{cases} \tag{5.4.51}$$

From the trace theorems, Theorems 5.4.1 and 5.4.2, it follows that the only non-bounded terms are, on one side $\|E_T\|_{TH^{1/2}(\Gamma)}$, and on the other side $\|E \cdot n\|_{H^{1/2}(\Gamma)}$. We estimate the products that contain these terms using an estimate of the form $\eta \|E_T\|_{TH^{1/2}(\Gamma)}^2 + (1/\eta) \|E_T\|_{TH^{-1/2}(\Gamma)}$, and then, it follows from the usual trace theorem that when η is small enough, this term is dominated by $\int_{\Omega} |\nabla E|^2 \, dx + \int_{\Omega} |E|^2 \, dx$. The term which contains $\|E \cdot n\|_{H^{1/2}(\Gamma)}$ is estimated in the same way.

We have proved that the norm in $(H^1(\Omega))^3$ can be estimated by the norm in X. Using the density of smooth functions in X and in $(H^1(\Omega))^3$, we deduce the theorem by a classical argument. We proceed in the same way for the space $\widetilde{X}$, using the estimate

$$\left| \int_\Gamma \operatorname{div}_\Gamma E_T \, (E \cdot n) d\gamma \right| \leq C \, \|E \cdot n\|_{H^{1/2}(\Gamma)} \, \|E_T\|_{TH^{1/2}(\Gamma)} \, . \tag{5.4.52}$$

∎

This theorem can be generalized in the following way. Let Ω be a domain whose smooth boundary is Γ_c, which can be divided into two domains Ω_1 and Ω_1 through a smooth interior boundary Γ_d (the subscript d stands for a dielectric interface) which does not intersect Γ_c. Let ε be a complex-valued function, bounded and non-vanishing which has bounded derivatives in Ω_1 and Ω_2, but is discontinuous when crossing Γ_d. We introduce the Hilbert spaces

$$X = \begin{cases} E \in \left(L^2(\Omega)\right)^3, \quad \operatorname{curl} E \in \left(L^2(\Omega)\right)^3, \\[2mm] \qquad \operatorname{div}(\varepsilon E) \in L^2(\Omega), \quad E_T \in H_{\operatorname{div}}^{-1/2}(\Gamma_c); \end{cases}$$

$$\widetilde{X} = \begin{cases} E \in \left(L^2(\Omega)\right)^3, \quad \operatorname{curl} E \in \left(L^2(\Omega)\right)^3, \\[2mm] \qquad \operatorname{div}(\varepsilon E) \in L^2(\Omega), \quad (E \cdot n) \in H^{1/2}(\Gamma_c). \end{cases}$$

Theorem 5.4.4 *The spaces X and $\widetilde{X}$ are included in $\left(H^1(\Omega_1)\right)^3 \times \left(H^1(\Omega_2)\right)^3$, if ε satisfies*

$$\Re\varepsilon(x) \geq \varepsilon_0 > 0. \tag{5.4.53}$$

Proof

For any smooth u and v in the space X

$$\operatorname{div}(\varepsilon v \nabla u) = \varepsilon(\nabla u \cdot \nabla v) + \varepsilon(\Delta u \cdot v) + \nabla\varepsilon \, v \, \nabla u, \tag{5.4.54}$$

$$\begin{cases} \operatorname{div}(\varepsilon \operatorname{curl} u \wedge v) \\[2mm] \quad = -\varepsilon(\operatorname{curl} u \cdot \operatorname{curl} v) + \varepsilon(\operatorname{curl} \operatorname{curl} u \cdot v) \\[2mm] \quad\quad + (\nabla\varepsilon \cdot (\operatorname{curl} u \wedge v)), \end{cases} \tag{5.4.55}$$

$$\begin{cases} \operatorname{div}(\varepsilon v \operatorname{div} u) \\[2mm] \quad = \varepsilon \operatorname{div} u \operatorname{div} v + \varepsilon(\nabla \operatorname{div} u \cdot v) + (\nabla\varepsilon \cdot v) \operatorname{div} u. \end{cases} \tag{5.4.56}$$

We integrate these three identities on Ω_1 and Ω_2 separately, and we add them in order to obtain $\varepsilon(\Delta - \nabla \operatorname{div} + \operatorname{curl} \operatorname{curl})$.

Using (5.4.49) and (5.4.50), it follows that

$$
\begin{cases}
\displaystyle\int_{\Omega_1\cup\Omega_2} \varepsilon\left[(\nabla u\cdot\nabla v)-(\operatorname{curl} u\cdot\operatorname{curl} v)-\operatorname{div} u\operatorname{div} v\right]dx \\[2mm]
\quad +\displaystyle\int_{\Omega_1\cup\Omega_2}(\nabla\varepsilon\cdot(v\nabla u+\operatorname{curl} u\wedge v-v\operatorname{div} u))\,dx \\[2mm]
\quad =\displaystyle\int_{\Gamma_e}\Big[-[\operatorname{div}_{\Gamma_e} u_T\,\varepsilon(v\cdot n)+\operatorname{div}_{\Gamma_e}(v_T)\varepsilon(u\cdot n)] \\[2mm]
\qquad\qquad -2H\varepsilon(u\cdot n)(v\cdot n)-\varepsilon\,(Ru_T\cdot u_T)\Big]d\gamma \\[2mm]
\quad +\displaystyle\int_{\Gamma_d}\Big[-[\operatorname{div}_{\Gamma_d} u_T\,(\varepsilon_1(v_1\cdot n)-\varepsilon_2(v_2\cdot n)) \\[2mm]
\qquad\qquad +\operatorname{div}_{\Gamma_d} v_T\,(\varepsilon_1(u_1\cdot n)-\varepsilon_2(u_2\cdot n)) \\[2mm]
\qquad\qquad +(\nabla_{\Gamma_d}\varepsilon_1\cdot v_T)(u_1\cdot n)-(\nabla_{\Gamma_d}\varepsilon_2\cdot v_T)(u_2\cdot n)\big] \\[2mm]
\qquad\qquad -2H\big[\varepsilon_1(u_1\cdot n)(v_1\cdot n)-\varepsilon_2(u_2\cdot n)(v_2\cdot n)\big] \\[2mm]
\qquad\qquad -\varepsilon_1\,(Ru_T\cdot v_T)+\varepsilon_2\,(R\,u_T\cdot v_T)\Big]d\gamma.
\end{cases}
\tag{5.4.57}
$$

The tangential values of u_T and v_T are continuous when crossing Γ_d. As $\operatorname{div}(\varepsilon u)\in L^2(\Omega)$ the values of $\varepsilon(u\cdot n)$ and $\varepsilon(v\cdot n)$ are continuous across Γ_d. Thus, the terms containing $\operatorname{div}_{\Gamma_d} u_T$ and $\operatorname{div}_{\Gamma_d} v_T$ disappear. The only remaining terms, on the boundary Γ_d, depend only on u_T, $(u\cdot n)$ and v_T, $(v\cdot n)$. They are estimated using the trace theorem.

From the Theorems 5.4.1 and 5.4.2, choosing $v=\bar u$, it follows that

$$
\begin{cases}
\displaystyle\int_\Omega |u|^2\,dx+\int_{\Omega_1}\varepsilon_1\,|\nabla u|^2\,dx+\int_{\Omega_2}\varepsilon_2\,|\nabla u|^2\,dx \\[2mm]
\quad \leq c\bigg[\displaystyle\int_\Omega |\operatorname{curl} u|^2\,dx+\int_\Omega |u|^2\,dx+\int_{\Omega_1}\varepsilon_1\,|\operatorname{div} u|^2\,dx \\[2mm]
\qquad +\displaystyle\int_{\Omega_2}\varepsilon_2\,|\operatorname{div} u|^2\,dx+\|u_T\|^2_{H^{-1/2}_{\mathrm{div}}(\Gamma_e)}\bigg].
\end{cases}
\tag{5.4.58}
$$

The real part of this identity gives the estimate. The proof for $\widetilde X$ is similar, using (5.4.52). ■

5.4.2 Variational formulations for the perfect conductor problem

We examine the exterior Maxwell problem in the domain Ω_e whose boundary is Γ. Let B_R be a ball of radius R, chosen large enough to contain Γ

and the part of the domain where ε and μ are not constant. Let us denote by S_R the sphere of radius R which is the boundary of B_R. We introduced in (5.3.88), the capacity operator for the Maxwell system, in the case of the unit sphere. The capacity operator for the sphere S_R is denoted by T_R. It can be deduced from T by changing k into kR, and thus this operator satisfies Theorems 5.3.5 and 5.3.6, except for the modification of the continuity and coercivity constants, which are divided by R.

Lemma 5.4.3 *An equivalent formulation of problem* (5.4.1) (5.4.5), *when* $z = 0$, *is*

$$
\begin{cases}
\operatorname{curl} E - i\omega\mu H = 0, & x \in \Omega_e \cap B_R, \\
\operatorname{curl} H + i\omega\varepsilon E = 0, & x \in \Omega_e \cap B_R, \\
E \wedge n|_\Gamma = g, \\
(H \wedge n - T_R E_T)|_{S_R} = 0.
\end{cases}
\tag{5.4.59}
$$

Proof

The solution of the Maxwell problem in the exterior of the sphere S_R, which has the trace $E \wedge n|_{S_R}$, is given by a sum of multipoles in the form described in Theorem 5.3.1. It satisfies the radiation conditions. We use it to extend our solution outside B_R. At the interface S_R, there is no jump of $E \wedge n$ and the link between $E \wedge n$ and $H \wedge n$ through the capacity operator shows that $H \wedge n$ has no jump either. Thus, $\operatorname{curl} E$ has no discontinuity and the extended solution is a solution of (5.4.1) and (5.4.5). Conversely, a solution of (5.4.1) which satisfies the radiation conditions can be expanded in the exterior of S_R as a sum of multipoles, and then the value on S_R of $H \wedge n$ is given by $T_R E_T$. ∎

By elimination of H or of E, we obtain two different formulations, one with E, the other with H. They are

$$
\begin{cases}
\operatorname{curl} \frac{1}{\mu} \operatorname{curl} E - \omega^2 \varepsilon E = 0, & x \in \Omega_e \cap B_R, \\
\operatorname{div}(\varepsilon E) = 0, & x \in \Omega_e \cap B_R, \\
E \wedge n|_\Gamma = g, \\
(\operatorname{curl} E \wedge n - i\omega\mu T_R E_T)|_{S_R} = 0,
\end{cases}
\tag{5.4.60}
$$

and

$$
\begin{cases}
\operatorname{curl} \frac{1}{\varepsilon} \operatorname{curl} H - \omega^2 \mu H = 0, & x \in \Omega_e \cap B_R, \\
\operatorname{div}(\mu H) = 0, & x \in \Omega_e \cap B_R, \\
\frac{i}{\omega\varepsilon}(\operatorname{curl} H \wedge n)\big|_\Gamma = g, \\
(\operatorname{curl} H \wedge n - i\omega\mu T_R H_T)|_{S_R} = 0.
\end{cases}
\tag{5.4.61}
$$

This last relation comes from the property (5.3.110) satisfied by the capacity operator.

We will now give two equivalent variational formulations of these problems. These variational formulations are of mixed type.

Let p denote a new unknown scalar function, which plays the role of a Lagrange multiplier for the constraint $\operatorname{div}(\varepsilon E) = 0$. We introduce a **decomposition of the field E** of the form

$$E = u + \nabla p. \tag{5.4.62}$$

It holds that

$$\operatorname{curl} u = \operatorname{curl} E. \tag{5.4.63}$$

We choose the function p which satisfies

$$\operatorname{div}(\varepsilon \nabla p) = 0, \tag{5.4.64}$$

and thus, u is such that

$$\operatorname{div}(\varepsilon u) = 0. \tag{5.4.65}$$

We choose the vector u which satisfies, on the surface Γ,

$$\operatorname{div}_\Gamma u_T = 0, \tag{5.4.66}$$

and thus, p is such that

$$\Delta_\Gamma p = -\operatorname{curl}_\Gamma g. \tag{5.4.67}$$

The operator Δ_Γ is an isomorphism of the subspace $H^{-3/2}(\Gamma)$, such that $\langle v, 1 \rangle = 0$, onto $H^{1/2}(\Gamma)/R$. Thus, for any $g \in TH^{-1/2}(\Gamma)$, equation (5.4.67) determines, up to a constant, the trace $p_\Gamma(g)$ on Γ of the function p in $H^{1/2}(\Gamma)$. The value of u_T on Γ, denoted by $u_T(g)$, is

$$u_T = n \wedge g - \nabla_\Gamma p_\Gamma(g) = u_T(g). \tag{5.4.68}$$

It holds that

$$\operatorname{curl}_\Gamma u_T = -\operatorname{div}_\Gamma g, \tag{5.4.69}$$

which, together with (5.4.66), proves that $u_T \in TH^{1/2}(\Gamma)$ when $g \in H_{\operatorname{div}}^{-1/2}(\Gamma)$.

Let V be the Hilbert space

$$V = \{v \in H(\operatorname{curl}, \Omega_e \cap B_R) ; v_T = 0 \operatorname{on} \Gamma\}.$$

From the trace Theorem 5.4.2, it is a closed subspace of $H(\operatorname{curl})$.

Let W be the Hilbert space

$$W = \{q \in H^1(\Omega_e \cap B_R) ; q|_\Gamma = 0\}.$$

We seek the function p in the space $H^1(\Omega_e \cap B_R)$.

Lemma 5.4.4 *The variational problem, whose unknowns are u and p with $u \in H(\mathrm{curl}, \Omega_e \cap B_R)$ and $p \in H^1(\Omega_e \cap B_R)$,*

$$
\begin{cases}
\displaystyle\int_{\Omega_e \cap B_R} \frac{1}{\mu}(\mathrm{curl}\, u \cdot \mathrm{curl}\, v)\, dx \\[2mm]
\displaystyle -\omega^2 \int_{\Omega_e \cap B_R} [\varepsilon(u \cdot v) + \varepsilon(\nabla p \cdot v)]\, dx \\[2mm]
\displaystyle -i\omega \int_{S_R} [(T_R u_T \cdot v_T) + (T_R \nabla_S p \cdot v_T)]\, d\gamma = 0, \quad \forall v \in V, \\[2mm]
\displaystyle -\omega^2 \int_{\Omega_e \cap B_R} \varepsilon(u \cdot \nabla q)\, dx \\[2mm]
\displaystyle -i\omega \int_{S_R} (T_R u_T \cdot \nabla_S q)\, d\gamma = 0, \quad \forall q \in W, \\[2mm]
p|_\Gamma = p_\Gamma(g), \\[2mm]
u_T|_\Gamma = u_T(g)
\end{cases}
\tag{5.4.70}
$$

is such that

$$
E = u + \nabla p
\tag{5.4.71}
$$

is a solution of (5.4.60). The field E associated with this solution belongs to $H(\mathrm{curl}, \Omega_e \cap B_R)$.

Proof

Using the Stokes formula and the expression (5.3.93), we interpret the second equation of (5.4.70), as

$$
\mathrm{div}(\varepsilon u) = 0, \quad x \in \Omega_e \cap B_R,
\tag{5.4.72}
$$

$$
\omega^2 \varepsilon(u \cdot n) - i\omega\, \mathrm{div}_S\, T_R u_T = 0, \quad \text{on } S_R.
\tag{5.4.73}
$$

Using the Stokes formula (5.4.47), we interpret the first equation of (5.4.70) as

$$
\mathrm{curl}\, \frac{1}{\mu}\, \mathrm{curl}\, u - \omega^2 \varepsilon(u + \nabla p) = 0,
\tag{5.4.74}
$$

$$
\begin{cases}
\displaystyle\int_{S_R} \left(\left(\frac{1}{\mu}\, \mathrm{curl}\, u \wedge n - i\omega T_R(u_T + \nabla_S p) \right) \cdot v_T \right) d\gamma = 0, \\[2mm]
\forall v_T \in H_{\mathrm{curl}}^{-1/2}(S_R).
\end{cases}
\tag{5.4.75}
$$

From the trace Theorem 5.4.2 and (5.4.75), it follows that

$$
\frac{1}{\mu}\, \mathrm{curl}\, u \wedge n - i\omega T_R(u_T + \nabla_S p) = 0.
\tag{5.4.76}
$$

Then, (5.4.62), (5.4.63) and (5.4.76) yield

$$\operatorname{curl} E \wedge n - i\omega\mu T_R E_T = 0, \quad \text{on } S_R. \tag{5.4.77}$$

The boundary condition on Γ results from the values of $p_\Gamma(g)$ and $u_T(g)$. ∎

We introduce the bilinear forms

$$\begin{cases} a(u,v) = \displaystyle\int_{\Omega, \cap B_R} \frac{1}{\mu} (\operatorname{curl} u \cdot \operatorname{curl} v)\, dx - \omega^2 \int_{\Omega, \cap B_R} \varepsilon(u \cdot v) dx \\ \qquad\qquad - i\omega \displaystyle\int_{S_R} (T_R u_T \cdot v_T)\, d\gamma. \end{cases} \tag{5.4.78}$$

$$b(\nabla p, v) = -\omega^2 \int_{\Omega, \cap B_R} \varepsilon(\nabla p \cdot v) dx - i\omega \int_{S_R} (T_R \nabla_S p \cdot v_T)\, d\gamma. \tag{5.4.79}$$

With this notation, the variational problem (5.4.70) takes the form

$$\begin{cases} a(u,v) + b(\nabla p, v) = 0, & \forall v \in V, \\ a(u, \nabla q) = 0, & \forall q \in W \end{cases} \tag{5.4.80}$$

or equivalently

$$\begin{cases} a(u,v) + b(\nabla p, v) = 0, & \forall v \in V, \\ b(\nabla p, \nabla q) = 0, & \forall q \in W. \end{cases} \tag{5.4.81}$$

From Theorem 5.4.2, there exists a lifting of the data $u_T(g)$ in the space $H(\operatorname{curl}, \Omega_e \cap B_R)$. This lifting $\mathcal{R}(u_T(g))$ belongs to the space $(H^1(\Omega_e \cap B_R))^3$ as $u_T(g)$ belongs to $TH^{1/2}(\Gamma)$. From the usual trace theorem, there exists a lifting of the data $p_\Gamma(g)$ in the space $H^1(\Omega_e \cap B_R)$. We change the unknowns using these liftings, introducing $\tilde{u} = u - \mathcal{R}(u_T(g))$ and $\tilde{p} = p - \mathcal{R}(p_\Gamma(g))$. The variational formulation (5.4.81) takes the form

$$\begin{cases} a(\tilde{u}, v) + b(\nabla \tilde{p}, v) = (g_1, v), & \forall v \in V, \\ b(\nabla \tilde{p}, \nabla q) = (g_2, \nabla q), & \forall q \in W, \end{cases} \tag{5.4.82}$$

where g_1 and g_2 depend on the chosen liftings. They belong to the dual spaces of V and W.

Let us prove the abstract theorem:

Theorem 5.4.5 (Fredholm alternative) *Let V be a Hilbert space. Let H be a Hilbert space which contains V. Let $a(u,v)$ be a continuous bilinear form on $V \times V$ which satisfies*

$$\Re[a(u,\bar{u})] \geq \alpha \|u\|_V^2 - c\|u\|_H^2, \quad \alpha > 0, \quad \forall u \in V. \tag{5.4.83}$$

Consider the variational problem

$$a(u,v) = (g,v); \quad \forall v \in V; \quad g \in V^*. \tag{5.4.84}$$

Suppose that the **injection of V into H is compact**. *Then, the variational problem* (5.4.83) *satisfies* **the Fredholm alternative** *i.e.,*

- *either it admits a unique solution in V,*
- *or it has a finite dimension kernel, and a unique solution up to any element in this kernel, when the duality product of the right-hand side g vanishes on every element in this kernel.*

Proof

We first prove the finite dimension of the kernel. Any element in the kernel satisfies

$$\|u\|_V^2 \leq \frac{1}{\alpha} \Re\left[a(u, \overline{u})\right] + c \|u\|_H^2 \leq c \|u\|_H^2 . \tag{5.4.85}$$

Thus, from the compact injection property, this linear space has a compact unit ball, and so has finite dimension.

We introduce, using the Galerkin method, an approximate solution for this problem. We denote by u_ε this solution which belongs to the quotient space V/N. Let us denote by V this quotient space.

We estimate the norm of this solution as

$$\begin{cases} \alpha \|u_\varepsilon\|_V^2 - c \|u_\varepsilon\|_H^2 \leq \Re\left[a\left(u_\varepsilon, \overline{u}_\varepsilon\right)\right] \\ \qquad\qquad \leq \|g\|_{V^*} \|u_\varepsilon\|_V , \end{cases} \tag{5.4.86}$$

and thus,

$$\|u_\varepsilon\|_V \leq c \left[\|u_\varepsilon\|_H + \|g\|_{V^*}\right]. \tag{5.4.87}$$

In order to show the convergence of this sequence, we use a contradiction argument.

- Either the sequence $\|u_\varepsilon\|_H$ is bounded and then u_ε is bounded in V, from (5.4.87). The weak convergence in V shows the convergence of the sequence.
- Or the sequence $\|u_\varepsilon\|_H$ is not bounded, but then $\widetilde{u}_\varepsilon = u_\varepsilon / \|u_\varepsilon\|_H$ is bounded in V, and its limit $\widetilde{u}$ satisfies

$$a(\widetilde{u}, v) = 0, \quad \forall v \in V. \tag{5.4.88}$$

Thus, $\widetilde{u}$ belongs to the kernel of the problem. In the quotient space, this kernel is reduced to zero. The contradiction comes from the fact that the injection of V into H is compact, which implies that $\widetilde{u}$ converges strongly in H toward zero. This is contradictory with $\|\widetilde{u}_\varepsilon\|_H = 1$. ∎

Remark

A closed variational formulation is: find $u \in V_g$ such that $a(u, v) = 0$; $\forall v \in V$; where V_g is an affine space. The space V_g is linked to the space V through the following hypothesis: there exists an element u_g such that any $u \in V_g$ admits the expression $u = u_g + v$, $v \in V$ and $a(u_g, v)$ is a continuous linear form on the space V. ∎

Theorem 5.4.6 *Let us suppose that ε and μ satisfy the hypothesis (5.4.2) and (5.4.3) where ε_0 and μ_0 have real values. The variational formulation (5.4.70) of the Maxwell problem (5.4.59) admits a unique solution when the* **data g belongs to $H_{\mathrm{div}}^{-1/2}(\Gamma)$.**

The electric field E is given by $E = u + \nabla p$ and it belongs to the space $H\,(\mathrm{curl}, \Omega_e \cap B_R)$, while $u \in H\,(\mathrm{curl}, \Omega_e \cap B_R)$ and $p \in H^1\,(\Omega_e \cap B_R)$. The magnetic field H belongs to $H(\mathrm{curl}, \Omega_e \cap B_R)$. **If moreover $g \in TH^{1/2}(\Gamma)$,** *then the electric field E belongs to $\left(H^1(\Omega_1) \cap H^1\,(\Omega_2 \cap B_R)\right)^3$.*

Proof

It relies on the variational formulation (5.4.70). The second equation of this variational formulation is an independent equation for the unknown $\tilde{p}$. From (5.4.79), it satisfies

$$\begin{cases} \Re\left[b(\nabla\tilde{p}, \nabla\overline{\tilde{p}})\right] \\ = -\omega^2 \int_{\Omega_c \cap B_R} \varepsilon |\nabla\tilde{p}|^2 dx + \omega\Im \int_{S_R} \left(\mathcal{T}_R\nabla_S p \cdot \nabla\overline{\tilde{p}}\right) d\gamma. \end{cases} \qquad (5.4.89)$$

From Poincaré's inequality and the estimate (5.3.108),

$$\|q\|^2_{H^1(\Omega_c \cap B_R)} \le c\,\Re\left[b(\nabla q, \nabla\overline{q})\right], \quad \forall\, q \in W. \qquad (5.4.90)$$

Thus, Theorem 5.4.5 proves the existence and uniqueness of $\tilde{p}$ in W. The vector $\tilde{u}$ satisfies

$$a(\tilde{u}, \nabla q) = (g_1 - g_2, \nabla q), \quad \forall q \in W, \qquad (5.4.91)$$

which yields

$$\mathrm{div}(\varepsilon\tilde{u}) = \mathrm{div}(\varepsilon\mathcal{R}(u_T(g))). \qquad (5.4.92)$$

We proved in Theorem 5.4.2 that we can choose a free divergence lifting $\mathcal{R}(u_T(g))$ without modifying (5.4.73). Equation (5.4.91) is thus

$$a(\tilde{u}, \nabla q) = 0, \quad \forall q \in W. \qquad (5.4.93)$$

We introduce the Hilbert space

$$V_0 = \Big\{ v \in H(\mathrm{curl}, \Omega_e \cap B_R),\ \mathrm{div}(\varepsilon v) = 0,\ v_T = 0, \quad \text{on } \Gamma,$$
$$\omega\varepsilon(v \cdot n) - i\,\mathrm{div}_S\,\mathcal{T}_R v_T = 0, \quad \text{on } S \Big\}.$$

The existence of $\tilde{u}$ depends on the first equation of (5.4.81). One part of this equation is (5.4.91), which is taken into account by looking for element V_0. Besides, the space ∇W is contained in the space V, which is spanned by the sum of V_0 and the space ∇W. Thus, it is sufficient to test (5.4.81) for elements of V_0. It holds that

$$a(\tilde{u}, v) = (g_1, v) - b(\nabla\tilde{p}, v), \quad \forall v \in V_0. \qquad (5.4.94)$$

We apply Theorem 5.4.5 in the space V_0 to the variational equation (5.4.94). We check all the hypotheses. We start from

$$\begin{cases} a(u, \bar{u}) = \displaystyle\int_{\Omega_r \cap B_R} \left[\frac{1}{\mu} |\operatorname{curl} u|^2 - \omega^2 \, \varepsilon \, |u|^2 \right] dx \\[2mm] \qquad\qquad - i\omega \displaystyle\int_{S_R} (\mathcal{T}_R u_T, \bar{u}_T) \, d\gamma, \end{cases} \tag{5.4.95}$$

which yields

$$\begin{cases} \Re\, a(u, \bar{u}) = \displaystyle\int_{\Omega_r \cap B_R} \left[\Re\left(\frac{1}{\mu}\right) |\operatorname{curl} u|^2 - \omega^2 \Re(\varepsilon) \, |u|^2 \right] dx \\[2mm] \qquad\qquad + \omega \Im \displaystyle\int_{S_R} (\mathcal{T}_R \, u_T \cdot \bar{u}_T) \, d\gamma, \end{cases} \tag{5.4.96}$$

and from the hypotheses on μ and ε, and inequality (5.3.90), it follows that

$$\begin{cases} \Re\, a(u, \bar{u}) \\[2mm] \geq \alpha \left[\displaystyle\int_{\Omega_r \cap B_R} \left(|\operatorname{curl} u|^2 + |u|^2 \right) dx + \|\operatorname{curl}_S \, u_T\|^2_{H^{-1/2}(S)} \right] \\[2mm] \qquad - c \left[\displaystyle\int_{\Omega_r \cap B_R} |u|^2 \, dx + \|\operatorname{curl}_S \, u_T\|^2_{H^{-1}(S)} \right. \\[2mm] \qquad\qquad \left. + \|\operatorname{div}_S \, u_T\|^2_{H^{-3/2}(S)} \right], \quad \alpha > 0. \end{cases} \tag{5.4.97}$$

This is (5.3.79) with V_0 as the space V, and the choice of the space

$$H = \left\{ u \in \left(L^2\left(\Omega_e \cap B_R\right)\right)^3, \operatorname{curl}_S u_T \in H^{-1}(S), \operatorname{div}_S u_T \in H^{-3/2}(S) \right\}.$$

It remains to check that the injection of V_0 into H is compact. This is the aim of

Lemma 5.4.5 *The Hilbert space V_0 is imbedded in the Hilbert space $\left(H^1\left(\Omega_1\right)\right)^3 \cap \left(H^1\left(\Omega_2 \cap B_R\right)\right)^3$, where the domains Ω_1 and Ω_2 are separated by the union of the discontinuity surfaces of the function ε.*

Proof

It is a consequence of the Green formula (5.4.57), as in Theorem 5.4.4. When choosing $v = \bar{u}$ in this inequality, we need to estimate all the negative terms in the left-hand side. The only new type of term is

$$\begin{cases} \displaystyle\int_{S_R} \left[\operatorname{div}_S u_T \varepsilon(\bar{u} \cdot n) + \operatorname{div}_S \bar{u}_T \varepsilon(u \cdot n)\right] d\gamma \\[2mm] = -2\Im \left[\displaystyle\int_{S_R} \frac{1}{\omega} \operatorname{div}_S \mathcal{T}_R u_T \operatorname{div}_S (\bar{u}_T) \, d\gamma \right]. \end{cases} \tag{5.4.98}$$

It can be computed using the expressions (5.3.91) and (5.3.93). The quantity $-\Re z_l(kR)-1$ which appears is positive. Therefore, this term is positive, which ends the proof of the lemma.   ∎

End of the proof of Theorem 5.4.6

It remains to show the uniqueness of the solution. It consists in proving that zero is the only solution when the data are zero.

Let us consider the null quantity

$$
\begin{cases}
\Im\left[a(u,\overline{u}+\nabla\overline{p})+b(\overline{u}+\nabla\overline{p},p)\right] \\[2mm]
= -\omega\Re\displaystyle\int_{S_R}(\mathcal{T}_R(u_T+\nabla_S p)\cdot(\overline{u}_T+\nabla_S\overline{p}))\,d\gamma \\[2mm]
+\displaystyle\int_{\Omega_e\cap B_R}\left[\Im\left(\frac{1}{\mu}\right)|\mathrm{curl}\,u|^2-\omega^2\,\Im(\varepsilon)\,|u+\nabla p|^2\right]dx.
\end{cases}
\tag{5.4.99}
$$

From Theorem 5.3.5 (formula (5.3.89)) and the properties of ε and μ, all the terms are negative.

Thus,

$$
E_T = 0, \quad \text{on } S, \tag{5.4.100}
$$

which, using the boundary condition on S, implies

$$
H_T = 0, \quad \text{on } S. \tag{5.4.101}
$$

It follows that $E = u + \nabla p$ satisfies

$$
\mathrm{curl}\,\frac{1}{\mu}\,\mathrm{curl}\,E - \omega^2\varepsilon E = 0 \tag{5.4.102}
$$

and that the following boundary conditions hold:

$$
\begin{cases}
E \wedge n|_\Gamma = 0, \\[2mm]
E \wedge n|_S = 0, \\[2mm]
\mathrm{curl}\,E \wedge n|_S = 0.
\end{cases}
\tag{5.4.103}
$$

Thus, the multipole solution, which solves the Maxwell equation in the exterior of the ball B_R, vanishes. The global solution has analytic components in each domain in Ω_e where ε and μ are analytic. As it vanishes in B_R, it is zero everywhere.

We have shown the existence and uniqueness in the space $V \times W$. Let us now prove the regularity result. The part u of the solution is already known to belong to $(H^1(\Omega_1)\cap H^1(\Omega_e\cap B_R))^3$. Thus, the field E will be in $\left(H^1(\Omega_1)\cap H^1(\Omega_e\cap B_R)\right)^3$, if p is in $H^2(\Omega_1)\cap H^2(\Omega_e\cap B_R)$. We use condition (5.4.76) and the normal component in the equation (5.4.74) to show that the function p satisfies

$$\begin{cases} \operatorname{div} \varepsilon \nabla p = 0, & \text{in } \Omega_e \cap B_R, \\ p|_\Gamma = p_\Gamma(g), \\ \operatorname{div}_S T_R \nabla_S p = i\omega \, \varepsilon_0 \dfrac{\partial p}{\partial n}, & \text{on } S. \end{cases} \tag{5.4.104}$$

When $g \in TH^{1/2}(\Gamma)$, $\operatorname{curl}_\Gamma g \in H^{-1/2}(\Gamma)$ and from (5.4.67), $p_\Gamma(g) \in H^{3/2}(\Gamma)$.

The operator $\operatorname{div}_S T_R \nabla_S p$ can be expanded on the spherical harmonics in the form (this is the expression on the unit sphere)

$$\operatorname{div}_S T \nabla_S p = \sum_{\ell=1}^{\infty} \sum_{m=-\ell}^{\ell} -i \sqrt{\frac{\varepsilon}{\mu}} \frac{k}{z_\ell(k)+1} \ell(\ell+1) p_\ell^m Y_\ell^m(\theta, \varphi). \tag{5.4.105}$$

It is easy to check in this expression that the operator $\operatorname{div}_S T_R \nabla_S p$ is an isomorphism of $H^s(S)$ onto $H^{s-2}(S)$ for any real s. The regularity results hold for an elliptic equation of the form (5.4.104) with a non-local condition and this ends the proof. ∎

The **fields E and H** play **symmetric roles** in the Maxwell equations, **except regarding the boundary conditions**. But in the variational formulation (5.4.61), this symmetry is broken. In the following variational formulation, the field H plays the central role.

Lemma 5.4.6 *A variational formulation for equation (5.4.61) is:*
Find H in the form

$$H = v + \nabla q, \quad v \in V, \quad q \in W,$$
$$V = H(\operatorname{curl}, \Omega_e \cap B_R),$$
$$W = H^1(\Omega_e \cap B_R)/P_0$$

such that

$$\begin{cases} \displaystyle\int_{\Omega_e \cap B_R} \frac{1}{\varepsilon} (\operatorname{curl} v \cdot \operatorname{curl} w)\, dx \\[2mm] \displaystyle - \omega^2 \int_{\Omega_e \cap B_R} \mu(v \cdot w)dx - i\omega \frac{\mu_0}{\varepsilon_0} \int_{S_R} (T_R v_T \cdot w_T)\, d\gamma \\[2mm] \displaystyle - \omega^2 \int_{\Omega_e \cap B_R} \mu(\nabla q \cdot w)dx - i\omega \frac{\mu_0}{\varepsilon_0} \int_{S_R} (T_R \nabla_S q \cdot w_T)\, d\gamma \\[2mm] \displaystyle \qquad\qquad = i\omega \int_{S_R} (g \cdot w_T)\, d\gamma, \quad \forall w \in V, \\[2mm] \displaystyle - \omega^2 \int_{\Omega_e \cap B_R} \mu(v \cdot \nabla p)dx - i\omega \frac{\mu_0}{\varepsilon_0} \int_{S_R} (T_R v_T \cdot \nabla_S p)\, d\gamma \\[2mm] \displaystyle \qquad\qquad = i\omega \int_{S_R} (g \cdot \nabla_S p)\, d\gamma, \quad \forall q \in W. \end{cases} \tag{5.4.106}$$

Proof

The boundary conditions on S associated with (5.4.106) are obtained in a way similar to that used in the proof of Lemma 5.4.4, using the relation (5.3.110) in Theorem 5.3.6.

The boundary conditions on Γ come from the Stokes fromula (5.4.47) using the first part of the equation. We obtain

$$\begin{cases} \int_\Gamma \left(\frac{1}{\varepsilon} (\operatorname{curl} v \wedge n) \cdot w_T \right) d\gamma \\ \qquad = -i\omega \int_\Gamma (g \cdot w_T) d\gamma, \quad \forall w_T \in H_{\operatorname{div}}^{-1/2}(\Gamma). \end{cases} \qquad (5.4.107)$$

Besides, choosing $w = \nabla p$ in the first equation, and substracting the second equation, it follows that

$$\begin{cases} -\omega^2 \int_{\Omega_e \cap B_R} \mu(\nabla q \cdot \nabla p) dx \\ \qquad = -i\omega \int_\Gamma p \operatorname{div}_\Gamma g d\gamma, \quad \forall p \in W, \text{ such that } p = 0 \text{ on } S. \end{cases} \qquad (5.4.108)$$

Thus, the function q satisfies

$$\operatorname{div}(\mu \nabla q) = 0, \quad \text{in } \Omega_e \cap B_R, \qquad (5.4.109)$$

and

$$\omega^2 \mu \frac{\partial}{\partial n} q = -i\omega \operatorname{div}_\Gamma g, \quad \text{on } \Gamma. \qquad (5.4.110)$$

Moreover, using the Green formula in the second part of (5.4.106), we obtain

$$\omega^2 \mu(v \cdot n) = 0, \quad \text{on } \Gamma. \qquad (5.4.111)$$

∎

Theorem 5.4.7 *Suppose that ε and μ satisfy the hypothesis (5.4.2) and (5.4.3), where ε_0 and μ_0 have real values. The Maxwell problem (5.4.58) admits the variational formulation (5.4.106). It has a unique solution H of the form $H = v + \nabla q$.*

If $g \in H_{\operatorname{div}}^{-1/2}(\Gamma)$, we have $v \in H(\operatorname{curl}, \Omega_e \cap B_R)$, $q \in H^1(\Omega_e \cap B_R)$, $H \in H(\operatorname{curl}, \Omega_e \cap B_R)$ and $E \in H(\operatorname{curl}, \Omega_e \cap B_R)$. If moreover g is such that $\operatorname{div}_\Gamma g \in H^{1/2}(\Gamma)$, H belongs to the space $\left(H^1(\Omega_1) \cap H^1(\Omega_2 \cap B_R) \right)^3$, while q is in $H^2(\Omega_1) \cap H^2(\Omega_2 \cap B_R)$ and v is in $\left(H^1(\Omega_1) \cap H^1(\Omega_2 \cap B_R) \right)^3$.

Proof

We apply Theorem 5.4.5 to the variational formulation (5.4.106). The proof is absolutely similar to that of Theorem 5.4.6, the bilinear forms a and b being similar, except the exchange of roles between μ and ε, and the factor μ_0/ε_0. But the space W has to be modified as the functions q do not vanish on Γ. We choose $H^1(\Omega_e \cap B_R)/P_0$, and the existence of q comes from the fact that $\|\nabla q\|_{L^2(\Omega, \cap B_R)}$ is a norm on this space. The space H is

$$H = \left\{ u \in (L^2(\Omega_e \cap B_R))^3, \operatorname{curl}_S u_T \in H^{-1}(S), \operatorname{div}_S u_T \in H^{-3/2}(S) \right\}.$$

The second equation (5.4.106) proves that v satisfies

$$\operatorname{div}(\mu v) = 0, \tag{5.4.112}$$

$$\omega \varepsilon_0 (v \cdot n) - i \operatorname{div}_S \mathcal{T}_R v_T = 0, \quad \text{on } S_R. \tag{5.4.113}$$

The Hilbert space V_0 is

$$V_0 = \Big\{ w \in H(\operatorname{curl}, \Omega_e \cap B_R), \ \operatorname{div}(\mu w) = 0, (w \cdot n)|_\Gamma = 0 \ ,$$

$$\omega \varepsilon_0 (w \cdot n) - i \operatorname{div}_S \mathcal{T}_R w_T = 0, \quad \text{on } S_R \Big\}.$$

Theorem 5.4.4 and Lemma 5.4.5 imply that V_0 is imbedded in the space $\left(H^1(\Omega_1) \cap H^1(\Omega_2 \cap B_R) \right)^3$, and that moreover $w_T \in TH^{1/2}(S)$. It is compactly imbedded in the space H, which yields the existence of v in the space V_0.

To prove uniqueness, we introduce the zero expression

$$\begin{cases} \Im\left[a(v, \overline{v} + \nabla \overline{q}) + b(\overline{v} + \nabla \overline{q}, q) \right] \\[2mm] = -\omega \frac{\mu_0}{\varepsilon_0} \Re \int_{S_R} (\mathcal{T}_R (v_T + \nabla_S q) \cdot (\overline{v}_T + \nabla_S \overline{q})) \, d\gamma \\[2mm] + \int_{\Omega_e \cap B_R} \left[\Im\left(\frac{1}{\varepsilon}\right) |\operatorname{curl} v|^2 - \omega^2 \Im(\mu) |v + \nabla q|^2 \right] dx. \end{cases} \tag{5.4.114}$$

From the properties of ε and μ, and Theorem 5.3.5, all its terms are negative. It follows that

$$H_T = 0, \quad \text{on } S. \tag{5.4.115}$$

The technique used in Theorem 5.4.6 then leads to uniqueness, and the existence follows from Theorem 5.4.5.

The right-hand side g is a continuous linear form on V, when $g \in H_{\operatorname{div}}^{-1/2}(\Gamma)$, and thus the duality $\int_\Gamma (g \cdot w_T) \gamma$ has a meaning using Theorem 5.4.2 and Lemma 5.4.1. The extra regularity of the field H, is linked to the H^2 regularity of the function q.

On the sphere S, as in Theorem 5.4.5, we have

$$\omega^2 \varepsilon_0 \frac{\partial q}{\partial n} - i\omega \operatorname{div}_S \mathcal{T}_R \nabla_S q = 0. \tag{5.4.116}$$

We also have

$$\begin{cases} \operatorname{div}(\mu \nabla q) = 0, \\[2mm] \dfrac{\partial q}{\partial n} = -\dfrac{i}{\omega \mu} \operatorname{div}_\Gamma g, \quad \text{on } \Gamma. \end{cases} \tag{5.4.117}$$

The H^2 regularity of the function q, in each of the domains where μ is regular, is satisfied when $\text{div}_\Gamma g \in H^{1/2}(\Gamma)$, which proves the theorem. ∎

Using the multipole expansion in the exterior of the ball B_R, we build a solution of the Maxwell exterior problem. We can combine the previous Theorems 5.3.3, 5.3.4 and 5.4.6, 5.4.7 to obtain the following results:

Theorem 5.4.8 *Let ε and μ be two functions of class C^1 in regular domains included in Ω_e, separated by analytic surfaces, real and constant outside a ball of radius R ($\varepsilon = \varepsilon_0$ and $\mu = \mu_0$). Suppose that they moreover satisfy, $\Re\varepsilon \geq 0$, $\Im\varepsilon \geq 0$, $\Re\mu \geq 0$, $\Im\mu \geq 0$. The solution of the exterior Maxwell system*

$$\begin{cases} \text{curl } E - i\omega\mu H = 0, & \text{in } \Omega_e, \\[2mm] \text{curl } E + i\omega\varepsilon E = 0, & \text{in } \Omega_e, \\[2mm] E \wedge n = g, & \text{on } \Gamma, \end{cases} \qquad (5.4.118)$$

which satisfies the Silver-Müller radiation condition, is unique in the space

$$X = \left\{ E/r \in \left(L^2(\Omega_e)\right)^3, H/r \in \left(L^2(\Omega_e)\right)^3, \left(E \cdot \frac{x}{r}\right) \in L^2(\Omega_e), \right.$$

$$\left. \left(H \cdot \frac{x}{r}\right) \in L^2(\Omega_e), r\left(\sqrt{\varepsilon_0}\, E_T - \sqrt{\mu_0}\, H \wedge \frac{x}{r}\right) \in \left(L^2(\Omega_e)\right)^3 \right\}$$

if the data $g \in H_{\text{div}}^{-1/2}(\Gamma)$.

If moreover g belongs to $TH^{1/2}(\Gamma)$, the solution E belongs to the spaces H_{loc}^1 constructed on the domains where ε has no discontinuity, and

$$\nabla_T E \in L^2\left(\Omega_e \cap B_R^c\right), \quad \frac{1}{r}\frac{\partial E}{\partial r} \in L^2\left(\Omega_e \cap B_R^c\right).$$

If moreover, g belongs to $H_{\text{div}}^{1/2}(\Gamma)$, the solution H belongs to the spaces H_{loc}^1 constructed on the domains where μ has no discontinuity, and

$$\nabla_T H \in L^2\left(\Omega_e \cap B_R^c\right), \quad \frac{1}{r}\frac{\partial H}{\partial r} \in L^2\left(\Omega_e \cap B_R^c\right).$$

5.4.3 Coupled variational formulations for impedance conditions

When the impedance z is different from zero, none of the previous variational formulations leads simply to an existence theorem. This is due to the fact that in these formulations, the fields E and H do not have the same role and the same regularity. The symmetry is broken. The coupled variational formulation respects this symmetry.

The equations are

$$\begin{cases} \operatorname{curl} E - i\omega\mu H = 0, & x \in \Omega_e \cap B_R, \\ \operatorname{curl} H + i\omega\varepsilon E = 0, & x \in \Omega_e \cap B_R, \\ (E \wedge n - zH_T)|_\Gamma = g, \\ (H \wedge n - T_R E_T)|_{S_R} = 0. \end{cases} \tag{5.4.119}$$

The variables E and H are the unknowns of the coupled variational formulations. We start from equation (5.4.119). For any test fields E^t and H^t, we have

$$\begin{cases} -\int_{\Omega_e \cap B_R} \left(\operatorname{curl} H \cdot \overline{E}^t \right) dx + \int_{\Omega_e \cap B_R} (\operatorname{curl} \overline{E} \cdot H^t) dx \\ \qquad + i\omega \int_{\Omega_e \cap B_R} \left(-\varepsilon(E \cdot \overline{E}^t) + \overline{\mu}(\overline{H} \cdot H^t) \right) dx = 0. \end{cases} \tag{5.4.120}$$

From the Stokes formula (5.4.33), it follows that

$$\begin{cases} \int_{\Omega_e \cap B_R} (E \cdot \operatorname{curl} H^t) \, dx - i\omega \int_{\Omega_e \cap B_R} \mu(H \cdot H^t) dx \\ \qquad - \int_S (H^t \cdot E \wedge n) \, d\gamma + \int_\Gamma (H^t \cdot E \wedge n) \, d\gamma = 0, \quad \forall H^t, \\ \int_{\Omega_e \cap B_R} (H \cdot \operatorname{curl} E^t) \, dx + i\omega \int_{\Omega_e \cap B_R} \varepsilon(E \cdot E^t) dx \\ \qquad - \int_S (E^t \cdot H \wedge n) \, d\gamma + \int_\Gamma (E^t \cdot H \wedge n) \, d\gamma = 0, \quad \forall E^t. \end{cases} \tag{5.4.121}$$

The boundary conditions on S admit the two forms

$$\begin{cases} H_T \wedge n = T_R E_T, \\ E_T \wedge n = -\frac{\mu_0}{\varepsilon_0} T_R H_T, \end{cases} \tag{5.4.122}$$

which are equivalent as a result of relation (5.3.110).

Let us suppose first that **neither z nor $1/z$ vanish**. Then, the boundary conditions on the surface Γ admit the two equivalent expressions

$$\begin{cases} E_T \wedge n = zH_T + g, \\ H_T \wedge n = -\frac{1}{z} E_T - \frac{1}{z}(g \wedge n). \end{cases} \tag{5.4.123}$$

Consider the Hilbert space

$$V = \left\{ v \in H\left(\operatorname{curl}, \Omega_e \cap B_R\right), v_T \in L^2\left(\Gamma\right) \right\}.$$

Replacing in equation (5.4.121) the quantities on Γ and S by their values extracted from the identities (5.4.122) and (5.4.123), we obtain a first variational formulation

$$
\begin{cases}
\displaystyle\int_{\Omega_c \cap B_R} \left(E \cdot \operatorname{curl} H^t\right) dx - i\omega \int_{\Omega_c \cap B_R} \mu(H \cdot H^t) dx \\[2mm]
\displaystyle\qquad + \int_S \frac{\mu_0}{\varepsilon_0} \left(H^t \cdot T_R H_T\right) d\gamma + \int_\Gamma \left(v^t \cdot z H_T\right) d\gamma \\[2mm]
\displaystyle\qquad\qquad = -\int_\Gamma \left(H^t \cdot g\right) d\gamma; \quad \forall H^t \in V, \\[3mm]
\displaystyle\int_{\Omega_c \cap B_R} \left(H \cdot \operatorname{curl} E^t\right) dx + i\omega \int_{\Omega_c \cap B_R} \varepsilon(E \cdot E^t) dx \\[2mm]
\displaystyle\qquad - \int_S \left(E^t \cdot T_R E_T\right) d\gamma - \int_\Gamma \left(E^t \cdot \frac{1}{z} E_T\right) d\gamma \\[2mm]
\displaystyle\qquad\qquad = \int_\Gamma \left(E^t \cdot \frac{1}{z}(g \wedge n)\right) d\gamma; \quad \forall E^t \in V.
\end{cases}
\tag{5.4.124}
$$

We have obtained two variational formulations which involve E and or H separately. They are the natural extensions of the ones used for perfect conductors. Using the above decompositions of the fields, these variational formulations can be used to prove some result of existence and uniqueness. Yet, in order to adapt to these formulations, the argument of compact injection, we need to specify the sign of the imaginary part of z. Moreover, it seems necessary to add quite restrictive hypotheses on the regularity of z. We introduce a **second variational formulation**, where the two fields E and H are linked through the boundary condition (5.4.123).

Let us consider the affine space

$$
V_g = \begin{cases}
E \in H(\operatorname{curl}, \Omega_e \cap B_R), \quad H \in H(\operatorname{curl}, \Omega_e \cap B_R), \\
\operatorname{div}(\varepsilon E) = 0, \quad \operatorname{div}(\mu H) = 0, & \text{on } \Omega_e \cap B_R, \\
\omega\varepsilon\,(E \cdot n) - i \operatorname{div}_S T_R E_T = 0, & \text{on } S, \\
\omega\mu(H \cdot n) - i \operatorname{div}_S T_R H_T = 0, & \text{on } S, \\
E_T \wedge n - z H_T = g, & \text{on } \Gamma.
\end{cases}
$$

We also introduce the associated Hilbert space

$$
V_g = \begin{cases}
E \in H(\operatorname{curl}, \Omega_e \cap B_R), \quad H \in H(\operatorname{curl}, \Omega_e \cap B_R), \\
\operatorname{div}(\varepsilon E) = 0, \quad \operatorname{div}(\mu H) = 0, & \text{on } \Omega_e \cap B_R, \\
\omega\varepsilon\,(E \cdot n) - i \operatorname{div}_S T_R E_T = 0, & \text{on } S, \\
\omega\mu(H \cdot n) - i \operatorname{div}_S T_R H_T = 0, & \text{on } S, \\
E_T \wedge n - z H_T = 0, & \text{on } \Gamma.
\end{cases}
$$

Lemma 5.4.7 *Suppose that z and $1/z$ act as multipliers in $H^{-1/2}(\Gamma)$, and that $\nabla_\Gamma(z)$ and $\nabla_\Gamma(1/z)$ act as multipliers from $TH^{1/2-s}(\Gamma)$ to $TH^{-1/2}(\Gamma)$, for some value $s > 0$. Suppose that $g \in TH^{1/2}(\Gamma)$. Then, the spaces V_0 and V_g are such that E belongs to $\left(H^1(\Omega_1)\right)^3 \cap \left(H^1(\Omega_2 \cap B_R)\right)^3$, where the domains Ω_1 and Ω_2 are separated by the discontinuity surfaces of the function ε, while H belongs to the Hilbert space $\left(H^1(\Omega_3)\right)^3 \cap \left(H^1(\Omega_4 \cap B_R)\right)^3$, where the domains Ω_3 and Ω_4 are separated by the discontinuity surfaces of the function μ. It follows that E_T and H_T are in the space $TH^{1/2}(\Gamma)$ which is compactly imbedded in the space $TL^2(\Gamma)$.*

Proof
As a consequence of Lemma 5.4.5 and Theorem 5.4.4, we need only to show that $\text{div}_\Gamma E_T$ and $\text{div}_\Gamma H_T$ belong to $H^{-1/2}(\Gamma)$. We have

$$\begin{cases} \text{div}_\Gamma(E_T) = -z\,\text{curl}_\Gamma H_T - (\nabla_\Gamma(z) \cdot (H \wedge n)) - \text{curl}_\Gamma g, \\ \text{div}_\Gamma(H_T) = \dfrac{1}{z}\,\text{curl}_\Gamma H_T + \left(\nabla_\Gamma\!\left(\dfrac{1}{z}\right) \cdot (E \wedge n)\right) - \text{div}_\Gamma \dfrac{1}{z}g. \end{cases} \tag{5.4.125}$$

From the properties of z, and these equalities, it follows that

$$\begin{cases} \|\text{div}_\Gamma E_T\|_{H^{-1/2}(\Gamma)} \leq c\Big(\|\text{curl}_\Gamma H_T\|_{H^{-1/2}(\Gamma)} \\ \qquad + \|H_T\|_{TH^{1/2-s}(\Gamma)} + \|\text{curl}_\Gamma g\|_{H^{-1/2}(\Gamma)}\Big), \end{cases} \tag{5.4.126}$$

$$\begin{cases} \|\text{div}_\Gamma H_T\|_{H^{-1/2}(\Gamma)} \leq c\Big(\|\text{curl}_\Gamma E_T\|_{H^{-1/2}(\Gamma)} \\ \qquad + \|E_T\|_{TH^{1/2-s}(\Gamma)} + \|\text{div}_\Gamma g\|_{H^{-1/2}(\Gamma)}\Big). \end{cases} \tag{5.4.127}$$

The lemma is now a consequence of the trace Theorem 5.4.2 and the estimates (5.4.4) and (5.4.3). ∎

We introduce the following **coupled variational formulation**:
Find $(E^t, H^t) \in V_g$ such that

$$\begin{cases} \displaystyle\int_{\Omega_\epsilon \cap B_R} \frac{1}{\mu}\left(\text{curl}\,E \cdot \text{curl}\,E^t\right)dx - \omega^2 \int_{\Omega_\epsilon \cap B_R} \varepsilon(E \cdot E^t)dx \\ \displaystyle - i\omega \int_S \left(E^t \cdot \mathcal{T}_R E_T\right)d\gamma - i\omega \int_\Gamma \left(E^t \cdot \frac{1}{z}E_T\right)d\gamma \\ \displaystyle + \int_{\Omega_\epsilon \cap B_R} \frac{1}{\varepsilon}\left(\text{curl}\,H \cdot \text{curl}\,H^t\right)dx - \omega^2 \int_{\Omega_\epsilon \cap B_R} \mu(H \cdot H^t)dx \\ \displaystyle - i\omega \int_S \frac{\mu_0}{\varepsilon_0}\left(H^t \cdot \mathcal{T}_R H_T\right)d\gamma - i\omega \int_\Gamma \left(H^t \cdot z H_T\right)d\gamma \\ \displaystyle \qquad\qquad\qquad = 0, \qquad \forall(E^t, H^t) \in V_0. \end{cases} \tag{5.4.128}$$

Theorem 5.4.9 *Suppose that ε and μ satisfy (5.4.2) and (5.4.3) with real positive value of ε_0 and μ_0. Suppose that the impedance z is such that z and $1/z$* **are bounded** *and satisfy*

$$\Re z(x) \geq 0. \tag{5.4.129}$$

Suppose that z and $1/z$ act as multipliers in $H^{-1/2}(\Gamma)$, and that $\nabla_\Gamma(z)$ and $\nabla_\Gamma(1/z)$ act as multipliers from $TH^{1/2-s}(\Gamma)$ to $TH^{-1/2}(\Gamma)$, for some value $s > 0$. Then, the Maxwell problem (5.4.119) admits a unique solution in the space V_g, when $g \in TH^{1/2}(\Gamma)$. This solution is such that E belongs to $\left(H^1(\Omega_1)\right)^3 \cap \left(H^1(\Omega_2 \cap B_R)\right)^3$, where the common boundary of Ω_1 and Ω_2 is the union of the discontinuity surfaces of the function ε, while H belongs to $\left(H^1(\Omega_3)\right)^3 \cap \left(H^1(\Omega_4 \cap B_R)\right)^3$, where the common boundary of Ω_3 and Ω_4 is the union of the discontinuity surfaces of the function μ. E_T and H_T are in $TH^{1/2}(\Gamma)$.

Proof

We obtain the coupled variational formulation (5.4.128) by multiplying the first equation of (5.4.119) by $H^t = \operatorname{curl} E^t / (i\omega\mu)$, then integrating and adding the second equation of (5.4.124). It follows that

$$\begin{cases} \displaystyle\int_{\Omega_e \cap B_R} \frac{1}{\mu}\left(\operatorname{curl} E \cdot \operatorname{curl} E^t\right) dx - \omega^2 \int_{\Omega_e \cap B_R} \varepsilon(E \cdot E^t) dx \\[2mm] \displaystyle - i\omega \int_S \left(E^t \cdot T_R E_T\right) d\gamma - i\omega \int_\Gamma \left(E^t \cdot \frac{1}{z} E_T\right) d\gamma \\[2mm] \displaystyle = i\omega \int_\Gamma \left(E^t \cdot \frac{1}{z}(g \wedge n)\right) d\gamma, \quad \forall E^t \in V. \end{cases} \tag{5.4.130}$$

We obtain a variational formulation in H in the same way, but exchanging the roles of E and H,

$$\begin{cases} \displaystyle\int_{\Omega_e \cap B_R} \frac{1}{\varepsilon}\left(\operatorname{curl} H \cdot \operatorname{curl} H^t\right) dx - \omega^2 \int_{\Omega_e \cap B_R} \mu(H \cdot H^t) dx \\[2mm] \displaystyle - i\omega \int_S \frac{\mu_0}{\varepsilon_0}\left(H^t \cdot T_R H_T\right) d\gamma - i\omega \int_\Gamma \left(H^t \cdot z H_T\right) d\gamma \\[2mm] \displaystyle = i\omega \int_\Gamma \left(H^t \cdot g\right) d\gamma, \quad \forall H^t \in V. \end{cases} \tag{5.4.131}$$

Moreover, from (5.4.119) we deduce the equations

$$\begin{cases} \operatorname{div} \varepsilon E = 0, & x \in \Omega_e \cap B_R, \\[1mm] \omega\varepsilon(E \cdot n) - i\frac{\mu_0}{\varepsilon_0} \operatorname{div}_S T_R E = 0, & x \in S, \\[1mm] \omega\varepsilon(E \cdot n) + i \operatorname{div}_\Gamma\left(\frac{1}{z} E\right) = 0, & x \in \Gamma. \end{cases} \tag{5.4.132}$$

$$\begin{cases} \operatorname{div} \mu H = 0, & x \in \Omega_e \cap B_R, \\[1mm] \omega\mu(H \cdot n) - i\frac{\mu_0}{\varepsilon_0} \operatorname{div}_S T_R H = 0, & x \in S, \\[1mm] \omega\mu(H \cdot n) + i \operatorname{div}_\Gamma (z H) = 0, & x \in \Gamma. \end{cases} \tag{5.4.133}$$

Adding (5.4.130) and (5.4.131) and using the boundary conditions (5.4.122) and (5.4.123), we obtain the coupled variational formulation.

From Lemma 5.4.7, Theorem 5.4.5 and the associated remark can be applied to the coupled variational formulation (5.4.128). The solution belongs to the space V_g. From the properties of the operator T_R, all the terms with a negative sign are compact. We need to build u_g, which is possible for example by solving a Maxwell equation (5.4.59) with $z = 0$. We obtain a field $(E_g, 0)$ which satisfies all the needed properties as a consequence of its regularity for a data $g \in TH^{1/2}(\Gamma)$.

In order to prove **uniqueness**, we introduce the null quantity extracted from (5.4.124),

$$
\begin{cases}
-\omega \Re \int_{S_R} (T_R E_T \cdot \overline{E}_T) \, d\gamma - \omega \int_\Gamma \Re\left(\frac{1}{z}\right) |E_T|^2 d\gamma \\[2mm]
+ \int_{\Omega_r \cap B_R} \left[\Im\left(\frac{1}{\mu}\right) |\text{curl } E|^2 - \omega^2 \Im(\varepsilon) |E|^2 \right] dx \\[2mm]
- \omega \frac{\mu_0}{\varepsilon_0} \Re \int_{S_R} (T_R H_T \cdot \overline{H}_T) \, d\gamma - \omega \int_\Gamma \left(\Re z \, |H_T|^2\right) d\gamma \\[2mm]
+ \int_{\Omega_r \cap B_R} \left[\Im\left(\frac{1}{\varepsilon}\right) |\text{curl } H|^2 - \omega^2 \Im(\mu) |H|^2 \right] dx.
\end{cases}
\tag{5.4.134}
$$

From Theorem 5.3.5 (formula (5.3.89)), the properties of z, and the properties of ε and μ, all its terms are negative, and thus,

$$E_T = 0, \quad \text{on } S \tag{5.4.135}$$

and

$$H_T = 0, \quad \text{on } S. \tag{5.4.136}$$

It follows that E satisfies

$$\text{curl } \frac{1}{\mu} \text{curl } E - \omega^2 \varepsilon E = 0, \tag{5.4.137}$$

and the boundary conditions

$$
\begin{cases}
E \wedge n|_\Gamma = 0, \\
E \wedge n|_S = 0, \\
\text{curl } E \wedge n|_S = 0.
\end{cases}
\tag{5.4.138}
$$

The rest of the proof is then similar to the one associated with the perfect conductor.   ∎

Remarks

– The regularity properties are different from those of theorem (5.4.7). When $g \in TH^{3/2}(\Gamma)$, E is probably in $\left(H^2(\Omega_1)\right)^3 \cap \left(H^2(\Omega_2 \cap B_R)\right)^3$, where the common boundary to Ω_1 and Ω_2 is the union of the discontinuity surfaces of the function ε, while H belongs to the space

$\left(H^2\left(\Omega_3\right)\right)^3 \cap \left(H^2\left(\Omega_4 \cap B_R\right)\right)^3$, where the common boundary to Ω_3 and Ω_4 is the union of the discontinuity surfaces of the function μ. This result cannot be true without imposing stronger hypotheses on z. We require that z and $1/z$ act as multipliers in $H^{1/2}(\Gamma)$, and that $\nabla_\Gamma(z)$ and $\nabla_\Gamma(1/z)$ act as multipliers from $TH^{3/2}(\Gamma)$ to $TH^{1/2}(\Gamma)$.

– The hypothesis $\Re z(x) \geq 0$ plays an essential role in the proof of uniqueness. When $\Re z(x)$ is negative, there probably exist eigenmodes, which correspond to surface waves. ∎

In the general case, z vanishes on a non-zero measure part Γ_0 of the surface Γ, while $1/z$ vanishes on a non-zero measure part Γ_∞ of Γ. We denote by Γ_1 the part of Γ where neither z nor $1/z$ are zero. The boundary conditions on S are unchanged and given by (5.4.122). The boundary conditions on Γ are

$$
\begin{cases}
E \wedge n = z H_T + g, & \text{on } \Gamma_1, \\
H \wedge n = -\frac{1}{z} E_T - \frac{1}{z}(g \wedge n), & \text{on } \Gamma_1, \\
g = 0, & \text{on } \Gamma_\infty, \\
E \wedge n = g, & \text{on } \Gamma_0, \\
H \wedge n = 0, & \text{on } \Gamma_\infty.
\end{cases}
\tag{5.4.139}
$$

We introduce the affine space

$$
V_g = \begin{cases}
E \in H(\text{curl}, \Omega_e \cap B_R), \quad H \in H(\text{curl}, \Omega_e \cap B_R), \\
\text{div}(\varepsilon E) = 0, \quad \text{div}(\mu H) = 0, & \text{on } \Omega_e \cap B_R, \\
\omega \varepsilon (E \cdot n) - i \, \text{div}_S T_R E_T = 0, & \text{on } S, \\
\omega \mu (H \cdot n) - i \, \text{div}_S T_R H_T = 0, & \text{on } S, \\
E \wedge n = z H_T + g, & \text{on } \Gamma_1, \\
E \wedge n = g, & \text{on } \Gamma_0, \\
H \wedge n = 0, & \text{on } \Gamma_\infty.
\end{cases}
$$

We introduce the associated Hilbert space

$$
V_g = \begin{cases}
E \in H(\text{curl}, \Omega_e \cap B_R), \quad H \in H(\text{curl}, \Omega_e \cap B_R), \\
\text{div}(\varepsilon E) = 0, \quad \text{div}(\mu H) = 0, & \text{on } \Omega_e \cap B_R, \\
\omega \varepsilon (E \cdot n) - i \, \text{div}_S T_R E_T = 0, & \text{on } S, \\
\omega \mu (H \cdot n) - i \, \text{div}_S T_R H_T = 0, & \text{on } S, \\
E \wedge n = z H_T, & \text{on } \Gamma_1, \\
E \wedge n = 0, & \text{on } \Gamma_0, \\
H \wedge n = 0, & \text{on } \Gamma_\infty.
\end{cases}
$$

Lemma 5.4.8 *Suppose that z and $1/z$ act as multipliers in $H^{-1/2}(\Gamma)$, and that $\nabla_\Gamma(z)$ and $\nabla_\Gamma(1/z)$ act as multipliers from $TH^{1/2-s}(\Gamma)$ to $TH^{-1/2}(\Gamma)$, for some value $s > 0$. Suppose that $g \in TH^{1/2}(\Gamma)$. Then, the* *spaces V_0 and V_g are such that E belongs to $\left(H^1(\Omega_1)\right)^3 \cap \left(H^1(\Omega_2 \cap B_R)\right)^3$, where the common boundary of Ω_1 and Ω_2 is the union of the discontinuity surfaces of the function ε, while H belongs to $\left(H^1(\Omega_3)\right)^3 \cap \left(H^1(\Omega_4 \cap B_R)\right)^3$, where the common boundary of Ω_3 and Ω_4 is the union of the discontinuity surfaces of the function μ. Moreover, E_T and H_T belong to $TH^{1/2}(\Gamma)$ which is compactly imbedded in the space $TL^2(\Gamma)$.*

Proof

As a consequence of Lemma 5.4.5 and Theorem 5.4.4, we need only to estimate the terms $\int_\Gamma \left[\operatorname{div}_\Gamma E_T \, \varepsilon(\overline{E} \cdot n) + \operatorname{div}_\Gamma (H_T)\mu(\overline{H} \cdot n)\right]$.

On the subdomain Γ_1, it holds that

$$\begin{cases} \operatorname{div}_\Gamma (E_T) = -z \operatorname{curl}_\Gamma H_T - (\nabla_\Gamma(z) \cdot (H \wedge n)) - \operatorname{curl}_\Gamma g, \\ \operatorname{div}_\Gamma (H_T) = \dfrac{1}{z} \operatorname{curl}_\Gamma H_T + \left(\nabla_\Gamma\!\left(\dfrac{1}{z}\right) \cdot (E \wedge n)\right) - \operatorname{div}_\Gamma \dfrac{1}{z} g, \end{cases} \quad (5.4.140)$$

On the subdomain Γ_0,

$$(H \cdot n) = \frac{1}{i\omega\mu}\operatorname{div}_\Gamma g. \quad (5.4.141)$$

On the subdomain Γ_∞,

$$(E \cdot n) = 0. \quad (5.4.142)$$

Thus, the above corresponding integrals are restricted to Γ_0 and Γ_1. From the regularity of g, the linear terms which appear on Γ_0 are bounded. (5.4.140) gives an estimate for the terms on Γ_1, using the properties of z, exactly as was done before. The lemma is a result of the trace theorem (5.4.2) and the estimates of theorems (5.4.3) and (5.4.4). ∎

The new **coupled variational formulation** is:

$$\begin{cases} \text{Find } (E,\, H) \in V_g \quad \text{such that} \\[4pt] \displaystyle\int_{\Omega_e \cap B_R} \frac{1}{\mu}\left(\operatorname{curl} E \cdot \operatorname{curl} E^t\right) dx - \omega^2 \int_{\Omega_e \cap B_R} \varepsilon(E \cdot E^t)dx \\[8pt] \displaystyle - i\omega \int_S \left(E^t \cdot T_R E_T\right) d\gamma - i\omega \int_{\Gamma_1} \left(E^t \cdot \frac{1}{z}E_T\right) d\gamma \\[8pt] \displaystyle + \int_{\Omega_e \cap B_R} \frac{1}{\varepsilon}\left(\operatorname{curl} H \cdot \operatorname{curl} H^t\right) dx - \omega^2 \int_{\Omega_e \cap B_R} \mu(H \cdot H^t)dx \\[8pt] \displaystyle - i\omega \int_S \frac{\mu_0}{\varepsilon_0}\left(H^t \cdot T_R H_T\right) d\gamma - i\omega \int_{\Gamma_1} \left(H^t \cdot z H_T\right) d\gamma \\[8pt] \hspace{3cm} = 0, \quad \forall\, (E^t,\, H^t) \in V_0, \end{cases} \quad (5.4.143)$$

Theorem 5.4.10 *Suppose that ε and μ satisfy (5.4.2) and (5.4.3) with real positive ε_0 and μ_0. Suppose that the impedance z is such that*

$$\Re z(x) \geq 0. \quad (5.4.144)$$

Suppose that z and $1/z$ act as multipliers in $H^{-1/2}(\Gamma_1)$, and that $\nabla_{\Gamma_1}(z)$ and $\nabla_{\Gamma_1}(1/z)$ act as multipliers from $TH^{1/2-s}(\Gamma_1)$ to $TH^{-1/2}(\Gamma_1)$, for some value $s > 0$.

Then, the Maxwell problem (5.4.119) admits a unique solution in the space V_g, when $g \in TH^{1/2}(\Gamma)$. This solution is such that E belongs to $\left(H^1(\Omega_1)\right)^3 \cap \left(H^1(\Omega_2 \cap B_R)\right)^3$, where the common boundary of Ω_1 and Ω_2 is the union of the discontinuity surfaces of the function ε, while H belongs to $\left(H^1(\Omega_3)\right)^3 \cap \left(H^1(\Omega_4 \cap B_R)\right)^3$, where the common boundary of Ω_3 and Ω_4 is the union of the discontinuity surfaces of the function μ. E_T and H_T are in $TH^{1/2}(\Gamma)$.

Proof
The proof is exactly similar to those of theorem (5.4.9) using Lemma 5.4.8.

Remarks
– Regularity properties are quite different from those of Theorem 5.4.9. It is necessary to distinguish between the parts where z vanishes and the part where z is infinite. It is also known through some explicit examples that any discontinuity line of the impedance z creates a singularity of the fields E and H.

– The hypotheses of Lemma 5.4.8 on z are not explicit. It would be quite useful to exhibit simpler hypotheses in the neighbourhood of the zeros of z.

5.5   Integral Representations

In this section, we suppose that ε and μ are constant in regular domains, delimited by regular boundaries. We exhibit the integral representations of the Maxwell equation and give their principal properties.

Theorem 5.5.1 (Representation Theorem) *Let Ω_i be a bounded regular interior domain, which boundary is the regular surface Γ and denote by Ω_e the associated exterior domain. The exterior unit normal to Γ is denoted by n. Let E and H be regular solutions of the Maxwell equations*

$$\begin{cases} \operatorname{curl} E - i\omega\mu H = 0, \\ \operatorname{curl} H + i\omega\varepsilon E = 0, \end{cases} \quad \text{in } \Omega_i, \qquad (5.5.1)$$

$$\begin{cases} \operatorname{curl} E - i\omega\mu H = 0, \\ \operatorname{curl} H + i\omega\varepsilon E = 0, \\ E \text{ and } H \text{ satisfy the radiation} \\ \text{conditions of Silver-Müller .} \end{cases} \quad \text{in } \Omega_e, \qquad (5.5.2)$$

The values of the constants ε and μ are the same for the interior and the exterior problems.

Let us denote by j and m the tangent fields to the surface Γ,

$$\begin{cases} j = H_i \wedge n - H_e \wedge n, \\ \text{where } H_i \text{ and } H_e \text{ are respectively the interior and exterior} \\ \text{limits of the field } H \quad (j \text{ is the electric current).} \end{cases} \quad (5.5.3)$$

$$\begin{cases} m = E_i \wedge n - E_e \wedge n, \\ \text{where } E_i \text{ and } E_e \text{ are respectively the interior and exterior} \\ \text{limits of the field } E \quad (m \text{ is the magnetic current).} \end{cases} \quad (5.5.4)$$

We denote by $G(r)$ **the outgoing fundamental solution** *of the Helmholtz equation*

$$G(r) = \frac{1}{4\pi r} e^{ikr}, \quad k = \omega\sqrt{\varepsilon\mu}. \quad (5.5.5)$$

Then, **the fields E and H** *admit the integral representation*

$$\begin{cases} E(y) = i\omega\mu \int_\Gamma G(x-y)j(x)d\gamma(x) \\ \qquad + \frac{i}{\omega\varepsilon}\nabla\int_\Gamma G(x-y)\,\mathrm{div}_\Gamma\, j(x)d\gamma(x) \\ \qquad + \mathrm{curl}\int_\Gamma G(x-y)m(x)d\gamma(x), \quad y \notin \Gamma. \end{cases} \quad (5.5.6)$$

$$\begin{cases} H(y) = -i\omega\varepsilon \int_\Gamma G(x-y)m(x)d\gamma(x) \\ \qquad - \frac{i}{\omega\mu}\nabla\int_\Gamma G(x-y)\,\mathrm{div}_\Gamma\, m(x)d\gamma(x) \\ \qquad + \mathrm{curl}\int_\Gamma G(x-y)j(x)d\gamma(x), \quad y \notin \Gamma. \end{cases} \quad (5.5.7)$$

The **interior value of $E \wedge n$** *is*

$$\begin{cases} (E \wedge n)(y) = \frac{m(y)}{2} + \int_\Gamma \Big[\frac{\partial G}{\partial n_y}(x-y)m(x) \\ \qquad\qquad - \nabla_y G(x-y)\,(m(x) \cdot (n_y - n_x))\Big]d\gamma(x) \\ \qquad + i\omega\mu \int_\Gamma G(x-y)\,(j(x) \wedge n_y))\,d\gamma(x) \\ \qquad + \frac{i}{\omega\varepsilon}\int_\Gamma \Big[(\nabla_y G(x-y) \wedge (n_y - n_x))\,\mathrm{div}_\Gamma\, j(x) \\ \qquad\qquad + G(x-y)\overrightarrow{\mathrm{curl}}_\Gamma\,\mathrm{div}_\Gamma\, j(x)\Big]d\gamma(x). \end{cases} \quad (5.5.8)$$

The **interior value of** $H \wedge n$ *is*

$$
\left\{
\begin{aligned}
(H \wedge n)(y) = \frac{j(y)}{2} &+ \int_\Gamma \left[\frac{\partial G}{\partial n_y}(x-y)j(x) \right. \\
& \left. - \nabla_y G(x-y)\,(j(x)\cdot(n_y - n_x)) \right] d\gamma(x) \\
& - i\omega\varepsilon \int_\Gamma G(x-y)\,(m(x)\wedge n_y))\,d\gamma(x) \\
& - \frac{i}{\omega\mu} \int_\Gamma \left[(\nabla_y G(x-y)\wedge(n_y - n_x))\,\mathrm{div}_\Gamma\, m(x) \right. \\
& \left. + G(x-y)\overrightarrow{\mathrm{curl}}_\Gamma\,\mathrm{div}_\Gamma\, m(x) \right] d\gamma(x).
\end{aligned}
\right.
\tag{5.5.9}
$$

The **interior value of** $(E \cdot n)$ *is*

$$
\left\{
\begin{aligned}
(E \cdot n)(y) = \frac{i}{2\omega\varepsilon}\,\mathrm{div}_\Gamma\, j(y) & \\
& + i\omega\mu \int_\Gamma G(x-y)\,(j(x)\cdot n_y)\,d\gamma(x) \\
& + \frac{i}{\omega\varepsilon} \int \frac{\partial G}{\partial n_y}(x-y)\,\mathrm{div}_\Gamma\, j(x)d\gamma(x) \\
& + \int_\Gamma (((n_y - n_x)\wedge\nabla_y G(x-y))\cdot m(x))\,d\gamma(x) \\
& - \int_\Gamma G(x-y)\,\mathrm{curl}_\Gamma\, m(x)d\gamma(x).
\end{aligned}
\right.
\tag{5.5.10}
$$

The **interior value of** $(H \cdot n)$ *is*

$$
\left\{
\begin{aligned}
(H \cdot n)(y) = -\frac{i}{2\omega\mu}\,\mathrm{div}_\Gamma\, m(y) & \\
& - i\omega\varepsilon \int_\Gamma G(x-y)\,(m(x)\cdot n_y)\,d\gamma(x) \\
& - \frac{i}{\omega\mu} \int \frac{\partial G}{\partial n_y}(x-y)\,\mathrm{div}_\Gamma\, m(x)d\gamma(x) \\
& + \int_\Gamma (((n_y - n_x)\wedge\nabla_y G(x-y))\,j(x))\cdot d\gamma(x) \\
& - \int_\Gamma G(x-y)\,\mathrm{curl}_\Gamma\, j(x)d\gamma(x).
\end{aligned}
\right.
\tag{5.5.11}
$$

The **exterior value of** $E \wedge n$ *is*

$$
\begin{cases}
(E \wedge n)(y) = -\dfrac{m(y)}{2} + \displaystyle\int_{\Gamma} \left[\dfrac{\partial G}{\partial n_y}(x - y)m(x) \right. \\[2mm]
\qquad\qquad \left. - \nabla_y G(x - y)\left(m(x) \cdot (n_y - n_x)\right) \right] d\gamma(x) \\[3mm]
\qquad + i\omega\mu \displaystyle\int_{\Gamma} G(x - y)\left(j(x) \wedge n_y\right) d\gamma(x) \qquad\qquad (5.5.12) \\[3mm]
\qquad + \dfrac{i}{\omega\varepsilon} \displaystyle\int_{\Gamma} \left[\left(\nabla_y G(x - y) \wedge (n_y - n_x)\right) \operatorname{div}_{\Gamma} j(x) \right. \\[3mm]
\qquad\qquad \left. + G(x - y)\overrightarrow{\operatorname{curl}}_{\Gamma} \operatorname{div}_{\Gamma} j(x) \right] d\gamma(x).
\end{cases}
$$

The **exterior value of** $H \wedge n$ *is*

$$
\begin{cases}
(H \wedge n)(y) = -\dfrac{j(y)}{2} + \displaystyle\int_{\Gamma} \left[\dfrac{\partial G}{\partial n_y}(x - y)j(x) \right. \\[2mm]
\qquad\qquad \left. - \nabla_y G(x - y)\left(j(x) \cdot (n_y - n_x)\right) \right] d\gamma(x) \\[3mm]
\qquad - i\omega\varepsilon \displaystyle\int_{\Gamma} G(x - y)\left(m(x) \wedge n_y\right) d\gamma(x) \qquad\qquad (5.5.13) \\[3mm]
\qquad - \dfrac{i}{\omega\mu} \displaystyle\int_{\Gamma} \left[\left(\nabla_y G(x - y) \wedge (n_y - n_x)\right) \operatorname{div}_{\Gamma} m(x) \right. \\[3mm]
\qquad\qquad \left. + G(x - y)\overrightarrow{\operatorname{curl}}_{\Gamma} \operatorname{div}_{\Gamma} m(x) \right] d\gamma(x).
\end{cases}
$$

The **exterior value of** $(E \cdot n)$ *is*

$$
\begin{cases}
(E \cdot n)(y) = -\dfrac{i}{2\omega\varepsilon} \operatorname{div}_{\Gamma} j(y) \\[3mm]
\qquad + i\omega\mu \displaystyle\int_{\Gamma} G(x - y)\left(j(x) \cdot n_y\right) d\gamma(x) \\[3mm]
\qquad + \dfrac{i}{\omega\varepsilon} \displaystyle\int \dfrac{\partial G}{\partial n_y}(x - y) \operatorname{div}_{\Gamma} j(x) d\gamma(x) \qquad\qquad (5.5.14) \\[3mm]
\qquad + \displaystyle\int_{\Gamma} \left(\left((n_y - n_x) \wedge \nabla_y G(x - y)\right) \cdot m(x)\right) d\gamma(x) \\[3mm]
\qquad - \displaystyle\int_{\Gamma} G(x - y) \operatorname{curl}_{\Gamma} m(x) d\gamma(x).
\end{cases}
$$

The **exterior value of** $(H \cdot n)$ *is*

$$
\begin{cases}
(H \cdot n)(y) = \frac{i}{2\omega\mu} \operatorname{div}_\Gamma m(y) \\[2mm]
\quad - i\omega\varepsilon \int_\Gamma G(x - y)\,(m(x) \cdot n_y)\,d\gamma(x) \\[2mm]
\quad - \frac{i}{\omega\mu} \int_\Gamma \frac{\partial G}{\partial n_y}(x - y)\operatorname{div}_\Gamma m(x)\,d\gamma(x) \\[2mm]
\quad + \int_\Gamma \left(\left((n_y - n_x) \wedge \nabla_y G(x - y)\right) \cdot j(x)\right)\,d\gamma(x) \\[2mm]
\quad - \int_\Gamma G(x - y)\operatorname{curl}_\Gamma j(x)\,d\gamma(x).
\end{cases}
\tag{5.5.15}
$$

Proof

It is based on the known results for the integral representation of the scalar Helmholtz equation, and the use of some properties of distributions. The fields E and H, solutions of (5.5.1) and (5.5.2), satisfy in the sense of distributions in R^3 :

$$
\begin{cases}
\operatorname{curl} E - i\omega\mu H = m\delta_\Gamma, \\
\operatorname{curl} H + i\omega\varepsilon E = j\delta_\Gamma.
\end{cases}
\tag{5.5.16}
$$

Thus, E and H are the sum of two contributions, one associated with j, the other associated with m. Let us compute the contribution associated with j, which solves the equation

$$
\begin{cases}
\operatorname{curl} E - i\omega\mu H = 0, \\
\operatorname{curl} H + i\omega\varepsilon E = j\delta_\Gamma.
\end{cases}
\tag{5.5.17}
$$

It takes the form of the sum of a vector potential A and a scalar potential V:

$$
E = A + \nabla V.
\tag{5.5.18}
$$

Exactly as in the computation of the fundamental solution, these potentials must be linked by a gauge condition, which we choose to be the Lorentz gauge

$$
\operatorname{div} A - k^2 V = 0.
\tag{5.5.19}
$$

The potentials A and V are continuous across the surface Γ.

We have

$$
i\omega\varepsilon \operatorname{div} E = \operatorname{div}(j\delta_\Gamma) = (\operatorname{div}_\Gamma j)\,\delta_\Gamma,
\tag{5.5.20}
$$

which, in combination with the gauge condition yields

$$
\Delta V + k^2 V = -\frac{i}{\omega\varepsilon} \operatorname{div}_\Gamma j\delta_\Gamma.
\tag{5.5.21}
$$

Thus, the potential V is given by a single layer potential for the scalar Helmholtz equation, with density $(i \operatorname{div}_\Gamma j)/(\omega \varepsilon)$:

$$V(y) = \frac{i}{\omega \varepsilon} \int_\Gamma G(x - y) \operatorname{div}_\Gamma j(x) d\gamma(x). \tag{5.5.22}$$

From equation (5.5.17), it follows that

$$\begin{cases} \Delta A + k^2 A = \nabla \operatorname{div} A - \operatorname{curl} \operatorname{curl} A + k^2 A \\ \qquad = k^2 \nabla V - k^2 \nabla V + k^2 E - \operatorname{curl} \operatorname{curl} E \\ \qquad = -i\omega\mu j \delta_\Gamma. \end{cases} \tag{5.5.23}$$

Thus, the vector potential A is given by a single layer potential for the scalar Helmholtz equation, with density $i\omega\mu j$:

$$A(y) = i\omega\mu \int_\Gamma G(x - y) j(x) d\gamma(x).$$

Thus, the expression of this part of the field E is

$$\begin{cases} E(y) = i\omega\mu \int_\Gamma G(x - y) \, j(x) \, d\gamma(x) \\ \qquad + \frac{i}{\omega\varepsilon} \nabla \int_\Gamma G(x - y) \operatorname{div}_\Gamma j(x) d\gamma(x), \end{cases} \tag{5.5.24}$$

while the corresponding part of H is

$$H(y) = \operatorname{curl} \int_\Gamma G(x - y) j(x) d\gamma(x). \tag{5.5.25}$$

By an argument of symmetry, the part associated with m is

$$E(y) = \operatorname{curl} \int_\Gamma G(x - y) \, m(x) d\gamma(x), \tag{5.5.26}$$

$$\begin{cases} H(y) = -i\omega\varepsilon \int_\Gamma G(x - y) \, m(x) d\gamma(x) \\ \qquad - \frac{i}{\omega\mu} \nabla \int_\Gamma G(x - y) \operatorname{div}_\Gamma m(x) d\gamma(x). \end{cases} \tag{5.5.27}$$

The expression of the interior and exterior values are deduced from the properties of single layer potentials given in Theorems 3.1.1 and 3.1.2. The single layer potential is continuous across the surface Γ. The tangential part of its gradient is also continuous across the Γ. Its normal derivative is discontinuous. Besides, we have (see (5.4.50))

$$\operatorname{curl} u \wedge n = \frac{\partial}{\partial n} u_T - \nabla_T (u \cdot n) + \mathcal{R} u_T. \tag{5.5.28}$$

Thus, for a single layer potential u, the jump of $\operatorname{curl} u \wedge n$ is the same as the jump of $\partial u_T / \partial n$, where u_T denotes the tangential component. It follows that

$$
\begin{cases}
\lim_{y \to \Gamma^\pm} \left(\text{curl}_y \int_\Gamma G(x - y)j(x)d\gamma(x) \right) \wedge n_y \\
\quad = \pm \dfrac{j(y)}{2} + \int_\Gamma (\text{curl}_y(G(x - y)j(x)) \wedge n_y)\, d\gamma(x).
\end{cases}
\tag{5.5.29}
$$

We have

$$
\begin{cases}
\text{curl}_y(G(x - y)j(x)) \wedge n_y = (\nabla_y G(x - y) \wedge j(x)) \wedge n_y \\
\quad = \dfrac{\partial G}{\partial n_y}(x - y)j(x) - \nabla_y G(x - y)\,(j(x) \cdot (n_y - n_x)).
\end{cases}
\tag{5.5.30}
$$

In order to exhibit the limit of the gradient term, we use

$$
\begin{cases}
n_y \wedge \nabla \displaystyle\int_\Gamma G(x - y)\rho(x)d\gamma(x) \\[2mm]
\quad = \displaystyle\int_\Gamma n_y \wedge \nabla_y G(x - y)\rho(x)d\gamma(x) \\[2mm]
\quad = \displaystyle\int_\Gamma ((n_y - n_x) \wedge \nabla_y G(x - y))\, \rho(x)d\gamma(x) \\[2mm]
\qquad + \displaystyle\int_\Gamma \overrightarrow{\text{curl}}_{\Gamma_x} G(x - y)\rho(x)d\gamma(x) \\[2mm]
\quad = \displaystyle\int_\Gamma ((n_y - n_x) \wedge \nabla_y G(x - y))\, \rho(x)d\gamma(x) \\[2mm]
\qquad - \displaystyle\int_\Gamma G(x - y)\overrightarrow{\text{curl}}_\Gamma \rho(x)d\gamma(x),
\end{cases}
\tag{5.5.31}
$$

from which we obtain formulas (5.5.8), (5.5.9), (5.5.12) and (5.5.13).
In order to find the values of the normal components, we need the limit of
the normal component of the curl. We have

$$
\begin{cases}
\left(n_y \cdot \text{curl} \displaystyle\int_\Gamma G(x - y)m(x)d\gamma(x) \right) \\[2mm]
\quad = \displaystyle\int_\Gamma (n_y \cdot (\nabla_y G(x - y) \wedge m(x)))\, d\gamma(x) \\[2mm]
\quad = \displaystyle\int_\Gamma (((n_y - n_x) \wedge \nabla_y G(x - y)) \cdot m(x))\, d\gamma(x) \\[2mm]
\qquad + \displaystyle\int_\Gamma ((\nabla_x G(x - y) \wedge n_x) \cdot m(x))\, d\gamma(x) \\[2mm]
\quad = \displaystyle\int_\Gamma (((n_y - n_x) \wedge \nabla_y G(x - y)) \cdot m(x))\, d\gamma(x) \\[2mm]
\qquad - \displaystyle\int_\Gamma G(x - y)\,\text{curl}_\Gamma m(x)d\gamma(x),
\end{cases}
\tag{5.5.32}
$$

from which we obtain formulas (5.5.10), (5.5.11), (5.6.16) and (5.5.15). ∎

To the Maxwell integral representation are associated, exactly as for
the scalar equation, two **Calderon projectors**. These are the integral

operators which give the expressions, starting from the interior and exterior values of $E \wedge n$ and $H \wedge n$, the same quantities.

The **interior projector** correspond to null exterior values. This operator associates with m and j the expressions of $E_i \wedge n$ and $H_i \wedge n$ given by (5.5.8) and (5.5.9).

Three integral operators appear that we choose to write in their initial form, associated with formulas (5.5.6) and (5.5.7):

$$Sj(y) = -n_y \wedge \int_\Gamma G(x - y)j(x)d\gamma(x), \tag{5.5.33}$$

$$Tj(y) = -n_y \wedge \nabla \int_\Gamma G(x - y)\,\mathrm{div}_\Gamma\,j(x)d\gamma(x), \tag{5.5.34}$$

$$Rj(y) = -n_y \wedge \mathrm{curl} \int_\Gamma G(x - y)j(x)d\gamma(x). \tag{5.5.35}$$

The interior projector is the operator C_{int} which maps the couple (m, j) to the quantities

$$E_i \wedge n = \frac{m}{2} + Rm + \frac{i}{\omega\varepsilon}Tj + i\omega\mu Sj, \tag{5.5.36}$$

$$H_i \wedge n = -\frac{i}{\omega\mu}Tm - i\omega\varepsilon Sm + \frac{j}{2} + Rj. \tag{5.5.37}$$

The exterior projector is the operator C_{ext} which maps the couple (m, j) to the quantities

$$-E_e \wedge n = \frac{m}{2} - Rm - \frac{i}{\omega\varepsilon}Tj - i\omega\mu Sj, \tag{5.5.38}$$

$$-H_e \wedge n = \frac{i}{\omega\mu}Tm + i\omega\varepsilon Sm + \frac{j}{2} - Rj. \tag{5.5.39}$$

The property of projection for $(C \circ C = C)$ is equivalent to the two identities

$$R \circ R + \frac{1}{k^2}T \circ T + T \circ S + S \circ T + k^2 S \circ S = \frac{I}{4}, \tag{5.5.40}$$

$$R \circ T + T \circ R + k^2(R \circ S + S \circ R) = 0. \tag{5.5.41}$$

Let us examine the order of the operators R, S and T, to understand the meaning of these relations. From (5.5.8), R takes the form

$$\left\{ \begin{array}{l} Rj(y) = \int_\Gamma \left[\dfrac{\partial G}{\partial n_y}(x - y)j(x) \right. \\[2mm] \qquad\qquad \left. - \nabla_y G(x - y)\,(j(x) \cdot (n_y - n_x)) \right]d\gamma(x), \end{array} \right. \tag{5.5.42}$$

which shows that this operator is of order -1, i.e., it is continuous from $TH^s(\Gamma)$ into $TH^{s+1}(\Gamma)$. The operator S is also of order -1.

The operator T takes the form

$$
\begin{cases}
Tj(y) = \displaystyle\int_\Gamma \Big[(\nabla_y G(x-y) \wedge (n_y - n_x)) \operatorname{div}_\Gamma j(x) \\[2mm]
\hspace{3cm} - G(x-y) \overrightarrow{\operatorname{curl}}_\Gamma \operatorname{div}_\Gamma j(x) \Big] d\gamma(x).
\end{cases}
\tag{5.5.43}
$$

It is of order 1 and is continuous from $TH^{s+1}(\Gamma)$ into $TH^s(\Gamma)$. The expression (5.5.40) proves that, up to a compact operator K, we have

$$
T \circ T = k^2 \left(\frac{I}{4} - T \circ S - S \circ T + K \right),
$$

thus, the operator $T \circ T$ is not of order 2 as expected, but of order 0.

The most singular part of the operator T corresponds to its value when $k = 0$. It is the operator $\mathcal{T}$,

$$
\begin{cases}
\mathcal{T}j(y) = -\frac{1}{4\pi} n_y \wedge \nabla \displaystyle\int_\Gamma \frac{1}{|x-y|} \operatorname{div}_\Gamma j(x) d\gamma(x) \\[2mm]
\quad = \frac{1}{4\pi} \displaystyle\int_\Gamma \Big[\left(\nabla_y \frac{1}{|x-y|} \wedge (n_y - n_x) \right) \operatorname{div}_\Gamma j(x) \\[2mm]
\hspace{3cm} - \frac{1}{|x-y|} \overrightarrow{\operatorname{curl}}_\Gamma \operatorname{div}_\Gamma j(x) \Big] d\gamma(x).
\end{cases}
\tag{5.5.44}
$$

$\mathcal{T}$ is also of order 1.

When k tends to zero, the operators S and R have the limits

$$
\mathcal{S}j(y) = -\frac{1}{4\pi} n_y \wedge \int_\Gamma \frac{1}{|x-y|} j(x) d\gamma(x),
\tag{5.5.45}
$$

$$
\begin{cases}
\mathcal{R}j(y) = -\frac{1}{4\pi} n_y \wedge \operatorname{curl} \displaystyle\int_\Gamma \frac{1}{|x-y|} j(x) d\gamma(x), \\[2mm]
\quad = \frac{1}{4\pi} \displaystyle\int_\Gamma \Big[\frac{\partial}{\partial n_y} \left(\frac{1}{|x-y|} \right) j(x) \\[2mm]
\hspace{2cm} - \nabla_y \frac{1}{|x-y|} \left(j(x) \cdot (n_y - n_x) \right) \Big] d\gamma(x).
\end{cases}
\tag{5.5.46}
$$

$\mathcal{S}$ and $\mathcal{R}$ are of order -1. The principal terms of the expansion in k are the identities

$$
\mathcal{T} \circ \mathcal{T} = 0,
\tag{5.5.47}
$$

$$
\mathcal{R} \circ \mathcal{T} + \mathcal{T} \circ \mathcal{R} = 0.
\tag{5.5.48}
$$

The next term in the expansion of T is a factor of k^2,

$$
\begin{cases}
T = \mathcal{T} + k^2 \mathcal{U} + \cdots, \quad \left(\frac{e^{ikr}}{r} = \frac{1}{r} + ik - \frac{k^2}{2}r + \right), \\[2mm]
\mathcal{U}j(y) = \frac{1}{8\pi} n_y \wedge \nabla \int_\Gamma |x - y|\, \mathrm{div}_\Gamma\, j(x)\, d\gamma(x), \\[2mm]
\qquad = \frac{1}{8\pi} \int_\Gamma (n_y \wedge (y - x)) \frac{1}{|x - y|}\, \mathrm{div}_\Gamma\, j(x)\, d\gamma(x).
\end{cases}
\tag{5.5.49}
$$

$\mathcal{U}$ is of order -1. We thus obtain

$$
\mathcal{U} \circ \mathcal{T} + \mathcal{T} \circ \mathcal{U} + \mathcal{T} \circ \mathcal{S} + \mathcal{S} \circ \mathcal{T} + \mathcal{R} \circ \mathcal{R} = \frac{I}{4}.
\tag{5.5.50}
$$

It is possible to interpret a part of these identities, noticing that

$$
\begin{cases}
\mathrm{div}_\Gamma(\mathcal{T}j) = \mathrm{curl}_\Gamma\, (n \wedge \mathcal{T}j) \\[2mm]
\qquad = \mathrm{curl}_\Gamma \left(\nabla_\Gamma \int_\Gamma G(x - y)\, \mathrm{div}_\Gamma\, j(x) d\gamma(x) \right) = 0,
\end{cases}
\tag{5.5.51}
$$

and thus,

$$
\mathcal{T} \circ \mathcal{T} = 0.
\tag{5.5.52}
$$

Identity (5.5.40) is therefore equivalent to

$$
\mathcal{T} \circ \mathcal{S} = \frac{I}{2} + K, \qquad \text{where } K \text{ is compact,}
\tag{5.5.53}
$$

while $\mathcal{T} \circ \mathcal{R}$ has order -2 and not 0, as can be seen from identity (5.5.41). Another **significant rewriting** of identity (5.5.40) is

$$
\left(\frac{1}{k}T + kS \right) \circ \left(\frac{1}{k}T + kS \right) = \frac{I}{4} - \mathcal{R} \circ \mathcal{R},
\tag{5.5.54}
$$

which means that $2\,(T/k + kS)$ is its own inverse, up to a compact operator. Looking carefully at the expressions of T and S, we can see that this identity is in fact equivalent to those obtained in the scalar case for the single layer and double layer potentials, up to compact operators. It can be proved using these scalar identities, but through a quite difficult and technical computation.

5.6 Integral Equations

5.6.1 The perfect conductor

We examine in this section the specific properties and the variational formulation of the integral equation associated with the perfect conductor problem. We have seen that the scattering of a plane wave, by a perfect conducting object immersed in vacuum, leads to the exterior Maxwell problem

$$\begin{cases} \operatorname{curl} E - i\omega\mu_0 H = 0, & \text{in } \Omega_e, \\ \operatorname{curl} H + i\omega\varepsilon_0 E = 0, & \text{in } \Omega_e, \\ E \wedge n|_\Gamma = -E^{\text{inc}} \wedge n, & \text{on } \Gamma, \\ \left| \sqrt{\varepsilon_0} E - \sqrt{\mu_0} H \wedge n \right| \leq \frac{c}{r^2}. \end{cases} \tag{5.6.1}$$

We associate with it, the interior Maxwell problem

$$\begin{cases} \operatorname{curl} E - i\omega\mu_0 H = 0, & \text{in } \Omega_i, \\ \operatorname{curl} H + i\omega\varepsilon_0 E = 0, & \text{in } \Omega_i, \\ E \wedge n|_\Gamma = -E^{\text{inc}} \wedge n, & \text{on } \Gamma, \end{cases} \tag{5.6.2}$$

which always admits the solution

$$\begin{cases} E = -E^{\text{inc}}, \\ H = -H^{\text{inc}}. \end{cases} \tag{5.6.3}$$

This solution is not unique when the interior problem is not invertible, i.e., when k^2 is an eigenvalue of the interior Maxwell problem.

We apply the representation Theorem 5.5.1 to the solutions of the problems (5.6.1) and (5.6.2). We obtain

$$\begin{cases} E(y) = i\omega\mu_0 \int_\Gamma G(x - y)j(x)d\gamma(x) \\ \qquad + \frac{i}{\omega\varepsilon_0} \nabla \int_\Gamma G(x - y) \operatorname{div}_\Gamma j(x)d\gamma(x), \end{cases} \tag{5.6.4}$$

$$H(y) = \operatorname{curl} \int_\Gamma G(x - y)j(x)d\gamma(x), \tag{5.6.5}$$

$$j = -H^{\text{inc}} \wedge n - H \wedge n. \tag{5.6.6}$$

Remark that the current j is the tangential part of the total field H. The tangent field j is determined by the boundary condition on the surface Γ, which expression is given by (5.5.12).

We have obtained the **integral equation**

$$\begin{cases} -\left(E^{\text{inc}} \wedge n\right)(y) = i\omega\mu_0 \int_\Gamma G(x - y)\left(j(x) \wedge n_y\right)d\gamma(x) \\ \qquad + \frac{i}{\omega\varepsilon_0} \int_\Gamma \Big[(\nabla_y G(x - y) \wedge (n_y - n_x)) \operatorname{div}_\Gamma j(x) \\ \qquad\qquad + G(x - y)\overline{\operatorname{curl}}_\Gamma \operatorname{div}_\Gamma j(x) \Big] d\gamma(x). \end{cases} \tag{5.6.7}$$

In this first expression of the integral equation, only integrable kernels appear. But, linked to this property, second derivatives of the unknown j appear.

We introduce a variational formulation for this equation where appears only first derivatives of the unknown j, and yet no finite part or non-integrable kernels. It comes from equation (5.6.4), written on Γ, multiplying by a test vector j^t. Using the Stokes formula on the surface Γ, we obtain

$$
\begin{cases}
i\omega\mu_0 \displaystyle\int_\Gamma\int_\Gamma G(x-y)\left(j(x)\cdot j^t(y)\right)d\gamma(x)d\gamma(y) \\[3mm]
\quad -\dfrac{i}{\omega\varepsilon_0}\displaystyle\int_\Gamma\int_\Gamma G(x-y)\operatorname{div}_\Gamma j(x)\operatorname{div}_\Gamma j^t(y)d\gamma(x)d\gamma(y) \qquad (5.6.8)\\[3mm]
\quad = -\displaystyle\int_\Gamma \left(E^{\mathrm{inc}}\cdot j^t\right)d\gamma, \qquad \text{for any } j^t \text{ tangent to } \Gamma.
\end{cases}
$$

This variational formulation is called **the Rumsey principle**.

We replace the variational formulation (5.6.8), by a saddle-point formulation which is more suitable to prove existence and uniqueness, using the following abstract theorem.

Theorem 5.6.1 (Fredholm alternative) *Let V and W be two Hilbert spaces. Let $a(u,v)$ be a bilinear form continuous on $V \times V$ which satisfies*

$$\Re\left[a(u,\bar{u})\right] \geq \alpha\,\|u\|_V^2 - c\,\|u\|_H^2, \quad \alpha > 0, \quad \forall u \in V, \qquad (5.6.9)$$

where H is a Hilbert space containing V. Let $b(q,v)$ be a bilinear form continuous on $W \times V$ which satisfies:

$$\sup_{\|u\|_V=1} |b(q,u)| \geq \beta\,\|q\|_W - c\,\|q\|_L, \quad \beta > 0, \quad \forall q \in W, \qquad (5.6.10)$$

where L is a Hilbert space containing W.

Consider the following variational problem, with $g_1 \in V^*$ and $g_2 \in W^*$:

$$
\begin{cases}
a(u,v) + b(p,v) = (g_1, v), & \forall v \in V, \\[2mm]
b(q,u) = (g_2, q), & \forall q \in W.
\end{cases}
\qquad (5.6.11)
$$

Denote by V_0 the kernel of the bilinear form b in V, i.e.,

$$V_0 = \{u \in V,\ b(q,u) = 0\ ,\quad \forall q \in W\}.$$

Suppose that **the injection from V_0 into H is compact** *and that* **the injection from W into L is compact.** *Suppose that there exists an element $u_{g_2} \in V$ such that*

$$b(q, u_{g_2}) = (g_2, q), \quad \forall q \in V.$$

Then, the variational problem (5.6.11) satisfies **the Fredholm alternative**, *i.e.,*

- *either it admits a unique solution in $V \times W$,*
- *or it admits a finite dimension kernel, and a solution defined up to any element in this kernel, when the right-hand side (g_1, g_2) vanishes on any element in this kernel.*

Proof

Let us first prove that this kernel has finite dimension. From (5.6.9) and (5.6.10), any element in this kernel satisfies

$$\|u\|_V^2 \leq \frac{1}{\alpha} \Re\left[a(u, \overline{u})\right] + c\|u\|_H^2 \leq c\|u\|_H^2 + c\frac{1}{\beta}\|p\|_L \|u\|_V,$$

or else

$$\|u\|_V + \|p\|_W \leq c\|u\|_H + c\|p\|_L. \tag{5.6.12}$$

Moreover, $u \in V_0$. From the compact injection hypothesis, it follows that this linear space has a compact unit ball and thus has finite dimension. From the property of g_2, it is equivalent to consider the case where g_2 is zero.

We then introduce, using the Galerkin technique, an approximated solution for our problem, denoted $(u_\varepsilon, p_\varepsilon)$, in the quotient space $(V \times W)/N$. We continue to denote this new space by $V \times W$. Notice that u_ε belongs to V_0. Estimates of this solution are

$$\|u_\varepsilon\|_V + \|p_\varepsilon\|_W \leq c\|u_\varepsilon\|_H + c\|p_\varepsilon\|_L + c\|g_1\|_{V*} + c\|g_2\|_{W*}. \tag{5.6.13}$$

We proceed by contradiction to show the convergence of this sequence:

– either the sequence $\|u_\varepsilon\|_H + \|p_\varepsilon\|_L$ is bounded, and then from (5.6.13), we deduce that $(u_\varepsilon, p_\varepsilon)$ is bounded in $V \times W$. We then have a weak convergence in $V \times W$.

– or the sequence $\|u_\varepsilon\|_H + \|p_\varepsilon\|_L$ is not bounded, and then $\widetilde{u}_\varepsilon = u_\varepsilon/(\|u_\varepsilon\|_H + \|p_\varepsilon\|_L)$ is bounded in V, while $\widetilde{p}_\varepsilon = p_\varepsilon/(\|u_\varepsilon\|_H + \|p_\varepsilon\|_L)$ is bounded in W, and their weak limits $(\widetilde{u}, \widetilde{p})$ are such that

$$\begin{cases} a(\widetilde{u}, v) + b(\widetilde{p}, v) = 0, & \forall v \in V, \\ \\ b(q, \widetilde{u}) = 0, & \forall q \in W. \end{cases} \tag{5.6.14}$$

Thus $(\widetilde{u}, \widetilde{p})$ are in the kernel of this problem.

As this kernel is reduced to zero in the quotient space, the quantity $\|\widetilde{u}_\varepsilon\|_H + \|\widetilde{p}_\varepsilon\|_L$, whose value is 1, cannot tend to zero. But, from the compact injections of V_0 into H and of W into L, there exists a subsequence which converges strongly. This is contradictory. ∎

We **decompose the vector** j in the form

$$j = g + \overrightarrow{\mathrm{curl}}_\Gamma p. \tag{5.6.15}$$

The **saddle-point formulation** associated with (5.6.8) is

$$
\begin{cases}
-k^2 \displaystyle\int_\Gamma \int_\Gamma G(x-y)\left(g(x)\cdot g^t(y)\right) d\gamma(x)d\gamma(y) \\[2mm]
\quad + \displaystyle\int_\Gamma \int_\Gamma G(x-y)\operatorname{div}_\Gamma g(x)\operatorname{div}_\Gamma g^t(y)d\gamma(x)d\gamma(y) \\[2mm]
\qquad - k^2 \displaystyle\int_\Gamma \int_\Gamma G(x-y)\left(\overrightarrow{\operatorname{curl}}_\Gamma p(x)\cdot g^t(y)\right)d\gamma(x)d\gamma(y) \quad (5.6.16) \\[2mm]
\qquad\qquad = -i\omega\varepsilon_0 \displaystyle\int_\Gamma \left(E^{\mathrm{inc}}\cdot g^t\right)d\gamma, \qquad \forall\, g^t, \\[2mm]
-k^2 \displaystyle\int_\Gamma \int_\Gamma G(x-y)\left(g(x)\cdot \overrightarrow{\operatorname{curl}}_\Gamma q(y)\right)d\gamma(x)d\gamma(y) = 0, \quad \forall\, q.
\end{cases}
$$

It is of the form (5.6.11) where a is given by

$$
\begin{cases}
a\,(g,g^t) = \displaystyle\int_\Gamma \int_\Gamma G(x-y)\operatorname{div}_\Gamma g(x)\operatorname{div}_\Gamma g^t(y)d\gamma(x)d\gamma(y) \\[2mm]
\qquad - k^2 \displaystyle\int_\Gamma \int_\Gamma G(x-y)\left(g(x)\cdot g^t(y)\right)d\gamma(x)d\gamma(y),
\end{cases}
\quad (5.6.17)
$$

and b is given by

$$
b(g,q) = -k^2 \int_\Gamma \int_\Gamma G(x-y)\left(g(x)\cdot \overrightarrow{\operatorname{curl}}_\Gamma q(y)\right)d\gamma(x)d\gamma(y). \quad (5.6.18)
$$

Theorem 5.6.2 *If k^2 is not an eigenvalue of the interior problem (5.6.2) and if $E_T^{\mathrm{inc}} \in H_{\mathrm{curl}}^{-1/2}(\Gamma)$, the integral equation (5.6.16) (and thus (5.6.8)) admits a unique solution such that $g \in TH^{1/2}(\Gamma)$ and $p \in H^{1/2}(\Gamma)$, and thus $j \in H_{\mathrm{div}}^{-1/2}(\Gamma)$. If, moreover $\operatorname{curl}_\Gamma E_T^{\mathrm{inc}} \in H^{1/2}(\Gamma)$, $p \in H^{3/2}(\Gamma)$ and $j \in TH^{1/2}(\Gamma)$. If, moreover $\operatorname{div}_\Gamma E_T^{\mathrm{inc}} \in H^{-1/2}(\Gamma)$ (i.e. $E_T^{\mathrm{inc}} \in TH^{1/2}(\Gamma)$), $\operatorname{div}_\Gamma j \in H^{1/2}(\Gamma)$, and in particular if $E_T^{\mathrm{inc}} \in H_{\mathrm{curl}}^{1/2}(\Gamma)$, $j \in H_{\mathrm{div}}^{1/2}(\Gamma)$.*

Proof

We use Theorem 5.6.1. From the properties of the single layer potential, it follows that

$$
\Re a(g,\bar g) \geq \alpha\,\|\operatorname{div}_\Gamma g\|_{H^{-1/2}(\Gamma)}^2 - c\,k^2\,\|g\|_{TH^{-1/2}(\Gamma)}^2, \qquad \alpha > 0. \quad (5.6.19)
$$

Thus, the right choice for the space V is

$$
V = \left\{g \in TH^{-1/2}(\Gamma), \operatorname{div}_\Gamma g \in H^{-1/2}(\Gamma)\right\} = H_{\mathrm{div}}^{-1/2}(\Gamma).
$$

The space H is $TH^{-1/2}(\Gamma)$. We thus have

$$
\begin{cases}
\left| b\left(\overrightarrow{\operatorname{curl}}_\Gamma q, \overrightarrow{\operatorname{curl}}_\Gamma \bar q\right)\right| \\[2mm]
\quad \geq \beta k^2 \left\|\overrightarrow{\operatorname{curl}}_\Gamma q\right\|_{TH^{-1/2}(\Gamma)}^2 - c\,k^4 \left\|\overrightarrow{\operatorname{curl}}_\Gamma q\right\|_{TH^{-3/2}(\Gamma)}^2 \quad (5.6.20) \\[2mm]
\quad \geq k^2 \left(\beta\,\|q\|_{H^{1/2}(\Gamma)}^2 - c\,k^2\,\|q\|_{H^{-1/2}(\Gamma)}^2\right), \qquad \beta > 0.
\end{cases}
$$

The right choice for the space W is thus $H^{1/2}(\Gamma)$, and the space L is $H^{-1/2}(\Gamma)$.

It remains to exhibit the space V_0 which is defined by the identity

$$\int_\Gamma \int_\Gamma G(x-y)\left(g(x) \cdot \overrightarrow{\mathrm{curl}}_\Gamma q(y)\right) d\gamma(x)d\gamma(y) = 0, \quad \forall q \in H^{1/2}(\Gamma). \quad (5.6.21)$$

From the Stokes formula on the surface Γ, this is equivalent to

$$\mathrm{curl}_\Gamma \int_\Gamma G(x-y)g(x)d\gamma(x) = 0.$$

Using (5.5.32), we can rewrite this equation in the form

$$\begin{cases} \displaystyle\int_\Gamma G(x-y)\,\mathrm{curl}_\Gamma\, g(x)d\gamma(x) \\[4mm] \displaystyle\quad = \int_\Gamma \left(\left((n_y - n_x)\wedge \nabla_y G(x-y)\right) \cdot g(x)\right)d\gamma(x). \end{cases} \quad (5.6.22)$$

From the results of Chapter 4, the operator on the right-hand side in (5.6.22) is pseudo-homogeneous of class 1. Thus, the right-hand side belongs to $H^{1/2}(\Gamma)$. The operator on the left-hand side is also pseudo-homogeneous of class 1 and invertible, from the properties of the single layer potential, since k^2 is not an eigenvalue of the interior problem (5.6.2). It follows that $\mathrm{curl}_\Gamma\, g \in H^{-1/2}(\Gamma)$. Thus, the space V_0 is

$$V_0 = \left\{g \in H_{\mathrm{div}}^{-1/2}(\Gamma), \quad \mathrm{curl}_\Gamma\, g \in H^{-1/2}(\Gamma)\right\}.$$

We have proved that this space is $TH^{1/2}(\Gamma)$ which is compactly imbedded in $TH^{-1/2}(\Gamma)$.

Theorem 5.6.1 shows that the Fredholm alternative holds. The data E_T^{inc} has to be in the space V^* which, from Lemma 5.4.1, is $H_{\mathrm{curl}}^{-1/2}(\Gamma)$.

The part $\overrightarrow{\mathrm{curl}}p$ is a solution to the equation

$$\begin{cases} \displaystyle\int_\Gamma\int_\Gamma G(x-y)\left(\overrightarrow{\mathrm{curl}}_\Gamma p(x) \cdot \overrightarrow{\mathrm{curl}}_\Gamma q(y)\right) d\gamma(x)d\gamma(y) \\[4mm] \displaystyle\quad = \frac{i}{\omega\mu_0}\int_\Gamma \mathrm{curl}_\Gamma\, E_T^{\mathrm{inc}}q\,d\gamma \\[4mm] \displaystyle\quad - \int_\Gamma\int_\Gamma G(x-y)\left(g(x) \cdot \overrightarrow{\mathrm{curl}}_\Gamma q(y)\right) d\gamma(x)d\gamma(y). \end{cases} \quad (5.6.23)$$

The operator on the left-hand side is invertible and of class -1. The second part of the right-hand side belongs to $H^{1/2}(\Gamma)$. Thus, the solution belongs to $H^{3/2}(\Gamma)$ when $\mathrm{curl}_\Gamma\, E_T^{\mathrm{inc}} \in H^{1/2}(\Gamma)$ and then $\overrightarrow{\mathrm{curl}}q \in TH^{1/2}(\Gamma)$ and $j \in TH^{1/2}(\Gamma)$. If, moreover $E_T^{\mathrm{inc}} \in TH^{1/2}(\Gamma)$, equation (5.6.16) shows that $\mathrm{div}_\Gamma\, g \in H^{1/2}(\Gamma)$, which is $\mathrm{div}_\Gamma\, j$.

Uniqueness is deduced from the integral representation theorem, which shows the equivalence between the integral equation and the set of the interior and exterior problems, which are known to have a unique solution.

Comments

The two above variational formulations are used to build a **numerical approximation** of the electric surface currents in the scattering problem for a perfect conductor. In both cases, it is convenient to use a finite element on the boundary, which generates a subspace of $H(\mathrm{div})$ for the variables j and g and a subspace of $H^{1/2}(\Gamma)$ for the variable p. The first works on this subject are due to A. Bendali [46, 47] for the Rumsey formulation (5.6.8).

We introduce a **third formulation** that is quite close to the second one. It was first introduced by A. De La Bourdonnaye [68] in order to analyse the spectrum of the associated operator. We look for the vector j in the form of its **Helmholtz decomposition.** If we suppose that the surface Γ is simply connected, it has the expression

$$j = \overrightarrow{\mathrm{curl}}_\Gamma p + \nabla_\Gamma q. \tag{5.6.24}$$

It follows from the variational equation (5.6.8) that

$$
\left\{
\begin{aligned}
&\int_\Gamma \int_\Gamma G(x-y) \Delta_\Gamma q(x) \Delta_\Gamma q^t(y) d\gamma(x) d\gamma(y) \\
&\quad - k^2 \int_\Gamma \int_\Gamma G(x-y) \left(\nabla_\Gamma q(x) \cdot \nabla_\Gamma q^t(y) \right) d\gamma(x) d\gamma(y) \\
&\quad - k^2 \int_\Gamma \int_\Gamma G(x-y) \left(\overrightarrow{\mathrm{curl}}_\Gamma p(x) \cdot \nabla_\Gamma q^t(y) \right) d\gamma(x) d\gamma(y) \\
&\quad = i\omega\varepsilon \int_\Gamma \mathrm{div}_\Gamma\, E_T^{\mathrm{inc}} q^t d\gamma, \\
&\qquad \text{for any } q^t \text{ in } H^{3/2}(\Gamma).
\end{aligned}
\right. \tag{5.6.25}
$$

$$
\left\{
\begin{aligned}
&-k^2 \int_\Gamma \int_\Gamma G(x-y) \left(\overrightarrow{\mathrm{curl}}_\Gamma p(x) \cdot \overrightarrow{\mathrm{curl}}_\Gamma p^t(y) \right) d\gamma(x) d\gamma(y) \\
&\quad - k^2 \int_\Gamma \int_\Gamma G(x-y) \left(\nabla_\Gamma q(x) \cdot \overrightarrow{\mathrm{curl}}_\Gamma p^t(y) \right) d\gamma(x) d\gamma(y) \\
&\quad = -i\omega\varepsilon \int_\Gamma \mathrm{curl}_\Gamma\, E_T^{\mathrm{inc}} p^t d\gamma, \\
&\qquad \text{for any } p^t \text{ in } H^{1/2}(\Gamma).
\end{aligned}
\right. \tag{5.6.26}
$$

These two equations are coupled. The principal part of the first equation is

$$\int_\Gamma \int_\Gamma G(x-y) \Delta_\Gamma q(x) \Delta_\Gamma q^t(y) d\gamma(x) d\gamma(y),$$

which, up to a compact operator, is coercive in $H^{3/2}(\Gamma)$, due to the properties of the single layer potential. The principal part of the second equation is

$$\int_\Gamma \int_\Gamma G(x-y) \left(\overrightarrow{\mathrm{curl}}_\Gamma p(x) \cdot \overrightarrow{\mathrm{curl}}_\Gamma p^t(y) \right) d\gamma(x) d\gamma(y),$$

which, up to a compact operator, is coercive in $H^{1/2}(\Gamma)$, due to the properties of the double layer potential.

We prove now that the coupling terms are also compact operators. We change the corresponding integrals using

$$\begin{cases} \int_\Gamma \int_\Gamma G(x-y) \left(\nabla_\Gamma q(x) \cdot \overrightarrow{\mathrm{curl}}_\Gamma p(y) \right) d\gamma(x) d\gamma(y) \\[2mm] \qquad = \int_\Gamma p(y) \, \mathrm{curl}_\Gamma \int_\Gamma G(x-y) \nabla_\Gamma q(x) d\gamma(x) d\gamma(y). \end{cases} \tag{5.6.27}$$

From identity (5.5.32), we obtain

$$\begin{cases} \int_\Gamma \int_\Gamma G(x-y) \left(\nabla_\Gamma q(x) \cdot \overrightarrow{\mathrm{curl}}_\Gamma p(y) \right) d\gamma(x) d\gamma(y) \\[2mm] \qquad = \int_\Gamma \int_\Gamma (((n_y - n_x) \wedge \nabla_y G(x-y)) \cdot \nabla_\Gamma q(x)) \, p(y) d\gamma(x) d\gamma(y), \end{cases} \tag{5.6.28}$$

which, using the results on pseudo-homogeneous kernels, shows that this bilinear form is bounded by

$$\begin{cases} \left| \int_\Gamma \int_\Gamma G(x-y) \left(\nabla_\Gamma q(x) \cdot \overrightarrow{\mathrm{curl}}_\Gamma p(y) \right) d\gamma(x) d\gamma(y) \right| \\[2mm] \qquad\qquad \leq c \|q\|_{H^{1/2}(\Gamma)} \|p\|_{H^{-1/2}(\Gamma)} \, . \end{cases} \tag{5.6.29}$$

It follows that the system of equations (5.6.25) (5.6.26) is of Fredholm type. It is also equivalent to equation (5.6.8), and thus it satisfies the existence and uniqueness properties, when k^2 is not an eigenvalue of the interior problem (5.6.2).

This gives a new proof of Theorem 5.6.1. This new proof allows us to extend the regularity properties to the spaces H^s for any real s, and to show that equation (5.6.8) admits a solution j in $H^s_{\mathrm{div}}(\Gamma)$ when E^{inc}_T belongs to $H^s_{\mathrm{curl}}(\Gamma)$. Moreover, it can be used to exhibit solutions in spaces of the type $W^{s,p}$.

5.6.2   The zero frequency limit

We can use the previous results, and in particular the formulation (5.6.25) and (5.6.26), to study the limit of the perfect conductor when the pulsation ω tends to zero. When the surface Γ is not simply connected, the vector j

admits the decomposition

$$\begin{cases} j = \overrightarrow{\mathrm{curl}}_\Gamma p + \nabla_\Gamma q + \alpha, & \text{where the fields } \alpha \\ \text{are such that:} \quad \Delta_\Gamma \alpha = 0. \end{cases} \qquad (5.6.30)$$

The linear space $\mathcal{N}$ of the fields α such that $\Delta_\Gamma \alpha = 0$, is finite dimensional. In this case, equations (5.6.25) and (5.6.26) can be written in the form

$$\begin{cases} \Delta_\Gamma \int_\Gamma G(x-y)\Delta_\Gamma q(x) d\gamma(x) \\[2mm] + k^2 \, \mathrm{div}_\Gamma \int_\Gamma G(x-y)\nabla_\Gamma q(x) d\gamma(x) \\[2mm] + k^2 \, \mathrm{div}_\Gamma \int_\Gamma G(x-y)\left(\overrightarrow{\mathrm{curl}}_\Gamma p + \alpha\right) d\gamma(x) = ik\sqrt{\dfrac{\varepsilon}{\mu}}\,\mathrm{div}_\Gamma\, E_T^{\mathrm{inc}}, \\[2mm] - k^2 \, \mathrm{curl}_\Gamma \int_\Gamma G(x-y)\left(\overrightarrow{\mathrm{curl}}_\Gamma p + \alpha\right) d\gamma(x) \qquad\qquad (5.6.31) \\[2mm] - k^2 \, \mathrm{curl}_\Gamma \int_\Gamma G(x-y)\nabla_\Gamma q(x) d\gamma(x) = -ik\sqrt{\dfrac{\varepsilon}{\mu}}\,\mathrm{curl}_\Gamma\, E_T^{\mathrm{inc}}, \\[2mm] - k^2 \int_\Gamma \int_\Gamma G(x-y)\left(\left(\overrightarrow{\mathrm{curl}}_\Gamma p + \nabla_\Gamma q + \alpha\right)\cdot \alpha^t\right) d\gamma(x) d\gamma(y) \\[2mm] \qquad = -ik\sqrt{\dfrac{\varepsilon}{\mu}}\int_\Gamma \left(E_T^{\mathrm{inc}}\cdot \alpha^t\right) d\gamma, \quad \text{for any } \alpha^t \text{ in } \mathcal{N}. \end{cases}$$

In order to obtain the limit equations, we expand this system in the small parameter k. We start from a Taylor expansion of the fundamental solution in the form

$$G(r) = \frac{1}{4\pi}\left(\frac{1}{r} + ik - k^2 r + \cdots\right). \qquad (5.6.32)$$

We seek for an expansion of the scalar unknowns p, q, α using the a priori ansatz

$$p = \frac{1}{k}\, p_{-1} + p_0 + \cdots, \qquad (5.6.33)$$

$$q = k\, q_1 + k^2\, q_2 + \cdots, \qquad (5.6.34)$$

$$\alpha = \frac{1}{k}\alpha_{-1} + \alpha_0 + \cdots. \qquad (5.6.35)$$

Carrying all these expansions of the kernel and the unknowns into equation (5.6.31), the terms in factor of k^2 provide the limit equation for the new unknowns which is

$$
\begin{cases}
\Delta_\Gamma \displaystyle\int_\Gamma \frac{1}{4\pi |x-y|} \Delta_\Gamma q_1 d\gamma(x) \\[2mm]
\qquad + \mathrm{div}_\Gamma \displaystyle\int_\Gamma \frac{1}{4\pi |x-y|} \left(\mathrm{curl}_\Gamma\, p_{-1} + \alpha_{-1} \right) d\gamma(x) \\[2mm]
\qquad\qquad = \lim_{k\to 0} i\sqrt{\frac{\varepsilon}{\mu}}\, \mathrm{div}_\Gamma\, E_T^{\mathrm{inc}}, \\[3mm]
- \mathrm{curl}_\Gamma \displaystyle\int_\Gamma \frac{1}{4\pi |x-y|} \left(\overrightarrow{\mathrm{curl}}_\Gamma p_{-1} + \alpha_{-1} \right) d\gamma(x) \\[2mm]
\qquad\qquad = \lim_{k\to 0} -i\sqrt{\frac{\varepsilon}{\mu}}\, \mathrm{curl}_\Gamma\, E_T^{\mathrm{inc}}, \\[3mm]
- \displaystyle\int_\Gamma\int_\Gamma \frac{1}{4\pi |x-y|} \left(\left(\overrightarrow{\mathrm{curl}}_\Gamma p_{-1} + \alpha_{-1} \right) \cdot \alpha^t \right) d\gamma(x) d\gamma(y) \\[2mm]
\qquad\qquad = \lim_{k\to 0} -i\sqrt{\frac{\varepsilon}{\mu}} \displaystyle\int_\Gamma \left(E_T^{\mathrm{inc}} \cdot \alpha^t \right) d\gamma, \quad \text{for any } \alpha^t \text{ in } \mathcal{N}.
\end{cases}
\tag{5.6.36}
$$

We made several choices and some of them may seem arbitrary. Let us explain some of them.

Most often, the data E^{inc} is the trace on the surface Γ of a field which satisfies the Maxwell equations. Then

$$
\mathrm{curl}_\Gamma\, E_T^{\mathrm{inc}} = \left(n \cdot \mathrm{curl}\, E^{\mathrm{inc}} \right) = ik\sqrt{\frac{\mu}{\varepsilon}} \left(H^{\mathrm{inc}} \cdot n \right),
\tag{5.6.37}
$$

which shows that the limit of $\mathrm{curl}_\Gamma\, E_T^{\mathrm{inc}}$ is zero. Moreover, we have

$$
0 = \mathrm{div}\, E^{\mathrm{inc}} = \mathrm{div}_\Gamma\, E_T^{\mathrm{inc}} + 2H \left(E^{\mathrm{inc}} \cdot n \right) + \frac{\partial}{\partial n} \left(E^{\mathrm{inc}} \cdot n \right),
\tag{5.6.38}
$$

which shows that the limit of $\mathrm{div}_\Gamma\, E_T^{\mathrm{inc}}$ is also that of $-2H \left(E^{\mathrm{inc}} \cdot n \right)$ which is zero. It is also the case of the third term.

The terms on the right-hand side of (5.6.31) are of order k, which implies that the first non-zero term in the unknown q is the coefficient of k, which implies first terms of order $1/k$ for p and α. Two slightly different situations appear, depending on whether the set $\mathcal{N}$ is 0 or not.

In the second case, the terms p_{-1} and α_{-1} are non-zero and satisfy the equations

$$
\begin{cases}
\frac{1}{4\pi} \displaystyle\int_\Gamma\int_\Gamma \frac{1}{|x-y|} \left(\left(\overrightarrow{\mathrm{curl}}_\Gamma p_{-1} + \alpha_{-1} \right)(x) \right. \\[2mm]
\qquad\qquad \left. \cdot\, \overrightarrow{\mathrm{curl}}_\Gamma p_{-1}^t(y) \right) d\gamma(x) d\gamma(y) = 0, \\[3mm]
\frac{1}{4\pi} \displaystyle\int_\Gamma\int_\Gamma \frac{1}{|x-y|} \left(\left(\mathrm{curl}_\Gamma\, p_{-1} + \alpha_{-1} \right)(x) \cdot \alpha^t(y) \right) d\gamma(x) d\gamma(y) \\[2mm]
\qquad\qquad = \displaystyle\int_\Gamma \lim_{k\to 0} i\sqrt{\frac{\varepsilon}{\mu}} \left(E_T^{\mathrm{inc}} \cdot \alpha^t \right) d\gamma.
\end{cases}
\tag{5.6.39}
$$

In the first case, the dominant term is p_0, solution of

$$\begin{cases} \frac{1}{4\pi} \int_\Gamma \int_\Gamma \frac{1}{|x-y|} \left(\overrightarrow{\mathrm{curl}}_\Gamma p_0(x) \cdot \overrightarrow{\mathrm{curl}}_\Gamma p_0^t(y) \right) d\gamma(x) d\gamma(y) \\ \qquad = i \int_\Gamma \sqrt{\frac{\varepsilon}{\mu}} \lim_{k \to 0} \left(\frac{1}{k} \mathrm{curl}_\Gamma E_T^{\mathrm{inc}} \right) p_0 d\gamma. \end{cases} \qquad (5.6.40)$$

Besides, both cases can be put into the form

$$\begin{cases} \frac{1}{4\pi} \int_\Gamma \int_\Gamma \frac{1}{|x-y|} \left(j(x) \cdot j^t(y) \right) d\gamma(x) d\gamma(y) = \int_\Gamma \left(g \cdot j^t \right) d\gamma, \\ \qquad \text{for any } j^t \text{ such that } \mathrm{div}_\Gamma \, j^t = 0. \end{cases} \qquad (5.6.41)$$

The right-hand sides g are different in the two cases.

We have already studied this equation which admits a unique solution j in $H_{\mathrm{div}}^{-1/2}(\Gamma)$ (p_{-1} or $p_0 \in H^{1/2}(\Gamma)$). From the properties of equation (5.6.31), it is possible to show in both cases the convergence with k of the solution j towards the solution associated with the limit problem. It is also possible to compute the next terms. Starting from the order 2, the coupling with the terms in q reappears.

5.6.3 The dielectric case

We will study here the case where the domain Ω_i is a dielectric medium characterized by the constants ε_i, μ_i while the domain Ω_e is a dielectric medium characterized by the constants ε_e, μ_e. Suppose that the field is created by an incident wave, a solution of the Maxwell equations, and that the exterior unknown is the scattered field. Then, the equations are

$$\begin{cases} \mathrm{curl}\, E_i - i\omega\mu_i H_i = 0, \\ \mathrm{curl}\, H_i + i\omega\varepsilon_i E_i = 0, \end{cases} \qquad \text{in } \Omega_i; \qquad (5.6.42)$$

$$\begin{cases} \mathrm{curl}\, E_e - i\omega\mu_e H_e = 0, \\ \mathrm{curl}\, H_e + i\omega\varepsilon_e E_e = 0, \\ \left| \sqrt{\varepsilon_e}\, E_e - \sqrt{\mu_e} H_e \wedge n \right| \leq \frac{c}{r^2}; \end{cases} \qquad \text{in } \Omega_e; \qquad (5.6.43)$$

$$\begin{cases} E_i \wedge n - \left(E_e + E^{\mathrm{inc}} \right) \wedge n = 0, \qquad \text{on } \Gamma, \\ H_i \wedge n - \left(H_e + H^{\mathrm{inc}} \right) \wedge n = 0, \qquad \text{on } \Gamma. \end{cases} \qquad (5.6.44)$$

In contrast with the case of the perfect conductor, there is not a unique integral representation for this problem. We have chosen to introduce here the most commonly used representation, which consists in extending by zero to the complementary domain, the solution in each of the considered domains.

Let us introduce the notation

$$\begin{cases} k_i = \omega\sqrt{\varepsilon_i\mu_i}, \\[2mm] G_i(r) = \frac{1}{4\pi r}e^{ik_ir}. \end{cases} \tag{5.6.45}$$

$$\begin{cases} k_e = \omega\sqrt{\varepsilon_e\mu_e}, \\[2mm] G_e(r) = \frac{1}{4\pi r}e^{ik_er}. \end{cases} \tag{5.6.46}$$

Let us then introduce the electric and magnetic interior currents

$$\begin{cases} j_i = H_i \wedge n, \\[2mm] m_i = E_i \wedge n, \end{cases} \tag{5.6.47}$$

as the electric and magnetic exterior currents

$$\begin{cases} j_e = -H_e \wedge n, \\[2mm] m_e = -E_e \wedge n. \end{cases} \tag{5.6.48}$$

The unit normal is oriented toward the exterior of Ω_i and thus toward the interior of Ω_e.

From Theorem 5.5.1, the electric and magnetic **interior field** has the representation

$$\begin{cases} E_i(y) = i\omega\mu_i \displaystyle\int_\Gamma G_i(x-y)j_i(x)d\gamma(x) \\[4mm] \qquad + \frac{i}{\omega\varepsilon_i}\nabla\displaystyle\int_\Gamma G_i(x-y)\,\mathrm{div}_\Gamma\, j_i(x)d\gamma(x) \\[4mm] \qquad\qquad + \mathrm{curl}\displaystyle\int_\Gamma G_i(x-y)m_i(x)d\gamma(x); \end{cases} \tag{5.6.49}$$

$$\begin{cases} H_i(y) = -i\omega\varepsilon_i \displaystyle\int_\Gamma G_i(x-y)m_i(x)d\gamma(x) \\[4mm] \qquad - \frac{i}{\omega\mu_i}\nabla\displaystyle\int_\Gamma G_i(x-y)\,\mathrm{div}_\Gamma\, m_i(x)d\gamma(x) \\[4mm] \qquad\qquad + \mathrm{curl}\displaystyle\int_\Gamma G_i(x-y)j_i(x)d\gamma(x); \end{cases} \tag{5.6.50}$$

while the **exterior field** has the representation

$$\begin{cases} E_e(y) = i\omega\mu_e \int_\Gamma G_e(x-y)j_e(x)d\gamma(x) \\ \qquad + \frac{i}{\omega\varepsilon_e}\nabla\int_\Gamma G_e(x-y)\,\mathrm{div}_\Gamma\, j_e(x)d\gamma(x) \qquad (5.6.51) \\ \qquad + \mathrm{curl}\int_\Gamma G_e(x-y)m_e(x)d\gamma(x); \end{cases}$$

$$\begin{cases} H_e(y) = -i\omega\varepsilon_e \int_\Gamma G_e(x-y)m_e(x)d\gamma(x) \\ \qquad - \frac{i}{\omega\mu_e}\nabla\int_\Gamma G_e(x-y)\,\mathrm{div}_\Gamma\, m_e(x)d\gamma(x) \qquad (5.6.52) \\ \qquad + \mathrm{curl}\int_\Gamma G_e(x-y)j_e(x)d\gamma(x). \end{cases}$$

In order to obtain the associated integral equations for the corresponding unknowns j and m, it is sufficient to express the continuity of the tangential traces, i.e., the jump equations (5.6.44). We use the boundary values of the representation given by the expressions (5.5.8) (5.5.9) (5.5.12) and (5.5.13). We obtain on one side

$$\begin{cases} m_i + m_e = E^{\mathrm{inc}} \wedge n, \\ \\ j_i + j_e = H^{\mathrm{inc}} \wedge n; \end{cases} \qquad (5.6.53)$$

and on the other side

$$\begin{cases} \frac{1}{2}E^{\mathrm{inc}} \wedge n = \int_\Gamma \Big[\frac{\partial G_i}{\partial n_y}(x-y)m_i(x) - \frac{\partial G_e}{\partial n_y}(x-y)m_e(x) \\ \qquad - \nabla_y G_i(x-y)\,(m_i(x)\cdot(n_y - n_x)) \\ \qquad + \nabla_y G_e(x-y)\,(m_e(x)\cdot(n_y - n_x))\Big]d\gamma(x) \\ \quad + i\omega \int_\Gamma \Big[\mu_i G_i(x-y)\,(j_i(x)\wedge n_y) \\ \qquad - \mu_e G_e(x-y)\,(j_e(x)\wedge n_y)\Big]d\gamma(x) \qquad (5.6.54) \\ \quad + \frac{i}{\omega}\int_\Gamma \Big[\frac{1}{\varepsilon_i}\Big[(\nabla_y G_i(x-y)\wedge(n_y - n_x))\,\mathrm{div}_\Gamma\, j_i(x) \\ \qquad + G_i(x-y)\overrightarrow{\mathrm{curl}}_\Gamma\,\mathrm{div}_\Gamma\, j_i(x)\Big] \\ \qquad - \frac{1}{\varepsilon_e}\Big[(\nabla_y G_e(x-y)\wedge(n_y - n_x))\,\mathrm{div}_\Gamma\, j_e(x) \\ \qquad + G_e(x-y)\overrightarrow{\mathrm{curl}}_\Gamma\,\mathrm{div}_\Gamma\, j_e(x)\Big]\Big]d\gamma(x); \end{cases}$$

$$\begin{cases}
\frac{1}{2}H^{\text{inc}} \wedge n = \int_\Gamma \Big[\frac{\partial G_i}{\partial n_y}(x - y)j_i(x) - \frac{\partial G_e}{\partial n_y}(x - y)j_e(x) \\
\qquad\qquad - \nabla_y G_i(x - y)\,(j_i(x) \cdot (n_y - n_x)) \\
\qquad\qquad + \nabla_y G_e(x - y)\,(j_e(x) \cdot (n_y - n_x)) \Big] d\gamma(x) \\
\qquad - i\omega \int_\Gamma \Big[\varepsilon_i G_i(x - y)\,(m_i(x) \wedge n_y) \\
\qquad\qquad - \varepsilon_e G_e(x - y)\,(m_e(x) \wedge n_y) \Big] d\gamma(x) \qquad (5.6.55) \\
\qquad - \frac{i}{\omega} \int_\Gamma \Big[\frac{1}{\mu_i}\Big[(\nabla_y G_i(x-y) \wedge (n_y - n_x))\,\mathrm{div}_\Gamma\, m_i(x) \\
\qquad\qquad + G_i(x - y)\overrightarrow{\mathrm{curl}}_\Gamma\,\mathrm{div}_\Gamma\, m_i(x) \Big] \\
\qquad\qquad - \frac{1}{\mu_e}\Big[(\nabla_y G_e(x-y) \wedge (n_y - n_x))\,\mathrm{div}_\Gamma\, m_e(x) \\
\qquad\qquad + G_e(x - y)\overrightarrow{\mathrm{curl}}_\Gamma\,\mathrm{div}_\Gamma\, m_e(x) \Big] \Big] d\gamma(x).
\end{cases}$$

This is a system of four equations for four unknowns.

We associate with these equations a variational formulation, based on the use of Stokes formulas on the surface Γ and analogous to the one introduced for the perfect conductor. Each domain contributes to a variational principle through the quantities

$$\int_\Gamma \big[(E \cdot j^t) - (H \cdot m^t) \big]\, d\gamma \qquad (j^t \text{ and } m^t \text{ are test currents}),$$

called the Rumsey integrals.

From (5.5.8) and (5.5.9), we obtain in the interior domain:

$$\begin{cases}
\int_\Gamma (E_i \cdot j^t)\, d\gamma \\
\quad = i\omega\mu_i \int_\Gamma \int_\Gamma G_i(x - y)\,(j_i(x) \cdot j^t(y))\, d\gamma(x)d\gamma(y) \\
\quad - \frac{i}{\omega\varepsilon_i} \int_\Gamma \int_\Gamma G_i(x - y)\,\mathrm{div}_\Gamma\, j_i(x)\,\mathrm{div}_\Gamma\, j^t(y) d\gamma(x)d\gamma(y) \\
\qquad\qquad\qquad\qquad\qquad\qquad\qquad\qquad (5.6.56) \\
\quad + \frac{1}{2} \int_\Gamma (m_i \cdot (j^t \wedge n))\, d\gamma \\
\quad + \int_\Gamma \int_\Gamma \Big[\frac{\partial G_i}{\partial n_y}(x - y)\,(m_i(x) \cdot (j^t(y) \wedge n_y)) \\
\qquad\qquad + \big(\overrightarrow{\mathrm{curl}}_{\Gamma_y} G_i(x-y) \cdot j^t(y)\big)(m_i(x) \cdot (n_y - n_x)) \Big] d\gamma(x)d\gamma(y);
\end{cases}$$

$$
\begin{cases}
\displaystyle \int_\Gamma \left(H_i \cdot m^t\right) d\gamma \\[2mm]
\displaystyle = -i\omega\varepsilon_i \int_\Gamma \int_\Gamma G_i(x-y)\left(m_i(x)\cdot m^t(y)\right) d\gamma(x)d\gamma(y) \\[2mm]
\displaystyle \quad + \frac{i}{\omega\mu_i}\int_\Gamma\int_\Gamma G_i(x-y)\,\mathrm{div}_\Gamma\,m_i(x)\,\mathrm{div}_\Gamma\,m^t(y)d\gamma(x)d\gamma(y) \\[2mm]
\displaystyle \quad + \frac12 \int_\Gamma \left(j_i\cdot(m^t\wedge n)\right)d\gamma \\[2mm]
\displaystyle \quad + \int_\Gamma\int_\Gamma\Big[\frac{\partial G_i}{\partial n_y}(x-y)\left(j_i(x)\cdot(m^t(y)\wedge n_y)\right) \\[2mm]
\displaystyle \qquad + \left(\overrightarrow{\mathrm{curl}}_{\Gamma_y}G_i(x-y)\cdot m^t(y)\right)\!\left(j_i(x)\cdot(n_y-n_x)\right)\Big]d\gamma(x)d\gamma(y).
\end{cases}
\tag{5.6.57}
$$

From (5.5.12) and (5.5.13), we obtain in the exterior domain:

$$
\begin{cases}
\displaystyle \int_\Gamma \left(E_e\cdot j^t\right)d\gamma \\[2mm]
\displaystyle = i\omega\mu_e \int_\Gamma\int_\Gamma G_e(x-y)\left(j_e(x)\cdot j^t(y)\right)d\gamma(x)d\gamma(y) \\[2mm]
\displaystyle \quad - \frac{i}{\omega\tilde{\varepsilon}_e}\int_\Gamma\int_\Gamma G_e(x-y)\,\mathrm{div}_\Gamma\,j_e(x)\,\mathrm{div}_\Gamma\,j^t(y)d\gamma(x)d\gamma(y) \\[2mm]
\displaystyle \quad - \frac12 \int_\Gamma \left(m_e\cdot(j^t\wedge n)\right)d\gamma \\[2mm]
\displaystyle \quad + \int_\Gamma\int_\Gamma\Big[\frac{\partial G_e}{\partial n_y}(x-y)\left(m_e(x)\cdot(j^t(y)\wedge n_y)\right) \\[2mm]
\displaystyle \qquad + \left(\overrightarrow{\mathrm{curl}}_{\Gamma_y}G_e(x-y)\cdot j^t(y)\right)\!\left(m_e(x)\cdot(n_y-n_x)\right)\Big]d\gamma(x)d\gamma(y);
\end{cases}
\tag{5.6.58}
$$

$$
\begin{cases}
\displaystyle \int_\Gamma \left(H_e\cdot m^t\right)d\gamma \\[2mm]
\displaystyle = -i\omega\varepsilon_e \int_\Gamma\int_\Gamma G_e(x-y)\left(m_e(x)\cdot m^t(y)\right)d\gamma(x)d\gamma(y) \\[2mm]
\displaystyle \quad + \frac{i}{\omega\mu_e}\int_\Gamma\int_\Gamma G_e(x-y)\,\mathrm{div}_\Gamma\,m_e(x)\,\mathrm{div}_\Gamma\,m^t(y)d\gamma(x)d\gamma(y) \\[2mm]
\displaystyle \quad - \frac12 \int_\Gamma \left(j_e\cdot(m^t\wedge n)\right)d\gamma \\[2mm]
\displaystyle \quad + \int_\Gamma\int_\Gamma\Big[\frac{\partial G_e}{\partial n_y}(x-y)\left(j_e(x)\cdot(m^t(y)\wedge n_y)\right) \\[2mm]
\displaystyle \qquad + \left(\overrightarrow{\mathrm{curl}}_{\Gamma_y}G_e(x-y)\cdot m^t(y)\right)\!\left(j_e(x)\cdot(n_y-n_x)\right)\Big]d\gamma(x)d\gamma(y).
\end{cases}
\tag{5.6.59}
$$

The variational formulation is obtained by the difference of these two previous contributions. In addition to the equations (5.6.53), we obtain the two equations

$$
\begin{cases}
i\omega \int_\Gamma \int_\Gamma ((\mu_i G_i(x-y)j_i(x) \\
\qquad - \mu_e G_e(x-y)j_e(x)) \cdot j^t(y))\, d\gamma(x)d\gamma(y) \\
\quad - \dfrac{i}{\omega} \int_\Gamma \int_\Gamma \Big(\dfrac{1}{\varepsilon_i} G_i(x-y)\,\mathrm{div}_\Gamma\, j_i(x) \\
\qquad - \dfrac{1}{\varepsilon_e} G_e(x-y)\,\mathrm{div}_\Gamma\, j_e(x)\Big)\,\mathrm{div}_\Gamma\, j^t(y)d\gamma(x)d\gamma(y) \\
\quad + \int_\Gamma \int_\Gamma \Big(\Big(\dfrac{\partial G_i}{\partial n_y}(x-y)m_i(x) \\
\qquad - \dfrac{\partial G_e}{\partial n_y}(x-y)m_e(x)\Big) \cdot \big(j^t(y) \wedge n_y\big)\Big)\,d\gamma(x)d\gamma(y) \\
\quad + \int_\Gamma \int_\Gamma \Big(\Big(\big(m_i(x)\cdot(n_y-n_x)\big)\overrightarrow{\mathrm{curl}}_{\Gamma_y} G_i(x-y) \\
\qquad -\big(m_e(x)\cdot(n_y-n_x)\big)\overrightarrow{\mathrm{curl}}_{\Gamma_y} G_e(x-y)\Big) \cdot j^t(y)\Big)d\gamma(x)d\gamma(y) \\
\qquad\qquad = \dfrac{1}{2}\int_\Gamma (E^{\mathrm{inc}}\cdot j^t)\, d\gamma, \quad \text{for any } j^t;
\end{cases}
\tag{5.6.60}
$$

$$
\begin{cases}
-i\omega \int_\Gamma \int_\Gamma ((\varepsilon_i G_i(x-y)m_i(x) \\
\qquad - \varepsilon_e G_e(x-y)m_e(x)) \cdot m^t(y))\, d\gamma(x)d\gamma(y) \\
\quad + \dfrac{i}{\omega} \int_\Gamma \int_\Gamma \Big(\dfrac{1}{\mu_i} G_i(x-y)\,\mathrm{div}_\Gamma\, m_i(x) \\
\qquad - \dfrac{1}{\mu_e} G_e(x-y)\,\mathrm{div}_\Gamma\, m_e(x)\Big)\,\mathrm{div}_\Gamma\, m^t(y)d\gamma(x)d\gamma(y) \\
\quad + \int_\Gamma \int_\Gamma \Big(\Big(\dfrac{\partial G_i}{\partial n_y}(x-y)j_i(x) \\
\qquad - \dfrac{\partial G_e}{\partial n_y}(x-y)j_e(x)\Big) \cdot (m^t(y) \wedge n_y)\, d\gamma(x)d\gamma(y)\Big) \\
\quad + \int_\Gamma \int_\Gamma \Big(\Big(\big(j_i(x)\cdot(n_y-n_x)\big)\overrightarrow{\mathrm{curl}}_{\Gamma_y} G_i(x-y) \\
\qquad -\big(j_e(x)\cdot(n_y-n_x)\big)\overrightarrow{\mathrm{curl}}_{\Gamma_y} G_e(x-y)\Big) \cdot m^t(y)\Big)d\gamma(x)d\gamma(y) \\
\qquad\qquad = \dfrac{1}{2}\int_\Gamma (H^{\mathrm{inc}}\cdot m^t)\, d\gamma, \quad \text{for any } m^t.
\end{cases}
\tag{5.6.61}
$$

In order to have more compact expressions, we introduce the following notations:

We set

$$s_i \left(j, j^t \right) = \int_\Gamma \int_\Gamma G_i(x - y) \left(j(x) \cdot j^t(y) \right) d\gamma(x) d\gamma(y); \qquad (5.6.62)$$

$$s_e \left(j, j^t \right) = \int_\Gamma \int_\Gamma G_e(x - y) \left(j(x) \cdot j^t(y) \right) d\gamma(x) d\gamma(y); \qquad (5.6.63)$$

$$t_i \left(j, j^t \right) = \int_\Gamma \int_\Gamma G_i(x - y) \operatorname{div}_\Gamma j(x) \operatorname{div}_\Gamma j^t(y) d\gamma(x) d\gamma(y); \qquad (5.6.64)$$

$$t_e \left(j, j^t \right) = \int_\Gamma \int_\Gamma G_e(x - y) \operatorname{div}_\Gamma j(x) \operatorname{div}_\Gamma j^t(y) d\gamma(x) d\gamma(y); \qquad (5.6.65)$$

$$\left\{ \begin{aligned} r_i \left(m, j^t \right) &= \int_\Gamma \int_\Gamma \left[\frac{\partial G_i}{\partial n_y}(x - y) \left(m(x) \cdot \left(j^t(y) \wedge n_y \right) \right) \right. \\ &\quad \left. + \left(\overrightarrow{\operatorname{curl}}_{\Gamma_y} G_i(x-y) \cdot j^t(y) \right) \left(m(x) \cdot \left(n_y - n_x \right) \right) \right] d\gamma(x) d\gamma(y); \end{aligned} \right. \qquad (5.6.66)$$

$$\left\{ \begin{aligned} r_e \left(m, j^t \right) &= \int_\Gamma \int_\Gamma \left[\frac{\partial G_e}{\partial n_y}(x - y) \left(m(x) \cdot \left(j^t(y) \wedge n_y \right) \right) \right. \\ &\quad \left. + \left(\overrightarrow{\operatorname{curl}}_{\Gamma_y} G_e(x-y) \cdot j^t(y) \right) \left(m(x) \cdot \left(n_y - n_x \right) \right) \right] d\gamma(x) d\gamma(y). \end{aligned} \right. \qquad (5.6.67)$$

From (5.5.30), we deduce that the terms r_i and r_e take the form

$$\left\{ \begin{aligned} r_i(m, j^t) &= \int_\Gamma \int_\Gamma \left[\frac{\partial G_i}{\partial n_y}(x - y) \left(m(x) \cdot \left(j^t(y) \wedge n_y \right) \right) \right. \\ &\quad \left. - \left(\nabla_y G_i(x-y) \cdot \left(j^t(y) \wedge n_y \right) \right) \left(m(x) \cdot \left(n_y - n_x \right) \right) \right] d\gamma(x) d\gamma(x) \\ &= \int_\Gamma \int_\Gamma \left(\nabla_y G_i(x - y) \cdot \left(m(x) \wedge j^t(y) \right) \right) d\gamma(x) d\gamma(y). \end{aligned} \right. \qquad (5.6.68)$$

$$\left\{ \begin{aligned} r_e(m, j^t) &= \int_\Gamma \int_\Gamma \left[\frac{\partial G_e}{\partial n_y}(x - y) \left(m(x) \cdot \left(j^t(y) \wedge n_y \right) \right) \right. \\ &\quad \left. - \left(\nabla_y G_e(x-y) \cdot \left(j^t(y) \wedge n_y \right) \right) \left(m(x) \cdot \left(n_y - n_x \right) \right) \right] d\gamma(x) d\gamma(x) \\ &= \int_\Gamma \int_\Gamma \left(\nabla_y G_e(x - y) \cdot \left(m(x) \wedge j^t(y) \right) \right) d\gamma(x) d\gamma(y). \end{aligned} \right. \qquad (5.6.69)$$

The variables j_e and m_e are herafter denoted by j and m. The systems (5.6.60) and (5.6.61) have the form

$$
\begin{cases}
i\omega \left(\mu_i s_i + \mu_e s_e \right) \left(j, j^t \right) - \frac{i}{\omega} \left(\frac{1}{\varepsilon_i} t_i + \frac{1}{\varepsilon_e} t_e \right) \left(j, j^t \right) \\[2mm]
\quad + \left(r_i + r_e \right) \left(m, j^t \right) \\[2mm]
\quad = -\frac{1}{2} \int_\Gamma \left(E^{\text{inc}} \cdot j^t \right) d\gamma + i\omega \mu_i s_i \left(H^{\text{inc}} \wedge n, j^t \right) \\[2mm]
\quad - \frac{i}{\omega} \frac{1}{\varepsilon_i} t_i \left(H^{\text{inc}} \wedge n, j^t \right) + r_i \left(E^{\text{inc}} \wedge n, j^t \right), \quad \forall j^t.
\end{cases}
\tag{5.6.70}
$$

$$
\begin{cases}
-i\omega \left(\varepsilon_i s_i + \varepsilon_e s_e \right) \left(m, m^t \right) + \frac{i}{\omega} \left(\frac{1}{\mu_i} t_i + \frac{1}{\mu_e} t_e \right) \left(m, m^t \right) \\[2mm]
\quad + \left(r_i + r_e \right) \left(j, m^t \right) \\[2mm]
\quad = -\frac{1}{2} \int_\Gamma \left(H^{\text{inc}} \cdot m^t \right) d\gamma - i\omega \varepsilon_i s_i \left(E^{\text{inc}} \wedge n, m^t \right) \\[2mm]
\quad + \frac{i}{\omega} \frac{1}{\mu_i} t_i \left(E^{\text{inc}} \wedge n, m^t \right) + r_i \left(H^{\text{inc}} \wedge n, m^t \right), \quad \forall m^t.
\end{cases}
\tag{5.6.71}
$$

In order to prove that the system of equations (5.6.70) (5.6.71) is of Fredholm type, we introduce the Helmholtz decomposition of the unknowns j and m. We write

$$
\begin{cases}
j = \overrightarrow{\text{curl}}_\Gamma p + \nabla_\Gamma q, & j^t = \overrightarrow{\text{curl}}_\Gamma p^t + \nabla_\Gamma q^t, \\[2mm]
m = \overrightarrow{\text{curl}}_\Gamma v + \nabla_\Gamma w, & m^t = \overrightarrow{\text{curl}}_\Gamma v^t + \nabla_\Gamma w^t.
\end{cases}
\tag{5.6.72}
$$

We rewrite the system of equations (5.6.68) and (5.6.69) with the help of the new scalar unknowns p, q, v, and w.

The equation for q is

$$
\begin{cases}
-\frac{i}{\omega} \int_\Gamma \int_\Gamma \left(\frac{1}{\varepsilon_i} G_i + \frac{1}{\varepsilon_e} G_e \right) \Delta_\Gamma q(x) \Delta_\Gamma q^t(y) d\gamma(x) d\gamma(y) \\[3mm]
+ i\omega \int_\Gamma \int_\Gamma \left(\mu_i G_i + \mu_e G_e \right) \left(\nabla_\Gamma q(x) \cdot \nabla_\Gamma q^t(y) \right) d\gamma(x) d\gamma(y) \\[3mm]
+ i\omega \int_\Gamma \int_\Gamma \left(\mu_i G_i + \mu_e G_e \right) \left(\overrightarrow{\text{curl}}_\Gamma p(x) \cdot \nabla_\Gamma q^t(y) \right) d\gamma(x) d\gamma(y) \\[3mm]
+ \left(r_i + r_e \right) \left(\overrightarrow{\text{curl}}_\Gamma v + \nabla_\Gamma w, \nabla_\Gamma q^t \right) = -\frac{1}{2} \int_\Gamma \left(E^{\text{inc}} \cdot \nabla_\Gamma q^t \right) d\gamma \\[3mm]
+ i\omega \mu_i s_i \left(H^{\text{inc}} \wedge n, \nabla_\Gamma q^t \right) + \frac{i}{\omega \varepsilon_i} t_i \left(H^{\text{inc}} \wedge n, \nabla_\Gamma q^t \right) \\[3mm]
+ r_i \left(E^{\text{inc}} \wedge n, \nabla_\Gamma q^t \right), \quad \forall q^t.
\end{cases}
\tag{5.6.73}
$$

Its dominant part is

$$
-\frac{i}{4\pi\omega} \int_\Gamma \int_\Gamma \left(\frac{1}{\varepsilon_i} + \frac{1}{\varepsilon_e} \right) \frac{1}{|x - y|} \Delta_\Gamma q(x) \Delta_\Gamma q^t(y) d\gamma(x) d\gamma(y), \tag{5.6.74}
$$

which is coercive in the space $H^{3/2}(\Gamma)$, and thus the unknown q belongs to $H^{3/2}(\Gamma)$. The second term of equation (5.6.73) and all the terms associated with the difference between $1/(4\pi|x-y|)$ and G_i or G_e, correspond to operators which are compact compared to the principal part.

The equation for w is

$$
\left\{
\begin{aligned}
&\frac{i}{\omega}\int_\Gamma\int_\Gamma\left(\frac{1}{\mu_i}G_i+\frac{1}{\mu_e}G_e\right)\Delta_\Gamma w(x)\Delta_\Gamma w^t(y)d\gamma(x)d\gamma(y)\\[2mm]
&-i\omega\int_\Gamma\int_\Gamma(\varepsilon_iG_i+\varepsilon_eG_e)\left(\nabla_\Gamma w(x)\cdot\nabla_\Gamma w^t(y)\right)d\gamma(x)d\gamma(y)\\[2mm]
&-i\omega\int_\Gamma\int_\Gamma(\varepsilon_iG_i+\varepsilon_eG_e)\left(\overrightarrow{\mathrm{curl}}_\Gamma v(x)\cdot\nabla_\Gamma w^t(y)\right)d\gamma(x)d\gamma(y)\quad(5.6.75)\\[2mm]
&+(r_i+r_e)\left(\overrightarrow{\mathrm{curl}}_\Gamma p+\nabla_\Gamma q,\nabla_\Gamma w^t\right)=-\frac{1}{2}\int_\Gamma\left(H^{\mathrm{inc}}\cdot\nabla_\Gamma w^t\right)d\gamma\\[2mm]
&-i\omega\varepsilon_is_i\left(E^{\mathrm{inc}}\wedge n,\nabla_\Gamma w^t\right)-\frac{i}{\omega\mu_i}t_i\left(E^{\mathrm{inc}}\wedge n,\nabla_\Gamma w^t\right)\\[2mm]
&+r_i\left(H^{\mathrm{inc}}\wedge n,\nabla_\Gamma w^t\right),\quad\forall w^t.
\end{aligned}
\right.
$$

Using the symmetry of the Maxwell system with respect to the unknowns E and H, the equation for w has the same structure. Thus, the unknown w belongs to $H^{3/2}(\Gamma)$.

The equation for p is

$$
\left\{
\begin{aligned}
&i\omega\int_\Gamma\int_\Gamma(\mu_iG_i+\mu_eG_e)\left(\overrightarrow{\mathrm{curl}}_\Gamma p(x)\cdot\overrightarrow{\mathrm{curl}}_\Gamma p^t(y)\right)d\gamma(x)d\gamma(y)\\[2mm]
&+i\omega\int_\Gamma\int_\Gamma(\mu_iG_i+\mu_eG_e)\left(\nabla_\Gamma q(x)\cdot\overrightarrow{\mathrm{curl}}_\Gamma p^t(y)\right)d\gamma(x)d\gamma(y)\\[2mm]
&+(r_i+r_e)\left(\overrightarrow{\mathrm{curl}}_\Gamma v+\nabla_\Gamma w,\overrightarrow{\mathrm{curl}}_\Gamma p^t\right)\qquad\qquad(5.6.76)\\[2mm]
&=-\frac{1}{2}\int_\Gamma\left(E^{\mathrm{inc}}\cdot\overrightarrow{\mathrm{curl}}_\Gamma p^t\right)d\gamma+i\omega\mu_is_i\left(H^{\mathrm{inc}}\wedge n,\overrightarrow{\mathrm{curl}}_\Gamma p^t\right)\\[2mm]
&+r_i\left(E^{\mathrm{inc}}\wedge n,\overrightarrow{\mathrm{curl}}_\Gamma p^t\right),\quad\forall p^t.
\end{aligned}
\right.
$$

Its dominant part is

$$
\frac{i\omega}{4\pi}\int_\Gamma\int_\Gamma(\mu_i+\mu_e)\frac{1}{|x-y|}\left(\overrightarrow{\mathrm{curl}}_\Gamma p(x)\cdot\overrightarrow{\mathrm{curl}}_\Gamma p^t(y)\right)d\gamma(x)d\gamma(y),
$$

which is coercive in the space $H^{1/2}(\Gamma)/R$, and thus the unknown p belongs to $H^{1/2}(\Gamma)$. All the terms associated to the difference between $1/(4\pi|x-y|)$ and G_i or G_e in equation (5.6.76), correspond to operators which are compact compared to the principal part.

The equation for v is

$$
\begin{cases}
-i\omega \int_\Gamma \int_\Gamma (\varepsilon_i G_i + \varepsilon_e G_e)\left(\overrightarrow{\mathrm{curl}}_\Gamma v(x) \cdot \overrightarrow{\mathrm{curl}}_\Gamma v^t(y)\right) d\gamma(x)d\gamma(y) \\
\quad - i\omega \int_\Gamma \int_\Gamma (\varepsilon_i G_i + \varepsilon_e G_e)\left(\nabla_\Gamma w(x) \cdot \overrightarrow{\mathrm{curl}}_\Gamma v^t(y)\right) d\gamma(x)d\gamma(y) \\
\quad + (r_i + r_e)\left(\overrightarrow{\mathrm{curl}}_\Gamma p + \nabla_\Gamma q, \overrightarrow{\mathrm{curl}}_\Gamma v^t\right) \qquad\qquad (5.6.77) \\
= -\tfrac{1}{2}\int_\Gamma \left(H^{\mathrm{inc}} \cdot \overrightarrow{\mathrm{curl}}_\Gamma v^t\right) d\gamma - i\omega\varepsilon_i s_i \left(E^{\mathrm{inc}} \wedge n, \overrightarrow{\mathrm{curl}}_\Gamma v^t\right) \\
\quad + r_i \left(H^{\mathrm{inc}} \wedge n, \overrightarrow{\mathrm{curl}}_\Gamma v^t\right), \quad \forall v^t.
\end{cases}
$$

From the argument of symmetry of the Maxwell system, the unknown v belongs to $H^{1/2}(\Gamma)$ and the associated equation has the same structure as (5.6.76).

We now examine the extra diagonal terms. Terms of the following form appear:

$$
\int_\Gamma \int_\Gamma G(x - y)\left(\overrightarrow{\mathrm{curl}}_\Gamma p(x) \cdot \nabla_\Gamma q(y)\right) d\gamma(x)d\gamma(y), \qquad (5.6.78)
$$

where the operator G is pseudo-homogeneous of class 1 and even. Such a term can be written

$$
\begin{cases}
\int_\Gamma \int_\Gamma G(x - y)\left(\overrightarrow{\mathrm{curl}}_\Gamma p(x) \cdot \nabla_\Gamma q(y)\right) d\gamma(x)d\gamma(y) \\
= \int_\Gamma \int_\Gamma (n_x \cdot \mathrm{curl}_x\,(G(x-y)\nabla_\Gamma q(y)))\, p(x)d\gamma(x)d\gamma(y) \\
= \int_\Gamma \int_\Gamma (n_x \cdot (\nabla_x G(x-y) \wedge \nabla_\Gamma q(y)))\, p(x)d\gamma(x)d\gamma(y) \\
= \int_\Gamma \int_\Gamma ((n_x - n_y) \wedge \nabla_x G(x-y) \cdot \nabla_\Gamma q(y))\, p(x)d\gamma(x)\gamma(y) \qquad (5.6.79) \\
\quad + \int_\Gamma \int_\Gamma (n_y \wedge (\nabla_x G + \nabla_y G) \cdot \nabla_\Gamma q(y))\, p(x)d\gamma(x)d\gamma(y) \\
\quad + \int_\Gamma \int_\Gamma \left(\overrightarrow{\mathrm{curl}}_{\Gamma_y} G(x-y) \cdot \nabla_\Gamma q(y)\right) p(x)d\gamma(x)\gamma(y) \\
= \int_\Gamma \int_\Gamma (((n_x - n_y) \wedge \nabla_x G(x-y)) \cdot \nabla_\Gamma q(y))\, p(x)d\gamma(x)d\gamma(y).
\end{cases}
$$

Thus, it satisfies the estimate

$$
\begin{cases}
\left| \int_\Gamma \int_\Gamma G(x - y)\left(\overrightarrow{\mathrm{curl}}_\Gamma p(x) \cdot \nabla_\Gamma q(y)\right) d\gamma(x)d\gamma(y)\right| \\
\qquad\qquad\qquad\qquad (5.6.80) \\
\leq c\,\|q\|_{H^{1/2}(\Gamma)}\,\|p\|_{H^{-1/2}(\Gamma)},
\end{cases}
$$

which proves that this extra diagonal operator is compact compared to the main diagonal part. The analogous term for v and w has the same structure.

It remains to examine the terms associated to the bilinear forms $r_i + r_e$. They are of several types. The first is

$$\int_\Gamma \int_\Gamma \left(\frac{\partial G_i}{\partial n_y} + \frac{\partial G_e}{\partial n_y} \right) \left(\overrightarrow{\mathrm{curl}}_\Gamma v \cdot \nabla_\Gamma p^t \right) d\gamma(x) d\gamma(y), \tag{5.6.81}$$

$$\int_\Gamma \int_\Gamma \left(\frac{\partial G_i}{\partial n_y} + \frac{\partial G_e}{\partial n_y} \right) \left(\overrightarrow{\mathrm{curl}}_\Gamma v \cdot \overrightarrow{\mathrm{curl}}_\Gamma q^t \right) d\gamma(x) d\gamma(y), \tag{5.6.82}$$

$$\int_\Gamma \int_\Gamma \left(\frac{\partial G_i}{\partial n_y} + \frac{\partial G_e}{\partial n_y} \right) \left(\nabla_\Gamma w \cdot \nabla_\Gamma p^t \right) d\gamma(x) d\gamma(y), \tag{5.6.83}$$

$$\int_\Gamma \int_\Gamma \left(\frac{\partial G_i}{\partial n_y} + \frac{\partial G_e}{\partial n_y} \right) \left(\nabla_\Gamma w \cdot \overrightarrow{\mathrm{curl}}_\Gamma q^t \right) d\gamma(x) d\gamma(y). \tag{5.6.84}$$

Term (5.6.84) is bounded by $c \|q^t\|_{H^{1/2}(\Gamma)} \|w\|_{H^{1/2}(\Gamma)}$ and is thus compact. Term (5.6.83) is bounded by $c \|w\|_{H^{1/2}(\Gamma)} \|p^t\|_{H^{1/2}(\Gamma)}$ and is also compact. Term (5.6.82) is bounded by $c \|v\|_{H^{1/2}(\Gamma)} \|q^t\|_{H^{1/2}(\Gamma)}$ and is also compact. The only term that is not obviously compact is (5.6.81).

The second type of term is

$$\int_\Gamma \int_\Gamma \left(\nabla_y (G_i + G_e) \cdot \overrightarrow{\mathrm{curl}}_\Gamma q^t \right) \left(\overrightarrow{\mathrm{curl}}_\Gamma v \cdot (n_y - n_x) \right) d\gamma(x) d\gamma(y), \tag{5.6.85}$$

$$\int_\Gamma \int_\Gamma \left(\nabla_y (G_i + G_e) \cdot \nabla_\Gamma p^t \right) \left(\overrightarrow{\mathrm{curl}}_\Gamma v \cdot (n_y - n_x) \right) d\gamma(x) d\gamma(y), \tag{5.6.86}$$

$$\int_\Gamma \int_\Gamma \left(\nabla_y (G_i + G_e) \cdot \overrightarrow{\mathrm{curl}}_\Gamma q^t \right) \left(\nabla_\Gamma w \cdot (n_y - n_x) \right) d\gamma(x) d\gamma(y), \tag{5.6.87}$$

$$\int_\Gamma \int_\Gamma \left(\nabla_y (G_i + G_e) \cdot \nabla_\Gamma p^t \right) \left(\nabla_\Gamma w \cdot (n_y - n_x) \right) d\gamma(x) d\gamma(y). \tag{5.6.88}$$

The only term which is not obviously compact is (5.6.86). It links the unknowns v and p.

Thus, to sum up the situation, the system has, up to compact operators, a block diagonal structure, where the unknowns q and w are linked together and to the unknowns p and v through compact operators.

Most of the extra diagonal terms correspond to compact operators. It remains to examine the terms (5.6.81) and (5.6.86), which couple the unknowns p and v, to conclude that the system is in fact block diagonally dominant. These two terms have integrable kernels and can be grouped in the form

$$\int_\Gamma \left(\overrightarrow{\text{curl}}_\Gamma p^t(y) \cdot \int_\Gamma \nabla_y (G_i + G_e) \wedge \overrightarrow{\text{curl}}_\Gamma v(x) d\gamma(x) \right) d\gamma(y). \qquad (5.6.89)$$

Let us denote by H the potential

$$H(y) = \text{curl} \int_\Gamma (G_i(x - y) + G_e(x - y)) \overrightarrow{\text{curl}}_\Gamma v(x) d\gamma(x). \qquad (5.6.90)$$

We have shown that this potential has a tangential discontinuity across the surface Γ. From the jump condition (5.5.29), it follows that

$$\begin{cases} \int_\Gamma \nabla_y (G_i(x - y) + G_e(x - y)) \wedge \overrightarrow{\text{curl}}_\Gamma v(x) d\gamma(x) \\ \\ = \lim_{\rho \to 0} \text{curl} \int_\Gamma (G_i(x - y - \rho n_y) - G_e(x - y - \rho n_y)) \overrightarrow{\text{curl}}_\Gamma v d\gamma. \end{cases} \qquad (5.6.91)$$

Using the Stokes formula, we modify each of these two contributions as

$$\begin{cases} \int_\Gamma \left(\overrightarrow{\text{curl}}_\Gamma p^t(y) \cdot \text{curl} \int_\Gamma G_i(x - y - \rho n_y) \overrightarrow{\text{curl}}_\Gamma v(x) d\gamma \right) d\gamma \\ \\ = \int_\Gamma p^t \left(n_y \cdot \text{curl curl} \int_\Gamma G_i(x - y - \rho n_y) \overrightarrow{\text{curl}}_\Gamma v d\gamma \right) d\gamma \\ \\ = \int_\Gamma p^t(y) \left(n_y \cdot \left[\nabla \text{div} \int_\Gamma G_i(x - y - \rho n_y) \overrightarrow{\text{curl}}_\Gamma v(x) d\gamma(x) \right. \right. \\ \\ \left. \left. - \Delta \int_\Gamma G_i(x - y - \rho n_y) \overrightarrow{\text{curl}}_\Gamma v(x) d\gamma(x) \right] \right) d\gamma(y). \end{cases} \qquad (5.6.92)$$

If $\mathcal{R}$ denotes the curvature tensor, we have

$$\begin{cases} \text{div} \int_\Gamma G_i(x - y - \rho n_y) \overrightarrow{\text{curl}}_\Gamma v(x) d\gamma(x) \\ \\ = \int_\Gamma \nabla_y G_i(x - y - \rho n_y)(I - \rho \mathcal{R}) \cdot \overrightarrow{\text{curl}} v(x) d\gamma(x), \end{cases} \qquad (5.6.93)$$

$$\begin{cases} \text{div} \int_\Gamma G_e(x - y + \rho n_y) \overrightarrow{\text{curl}}_\Gamma v(x) d\gamma(x) \\ \\ = \int_\Gamma \nabla_y G_e(x - y + \rho n_y)(I + \rho \mathcal{R}) \cdot \overrightarrow{\text{curl}} v(x) d\gamma(x). \end{cases} \qquad (5.6.94)$$

The terms where ρ appears correspond to operators with non-integrable kernels. Hopefully, they can be grouped by antisymmetry and their limit when $\rho \to 0$ is zero. That is, we would have

$$\begin{cases} \lim_{\rho \to 0} \int_\Gamma \left(\nabla_y \left(G_i(x - y - \rho n_y) \right. \right. \\ \qquad \left. \left. + G_e(x - y + \rho n_y) \right) \cdot \overrightarrow{\text{curl}}_\Gamma v(x) \right) d\gamma(x) \\ = - \lim_{\rho \to 0} \int_\Gamma \left(\nabla_x \left(G_i(x - y - \rho n_y) \right. \right. \\ \qquad \left. \left. + G_e(x - y + \rho n_y) \right) \cdot \overrightarrow{\text{curl}}_\Gamma v(x) \right) d\gamma(x) \\ = \lim_{\rho \to 0} \int_\Gamma \left(G_i(x - y - \rho n_y) \right. \\ \qquad \left. + G_e(x - y + \rho n_y) \right) \text{div}_\Gamma \overrightarrow{\text{curl}}_\Gamma v(x) d\gamma = 0, \end{cases} \tag{5.6.95}$$

$$\begin{cases} \Delta \int_\Gamma G_i(x - y - \rho n_y) \overrightarrow{\text{curl}}_\Gamma v(x) d\gamma(x) \\ = \int_\Gamma (\Delta G_i(x - y - \rho n_y)(I - \rho \mathcal{R})^2 \\ \qquad - \rho \nabla G_i(x - y - \rho n_y) \nabla \mathcal{R}) \overrightarrow{\text{curl}}_\Gamma v(x) d\gamma(x) \\ = -k_i^2 \int_\Gamma G_i(x - y - \rho n_y)(I - \rho \mathcal{R})^2 \overrightarrow{\text{curl}}_\Gamma v(x) d\gamma(x) \\ \qquad - \rho \int_\Gamma \nabla G_i(x - y - \rho n_y) \nabla \mathcal{R} \overrightarrow{\text{curl}}_\Gamma v(x) d\gamma(x). \end{cases} \tag{5.6.96}$$

Here, the contributions of the terms which are odd with respect to ρ, and have non-integrable kernels, cancel out by adding the two terms corresponding to G_i and G_e. Thus, we have proved that

$$\begin{cases} \int_\Gamma \int_\Gamma \left(\nabla_y (G_i + G_e) \cdot \left(\overrightarrow{\text{curl}}_\Gamma v(x) \wedge \overrightarrow{\text{curl}}_\Gamma p^t(y) \right) \right) d\gamma d\gamma \\ = \int_\Gamma \int_\Gamma (k_i^2 G_i + k_e^2 G_e) \, p^t(y) \left((n_y - n_x) \cdot \overrightarrow{\text{curl}}_\Gamma v(x) \right) d\gamma d\gamma. \end{cases} \tag{5.6.97}$$

Now, this term is bounded by

$$C \, \|v\|_{H^{-1/2}(\Gamma)} \, \|p^t\|_{H^{-1/2}(\Gamma)}$$

and is thus compact. Thus, we have proved that the system of equations (5.6.70) (5.6.71) is of Fredholm type.

Theorem 5.6.3 *The system of equations* (5.6.73) (5.6.75) (5.6.76) (5.6.77) *with unknowns* p, q, v, w, *admits a unique solution such that* p *and* v *belong to* $H^{1/2}(\Gamma)$, *whereas* q *and* w *belong to* $H^{3/2}(\Gamma)$.
The system of integral equations (5.6.53) (5.6.60) *and* (5.6.61) *admits a unique solution* j_i, j_e, m_i *and* m_e *which belong to* $H_{\text{div}}^{-1/2}(\Gamma)$, *if* $E_T^{\text{inc}} \in H_{\text{curl}}^{-1/2}(\Gamma)$ *and* $H_T^{\text{inc}} \in H_{\text{curl}}^{-1/2}(\Gamma)$.

Proof

The two integral formulations are equivalent. The system of equations (5.6.73) (5.6.75) (5.6.76) (5.6.77) with unknowns p, q, v, w being of Fredholm type, it is also the case for the system (5.6.53) (5.6.60) (5.6.61). Thus, the uniqueness of the solution will prove the theorem. The integral formulation was obtained using the Rumsey principle, and is thus equivalent to the equation

$$\int_{\Gamma} \left[((E_i - E_e) \cdot j^t) - ((H_i - H_e) \cdot m^t) \right] d\gamma = 0, \quad \forall j^t, \forall m^t. \quad (5.6.98)$$

The associated fields E and H, for a zero right-hand side, satisfy the Maxwell equations (5.6.42) (5.6.43) and the jump conditions

$$\begin{cases} [E \wedge n] = 0, \\ [H \wedge n] = 0. \end{cases} \quad (5.6.99)$$

From Theorem 5.4.6, it follows that they are zero. The currents j and m are the tangential traces of these fields and thus vanish. This proves uniqueness.

Remarks

– The above integral equations are pseudo-homogeneous operators of the type examined in Chapter 4. Thus, they satisfy the corresponding regularity properties in Hilbert spaces and L^p type spaces. We deduce similar regularity properties for the Maxwell equations. In particular, we find solutions such that p, $v \in H^s(\Gamma)$, q, $w \in H^{s+1}(\Gamma)$, i.e., such that j_i, j_e, m_i, $m_e \in H^s_{\text{div}}(\Gamma)$, for any $s \geq 3/2$ and for any real s using duality.

– The variational formulation (5.6.53) (5.6.60) (5.6.61) is an excellent setting for the introduction of an approximation, using mixed type finite elements similar to the ones used for the perfect conductor. The electric and the magnetic currents are approximated by finite element spaces, defined on the approximate surface Γ_h and included in $H(\text{div})$. Theorem (5.6.2) proves that the coupling terms between the electric and the magnetic currents are compact. Thus the proof of stability of this numerical technique and the estimation of the error estimates on the approximate solution are similar to the ones of the perfect conductor.

5.6.4   The infinite conductivity limit: The perfect conductor

We examine in this section the Maxwell problem in the case of an exterior domain with permittivity ε_e and permeability μ_e. The interior domain has permittivity ε_i and permeability μ_i, and is moreover conducting. Its conductivity constant is denoted by σ. We have seen that the associated harmonic problem has an equivalent permittivity $\widetilde{\varepsilon}$ given by

$$\widetilde{\varepsilon} = \varepsilon_i + i\frac{\sigma}{\omega}. \quad (5.6.100)$$

The system of equations is given by

$$\begin{cases} \operatorname{curl} E_i - i\omega\mu_i H_i = 0, \\ \operatorname{curl} H_i + i\omega\tilde{\varepsilon} E_i = 0, \end{cases} \quad \text{in } \Omega_i; \qquad (5.6.101)$$

$$\begin{cases} \operatorname{curl} E_e - i\omega\mu_e H_e = 0, \\ \operatorname{curl} H_e + i\omega\varepsilon_e E_e = 0, \qquad \text{in } \Omega_e; \\ \left| \sqrt{\varepsilon_e}\, E_e - \sqrt{\mu_e} H_e \wedge n \right| \le \frac{c}{r^2} \end{cases} \qquad (5.6.102)$$

$$\begin{cases} E_i \wedge n - \left(E_e + E^{\text{inc}} \right) \wedge n = 0, \quad \text{on } \Gamma; \\ H_i \wedge n - \left(H_e + H^{\text{inc}} \right) \wedge n = 0, \quad \text{on } \Gamma. \end{cases} \qquad (5.6.103)$$

We examine the limit of the unique solution of this problem when σ tends to infinity. The interior wave number k_i is then a complex number. We thus have

$$k_i = \omega\sqrt{\mu\tilde{\varepsilon}} \sim \sqrt{\frac{\omega\mu_i}{2}}\sqrt{\sigma}(1+i), \qquad (5.6.104)$$

and when σ tends to infinity, this interior wave number k_i also tends equally to infinity.

As a first step to understand this complicated limit process, we examine the properties of the **single layer and double layer potentials** associated to such complex wave numbers. We introduce two linked problems. In the first one, **the wave number is a pure imaginary number**. The associated fundamental solution is $e^{-k|x-y|}/(4\pi |x-y|)$ associated to the operator $-\Delta + k^2$ and then k is a real positive number.

The associated interior and exterior problems are

$$\begin{cases} -\Delta u + k^2 u = 0, \quad x \in \Omega_i \text{ or } \Omega_e; \\ u|_\Gamma = 0 \text{ or } \left. \frac{\partial u}{\partial n} \right|_\Gamma = g. \end{cases} \qquad (5.6.105)$$

They satisfy a coercivity inequality in the spaces $H^1(\Omega_i)$ and $H^1(\Omega_e)$ respectively. It is natural to use a norm depending on the wave number k in that case. We introduce the norms

$$\|u\|^2_{H^1_k(\Omega_i)} = \int_{\Omega_i} |\nabla u|^2\, dx + k^2 \int_{\Omega_i} |u|^2\, dx; \qquad (5.6.106)$$

$$\|u\|^2_{H^1_k(\Omega_e)} = \int_{\Omega_e} |\nabla u|^2\, dx + k^2 \int_{\Omega_e} |\nabla u|^2\, dx. \qquad (5.6.107)$$

Problems (5.6.105) are then coercive for these norms and the **coercivity constants are 1 and are thus independent** of k. We introduce also k

dependent norms for the associated trace spaces. In R^2, we define

$$\begin{cases} H_k^s(R^2) = \left\{ u \in L^2(R^2), \left(k^2 + |\xi|^2 \right)^{s/2} \hat{u} \in L^2(R^2) \right\}, \ s \text{ real}, \\ \|u\|_{H_k^s(R^2)}^2 = \left\| \left(k^2 + |\xi|^2 \right)^{s/2} \hat{u} \right\|_{L^2(R^2)}^2 . \end{cases} \tag{5.6.108}$$

Given a surface Γ and any real positive s, we define, using the charts introduced in Section 2.5.2, the following spaces and the associated norms:

$$\begin{cases} H_k^s(\Gamma) = \left\{ u \in L^2(\Gamma), \lambda_i u \circ \phi_i^{-1} \in H_k^1(R^2), \text{ for every }, \phi_i \right\} \\ \|u\|_{H_k^s(\Gamma)}^2 = \sum_{i=1}^{p} \left\| \lambda_i u \circ \phi_i^{-1} \right\|_{H_k^s(R^2)}^2 . \end{cases} \tag{5.6.109}$$

For a negative s, the space $H_k^s(\Gamma)$ is defined as the dual of $H_k^{-s}(\Gamma)$ with respect to the scalar product in $L^2(\Gamma)$.

Theorem 5.6.4 *The trace operator, which associates to u in $H_k^1(\Omega_i)$ or $H_k^1(\Omega_e)$ its trace on Γ, is continuous into $H_k^{1/2}(\Gamma)$ and it satisfies*

$$\|u\|_{H_k^{1/2}(\Gamma)} \leq c \|u\|_{H_k^1(\Omega_i)}, \tag{5.6.110}$$

$$\|u\|_{H_k^{1/2}(\Gamma)} \leq c \|u\|_{H_k^1(\Omega_e)}, \tag{5.6.111}$$

where the constants c are independent of k. There exists a lifting R which is a left inverse of the trace operator and satisfies

$$\|Rv_0\|_{H_k^1(\Omega_i)} \leq c \|v_0\|_{H_k^{1/2}(\Gamma)}, \tag{5.6.112}$$

$$\|Rv_0\|_{H_k^1(\Omega_e)} \leq c \|v_0\|_{H_k^{1/2}(\Gamma)}. \tag{5.6.113}$$

Proof
In the case of the halfspace, we modify the estimates of Lemma 2.5.2 and replace the lifting given by formula (2.5.63) and Lemma 2.5.3, by the expression

$$\hat{u}(\xi, z) = \hat{v}_0(\xi) e^{\left(k^2 + |\xi|^2 \right)^{1/2} z}. \tag{5.6.114}$$

The constants are 1 in this case. In the general case of a surface, the local charts do not depend on k and it is also the case of the continuity constants. ∎

We introduce the spaces $H_k^2(\Omega_i)$ and $H_k^2(\Omega_e)$:

$$\|u\|_{H_k^2(\Omega_i)}^2 = \left\| D^2 u \right\|_{L^2(\Omega_i)}^2 + k^2 \|\nabla u\|_{L^2(\Omega_i)}^2 + k^4 \|u\|_{L^2(\Omega_i)}^2, \tag{5.6.115}$$

$$\|u\|_{H_k^2(\Omega_e)}^2 = \left\| D^2 u \right\|_{L^2(\Omega_e)}^2 + k^2 \|\nabla u\|_{L^2(\Omega_e)}^2 + k^4 \|u\|_{L^2(\Omega_e)}^2. \tag{5.6.116}$$

In these spaces the two traces $u|_\Gamma$ and $\partial u/\partial n|_\Gamma$ are both well defined. In the same way as in the proof of the above lemma, we obtain

$$\|u\|_{H_k^{3/2}(\Gamma)} \leq c \|u\|_{H_k^2(\Omega_i)}, \tag{5.6.117}$$

$$\|u\|_{H_k^{3/2}(\Gamma)} \leq c \|u\|_{H_k^2(\Omega_r)}, \tag{5.6.118}$$

$$\left\|\frac{\partial u}{\partial n}\right\|_{H_k^{1/2}(\Gamma)} \leq c \|u\|_{H_k^2(\Omega_i)}, \tag{5.6.119}$$

$$\left\|\frac{\partial u}{\partial n}\right\|_{H_k^{1/2}(\Gamma)} \leq c \|u\|_{H_k^2(\Omega_r)}, \tag{5.6.120}$$

where the constants c are independent of k.

There exists a lifting R which is a left inverse of the trace operators and is continuous with constants independent of k. The tangential derivatives are continuous from $H_k^{s+1}(\Gamma)$ into $TH_k^s(\Gamma)$,

$$\|\nabla_\Gamma \varphi\|_{TH_k^s(\Gamma)} \leq c \|\varphi\|_{H_k^{s+1}(\Gamma)}. \tag{5.6.121}$$

The spaces $H_k^s(\Gamma)$ and the spaces $H^s(\Gamma)$ are the same, but have different norms. However, the equivalence constants between these norms depend on k. For $s > 0$,

$$\|u\|_{H^s(\Gamma)} \leq \|u\|_{H_k^s(\Gamma)} \leq k^s \|u\|_{H^s(\Gamma)}, \tag{5.6.122}$$

and for $s < 0$,

$$k^s \|u\|_{H^s(\Gamma)} \leq \|u\|_{H_k^s(\Gamma)} \leq \|u\|_{H^s(\Gamma)}. \tag{5.6.123}$$

The operator $-\Delta_\Gamma + k^2$ is an isomorphism from $H_k^{s+2}(\Gamma)$ onto $H_k^s(\Gamma)$ which norm is bounded independently of k.

Mimicking the proofs used in the case of the Laplacian, we can prove the following properties of the **single layer potential** given by

$$u(y) = \frac{1}{4\pi} \int_\Gamma \frac{e^{-k|x-y|}}{|x-y|} q(x) d\gamma(x). \tag{5.6.124}$$

Theorem 5.6.5 *The operator* (5.6.124) *is an isomorphism from* $H_k^{-1/2}(\Gamma)$ *onto* $H_k^{1/2}(\Gamma)$ *whose norm is bounded independently of* k. *Moreover it satisfies the coercivity inequality*

$$\left\{ \begin{array}{l} \dfrac{1}{4\pi} \int_\Gamma \int_\Gamma \dfrac{e^{-k|x-y|}}{|x-y|} q(x) q(y) d\gamma(x) d\gamma(y) \geq \alpha \|q\|^2_{H_k^{-1/2}(\Gamma)}, \\[2mm] \alpha > 0, \quad \text{where } \alpha \text{ is independent of } k. \end{array} \right. \tag{5.6.125}$$

Proof

It is based on the following remark: the quantity given by formula (5.6.125) is nothing but $\int_{R^3} \left(|\nabla u|^2 + k^2 |u|^2 \right) dx$, which is linked to the norm of the trace of u on Γ in $H_k^{1/2}(\Gamma)$. ∎

Let us consider the **double layer potential**

$$u(y) = -\frac{1}{4\pi} \int_\Gamma \frac{\partial}{\partial n_x} \left(\frac{e^{-k|x-y|}}{|x-y|} \right) \varphi(x) d\gamma(x). \tag{5.6.126}$$

It can be used to solve the Neumann problem through the introduction of the operator

$$N\varphi = \frac{1}{4\pi} \oint \frac{\partial^2}{\partial n_x \partial n_y} \left(\frac{e^{-k|x-y|}}{|x-y|} \right) \varphi(x) d\gamma(x), \tag{5.6.127}$$

and we then have

$$\begin{cases} \frac{1}{4\pi} \int_\Gamma \int_\Gamma \frac{e^{-k|x-y|}}{|x-y|} \left(\overrightarrow{\text{curl}}_\Gamma \varphi(x) \cdot \overrightarrow{\text{curl}}_\Gamma \varphi(y) \right) d\gamma(x) d\gamma(y) \\ + \frac{k^2}{4\pi} \int_\Gamma \int_\Gamma \frac{e^{-k|x-y|}}{|x-y|} \varphi(x)\varphi(y) \left(n_x \cdot n_y \right) d\gamma(x) d\gamma(y) \\ = \int_{\Omega_i} \left(|\nabla u|^2 + k^2 |u|^2 \right) dx + \int_{\Omega_e} \left(|\nabla u|^2 + k^2 |u|^2 \right) dx. \end{cases} \tag{5.6.128}$$

Thus, exactly as for Theorem 5.6.5, it follows that

$$\begin{cases} \frac{1}{4\pi} \int_\Gamma \int_\Gamma \frac{e^{-k|x-y|}}{|x-y|} \left(\overrightarrow{\text{curl}}_\Gamma \varphi(x) \cdot \overrightarrow{\text{curl}}_\Gamma \varphi(y) \right) d\gamma(x) d\gamma(y) \\ + \frac{k^2}{4\pi} \int_\Gamma \int_\Gamma \frac{e^{-k|x-y|}}{|x-y|} \varphi(x)\varphi(y) \left(n_x \cdot n_y \right) d\gamma(x) d\gamma(y) \\ \geq \alpha \, \|\varphi\|^2_{H_k^{1/2}(\Gamma)}, \quad \alpha > 0 \text{ independent of } k, \end{cases} \tag{5.6.129}$$

which proves that the operator N is an isomorphism from $H_k^{1/2}(\Gamma)$ onto $H_k^{-1/2}(\Gamma)$ whose norm is bounded independently of k.

We introduce **new atlases** on Γ which are **well adapted** to these new spaces. We build one in the following way. Let there be given a covering of R^3 by a collection of uniformly distributed balls of radius $1/k$. These balls are denote by B_i and their centers by b_i. Each ball intersects a finite number of neighbours.

Choose a partition of unity associated to this collection of balls. The associated functions λ_i have the value 1 on the ball B_i, and we choose this partition such that the support of λ_i is included in the union of the neighbouring balls that intersect B_i. Their gradients vanish in the complement of this domain and their norms are bounded by ck. Similarly, their derivatives of order m have norms bounded by ck^m.

The traces $\Gamma \cap B_i$ on the surface Γ of this collection of balls, and the associated partition of unity, constitute a covering of Γ by the open sets $\Gamma \cap B_i$, and a partition of unity, which we continue to denote by λ_i.

In each of these sets, we choose a "central" point y_i. The associated chart is the projection ϕ_i on the tangent plane to Γ at the point y_i. This mapping is an isomorphism of a neighbourhood of the point y_i (of size c/k) from $\Gamma_i = \Gamma \cap B_i$ into R^2. The new norm (N is the total number of charts and is of order ck^2) given by

$$\|u\|^2_{H^s_k(\Gamma)} = \sum_{i=1}^N \|\lambda_i u \circ \phi_i^{-1}\|^2_{H^s_k(R^2)} \qquad (5.6.130)$$

is a norm on the space $H^s_k(\Gamma)$ equivalent to the one introduced in (5.6.109). This can be easily checked using the properties of the new atlas. Although the number N is as large as ck^2, the **equivalence constants are independent of k**.

By localization and a dilation of ratio k, using the new system of charts, it is possible to show that the following operators are continuous:

$$\left\| \int_\Gamma \frac{(x - y \cdot n_x)}{|x-y|^3} e^{-k|x-y|} \varphi(x) d\gamma(x) \right\|_{H^{1/2}_k(\Gamma)} \le c \|\varphi\|_{H^{-1/2}_k(\Gamma)}, \qquad (5.6.131)$$

$$\left\{ \begin{array}{c} \left\| \int_\Gamma \frac{(x-y) \wedge (n_x - n_y)}{|x-y|^3} e^{-k|x-y|} \varphi(x) d\gamma(x) \right\|_{H^{1/2}_k(\Gamma)} \qquad (5.6.132) \\ \le c \|\varphi\|_{H^{-1/2}_k(\Gamma)}, \end{array} \right.$$

$$\left\{ \begin{array}{c} \left\| \int_\Gamma \frac{(x-y) \wedge (n_x - n_y)}{|x-y|^2} e^{-k|x-y|} \varphi(x) d\gamma(x) \right\|_{H^{3/2}_k(\Gamma)} \qquad (5.6.133) \\ \le c \|\varphi\|_{H^{-1/2}_k(\Gamma)}. \end{array} \right.$$

Any kernel which is the product of a pseudo-homogeneous kernel by e^{-kr} satisfies the same kind of continuity properties. The change of norm into the space H^s is taken into account by the dilation of ratio k. In particular, the function $e^{-kr}/(8\pi k)$ is the fundamental solution of the operator $(-\Delta + k^2)^2 = \Delta^2 - 2k^2\Delta + k^4$.

The integral operator associated to this kernel maps the space $H^s_k(\Gamma)$ into $H^{s+3}_k(\Gamma)$. It is moreover coercive in $H^{-3/2}_k(\Gamma)$ and satisfies

$$\frac{1}{k} \int_\Gamma \int_\Gamma e^{-k|x-y|} \varphi(x)\varphi(y) d\gamma d\gamma \ge \alpha \|\varphi\|^2_{H^{-3/2}_k(\Gamma)}, \quad \alpha > 0. \qquad (5.6.134)$$

The second problem that we examine now is associated to an **imaginary wave number** of the form $k_2 + ik_1$. Let

$$\tilde{k} = -k_1 + ik_2, k_1 > 0, k_2 > 0; k_2^2 = k_1^2 + \eta \quad (\eta \text{ bounded}). \qquad (5.6.135)$$

The associated differential operator is: $-\Delta + \left(k_1^2 - k_2^2\right) - 2ik_1k_2$, whose fundamental solution is $e^{(-k_1 + ik_2)\,r}/(4\pi r)$.

It results from the above equalities that the associated single layer potential satisfies

$$
\begin{cases}
\dfrac{1}{4\pi} \displaystyle\int_\Gamma \int_\Gamma \dfrac{e^{(-k_1 + ik_2)\,|x - y|}}{|x - y|} \varphi(x)\overline{\varphi}(y)d\gamma(x)d\gamma(y) \\[3mm]
\quad = \displaystyle\int_{R^3} \left(|\nabla u|^2 + \left(k_1^2 - k_2^2\right)|u|^2\right) dx - 2ik_1k_2 \int_{R^3} |u|^2\, dx, \qquad (5.6.136) \\[3mm]
u(y) = \dfrac{1}{4\pi} \displaystyle\int_\Gamma \dfrac{e^{(-k_1 + ik_2)\,|x - y|}}{|x - y|} \varphi(x)d\gamma(x).
\end{cases}
$$

The operator $-\Delta + \left(k_1^2 - k_2^2\right) - 2ik_1k_2$ is coercive in the spaces $H_{k_1}^1(\Omega_i)$ and $H_{k_1}^1(\Omega_e)$. It satisfies the following coercivity inequalities.

Lemma 5.6.1 *For k_1 large enough and η bounded, it holds*

$$
\begin{cases}
\Re\dfrac{1}{4\pi} \displaystyle\int_\Gamma \int_\Gamma \dfrac{e^{(-k_1 + ik_2)\,|x - y|}}{|x - y|} \varphi(x)\overline{\varphi}(y)d\gamma(x)d\gamma(y) \\[3mm]
\quad \geq \alpha \|\varphi\|^2_{H_{k_1}^{-1/2}(\Gamma)}, \text{ where } \alpha > 0 \text{ is independent of } k_1.
\end{cases} \qquad (5.6.137)
$$

$$
\begin{cases}
-\Im\dfrac{1}{4\pi} \displaystyle\int_\Gamma \int_\Gamma \dfrac{e^{(-k_1 + ik_2)\,|x - y|}}{|x - y|} \varphi(x)\overline{\varphi}(y)d\gamma(x)d\gamma(y) \\[3mm]
\quad \geq \alpha k_1^2 \|\varphi\|^2_{H_{k_1}^{-3/2}(\Gamma)}, \text{ where } \alpha > 0 \text{ is independent of } k_1.
\end{cases} \qquad (5.6.138)
$$

Moreover, these two bilinear forms are continuous on $H_{k_1}^{-1/2}(\Gamma)$ and $H_{k_1}^{-3/2}(\Gamma)$.

Proof
We use the above adapted system of charts, slightly modified by choosing the size of the balls to be bounded by $c\mathrm{Log}\,k_1/k_1$, for some large enough real c independent of k_1. Then, for any pair of points x and y in two different balls which do not intersect, and for **finite values of** m, linked to the choice of c, we have

$$k_1^m e^{-k_1 |x - y|} \leq 1 \quad \left(|x - y| \geq \dfrac{\mathrm{Log}\, k_1^m}{k_1}\right), \qquad (5.6.139)$$

and thus,

$$\left| D_y^{\ell_1} D_x^{\ell_2} \frac{e^{(-k_1 + ik_2)|x-y|}}{|x-y|} \right| \le \frac{c}{k_1^{m-1-\ell-\ell_2}}. \qquad (5.6.140)$$

We decompose the function φ with the help of the partition of unity (λ_i) in the form

$$\varphi = \sum_i \varphi_i, \quad \varphi_i = \lambda_i \varphi. \qquad (5.6.141)$$

It follows that

$$\begin{cases} \displaystyle\int_\Gamma \int_\Gamma \frac{e^{(-k_1 + ik_2)|x-y|}}{|x-y|} \varphi(x)\overline{\varphi}(y)d\gamma(x)d\gamma(y) \\[4mm] \displaystyle = \sum_{i,j} \int_{\Gamma_i} \int_{\Gamma_j} \frac{e^{(-k_1 + ik_2)|x-y|}}{|x-y|} \varphi_i(x)\varphi_j(y)d\gamma(x)d\gamma(y). \end{cases} \qquad (5.6.142)$$

Let us introduce

$$G(|x-y|) = \frac{e^{(-k_1 + ik_2)|x-y|}}{|x-y|}. \qquad (5.6.143)$$

For any pair Γ_i and Γ_j of sets with empty intersection,

$$\begin{cases} \displaystyle\left| \int_{\Gamma_i} \int_{\Gamma_j} G(|x-y|)\,\varphi_i(x)\varphi_j(y)d\gamma(x)d\gamma(y) \right| \\[4mm] \displaystyle\le \left\| \int_{\Gamma_i} G(|x-y|)\,\varphi_i(x)d\gamma(x) \right\|_{H_{k_1}^2(\Gamma_j)} \|\varphi_j\|_{H_{k_1}^{-2}(\Gamma_j)} \\[4mm] \displaystyle\le c \left[\int_{\Gamma_i} \int_{\Gamma_j} \left[k_1^4 |G|^2 + k_1^2 \left[|D_y G|^2 + |D_x G|^2 \right] \right. \right. \\[4mm] \displaystyle \left. \left. + |D_{xy}^2 G|^2 \right] d\gamma(x)d\gamma(y) \right]^{1/2} \|\varphi_i\|_{H_{k_1}^{-1}(\Gamma_i)} \|\varphi_j\|_{H_{k_1}^{-1}(\Gamma_j)}. \end{cases} \qquad (5.6.144)$$

From the estimates (5.6.140) and an ad hoc value of m, we obtain

$$\begin{cases} \displaystyle\left| \int_{\Gamma_i} \int_{\Gamma_j} G(|x-y|)\,\varphi_i(x)\varphi_j(y)d\gamma(x)d\gamma(y) \right| \\[4mm] \displaystyle\le \frac{c}{k_1^3} \|\varphi_i\|_{H_{k_1}^{-1/2}(\Gamma_i)} \|\varphi_j\|_{H_{k_1}^{-1/2}(\Gamma_j)}, \end{cases} \qquad (5.6.145)$$

and the contribution of these extra diagonal terms is dominated by the diagonal ones, due to the coercivity property. The same type of estimate with a larger m yields

$$\left\{ \left| \int_{\Gamma_i} \int_{\Gamma_j} G\left(|x-y|\right) \varphi_i(x)\varphi_j(y)d\gamma(x)d\gamma(y) \right| \right.$$

$$\left. \leq \frac{c}{k_1^3} \|\varphi_i\|_{H_{k_1}^{-3/2}(\Gamma_i)} \|\varphi_j\|_{H_{k_1}^{-3/2}(\Gamma_j)} . \right. \tag{5.6.146}$$

It remains to examine the diagonal or quasi-diagonal terms which are such that any pair of points x and y in the domain of these charts satisfies

$$|x - y| \leq c \frac{\text{Log } k_1^m}{k_1}. \tag{5.6.147}$$

The charts Γ_i are built using a regular covering of R^3 by balls B_i of center b_i.

To any chart Γ_i, we associate the p neighbouring charts, which are the ones such that the distance between the central points y_i and y_j is less than $p(\text{Log } k_1^m)/k_1$. The integer p is bounded and will be chosen later. Let us denote by Γ_i^p the union of Γ_i and its p neighbours, and $\sigma_p(i)$ the set of indexes j such that Γ_j is contained in Γ_i^p. We have

$$\left\{ \sum_i \int_\Gamma \int_\Gamma G\left(|x-y|\right) \left(\sum_{j \in \sigma_p(i)} \varphi_j(x) \right) \left(\sum_{\ell \in \sigma_p(i)} \overline{\varphi}_\ell(y) \right) d\gamma d\gamma \right.$$

$$= \frac{1}{N} \left(\sum_i \text{card}\sigma_p(i) \right) \int_\Gamma \int_\Gamma G\left(|x-y|\right) \varphi(x)\overline{\varphi}(y)d\gamma d\gamma + R,$$

$$R = \int_\Gamma \int_\Gamma G\left[\sum_i \left[\left(\sum_{j \in \sigma_p(i)} \varphi_j(x) \right) \left(\sum_{j \in \sigma_p(i)} \overline{\varphi}_\ell(x) \right) \right] \right.$$

$$\left. - \frac{1}{N} \left(\sum_i \text{card}\sigma_p(i) \right) \left[\left(\sum_j \varphi_j(x) \right) \left(\sum_\ell \overline{\varphi}_\ell(x) \right) \right] \right] d\gamma d\gamma. \tag{5.6.148}$$

The sum appearing in the expression of the residue R takes the form ($i \in \sigma_p(j) \Longleftrightarrow j \in \sigma_p(i)$)

$$\left\{ \sum_i \left(\sum_{j \in \sigma_p(i)} \varphi_j(x) \right) \left(\sum_{\ell \in \sigma_p(i)} \overline{\varphi}_\ell(y) \right) \right.$$

$$\left. = \sum_{j,\ell} \text{card}\left(\sigma_p(j) \cap \sigma_p(\ell)\right) \varphi_j(x)\overline{\varphi}_\ell(y), \right. \tag{5.6.149}$$

$$\begin{cases} R = \int_\Gamma \int_\Gamma G\left(|x-y|\right)\left[\sum_{\ell,k}\left[\text{card}\left(\sigma_p(j)\cap\sigma_p(\ell)\right)\right.\right. \\[2mm] \left.\left. -\frac{1}{N}\sum_i \text{card}\varphi_p(i)\right]\varphi_j(x)\overline{\varphi}_\ell(y)\right]d\gamma(x)d\gamma(y). \end{cases} \tag{5.6.150}$$

When k is large, the above system of charts is regular. Thus, the number $\text{card}\sigma_i(p)$ is bounded by cp^2. For any indexes j and k such that the points y_i and y_j satisfy (5.6.147), the difference

$$\left|\text{card}\sigma_p(j)\cap\text{card}\sigma_p(\ell) - \frac{1}{N}\sum_i \text{card}\sigma_p(i)\right|$$

is bounded by $c\,p$, for some large enough c independent of p.

For p large enough, using (5.6.145) and the continuity of the associated kernel in the corresponding spaces, we obtain

$$|\Re(R)| \le c\frac{p}{N}\sum_{j,\ell}\|\varphi_j\|_{H_{k_1}^{-1/2}(\Gamma_j)}\|\varphi\ell\|_{H_k^{-1/2}(\Gamma_j)}, \tag{5.6.151}$$

$$|\Im(R)| \le c\frac{p}{N}\sum_{j,\ell}\|\varphi_j\|_{H_{k_1}^{-3/2}(\Gamma_j)}\|\varphi\ell\|_{H_{k_1}^{-3/2}(\Gamma_j)}. \tag{5.6.152}$$

For p large enough, these terms are dominated by the coercive terms. It remains to show that the operators associated to the domains Γ_i^p are coercive.

We use for all the domain Γ_i^p, the chart associated to Γ_i, which is the projection on the tangent plane at the central point y_i. Let x and y be two points in Γ_i^p and let ξ, η be their projections on the tangent plane at the central point y_i. If we denote by n the normal to the tangent plane at the central point y_i, we have

$$\begin{cases} x = \xi + s(\xi)n, \\[2mm] y = \eta + s(\eta)n, \\[2mm] |x-y|^2 = |\xi-\eta|^2 + |s(\xi)-s(\eta)|^2. \end{cases} \tag{5.6.153}$$

As $\nabla s(y_i) = 0, D^2 s$ is bounded, and thus,

$$\begin{cases} |s(\xi)-s(\eta)| \le c|\nabla s(\lambda\xi + (1-\lambda)\eta)|\,|\xi-\eta| \\[2mm] \le c|\xi-\eta|\dfrac{p\,\text{Log}\,k_1^m}{k_1}. \end{cases} \tag{5.6.154}$$

The kernel G takes the form

$$\frac{e(-k_1+ik_2)|x-y|}{|x-y|} = \frac{e(-k_1+ik_2)|\xi-\eta|}{|\xi-\eta|}(1+\varphi(\xi,\eta)), \tag{5.6.155}$$

where the function φ and also its first derivatives are bounded by the quantity $(p \operatorname{Log} k_1^m)^2 / k_1$. The associated contribution is dominated by the one coming from the principal term.

It remains to examine the coercivity of the kernel G in a plane. There, we use the Fourier transform. The integral operator is a convolution, and thus its properties are related to those of the real and imaginary part of the Fourier transform of the kernel.

In R^3, the kernel G satisfies

$$(-\Delta - \eta - 2ik_1k_2)\, G(x, z) = 4\pi\delta_0 \tag{5.6.156}$$

and thus, its partial Fourier transform in the tangential variable X satisfies

$$\left(|\xi|^2 - \eta - 2ik_1k_2 - \frac{\partial^2}{\partial z^2}\right) \widehat{G}(\xi, z) = 2. \tag{5.6.157}$$

Choosing the solution exponentially decreasing in the variable z at infinity, it follows that

$$\widehat{G}(\xi) = \sqrt{|\xi|^2 - \eta - 2ik_1k_2}. \tag{5.6.158}$$

The determination of the square root corresponds to evanescent solutions of (5.6.157), which have a negative imaginary part.

We have

$$\begin{cases} \rho = \left(|\xi|^4 - 2\eta\,|\xi|^2 + \left(2k_1^2 - \eta\right)^2\right)^{1/2}, \\[2mm] \Re\widehat{G} = \dfrac{\sqrt{\xi^2 - \eta + \rho}}{\sqrt{2\rho}}, \\[2mm] \Im\widehat{G} = -\dfrac{-2\sqrt{2}k_1k_2}{\rho\sqrt{\xi^2 - \eta + \rho}}. \end{cases} \tag{5.6.159}$$

For $k_1^2 \geq \eta$,

$$\begin{cases} \frac{1}{2}\left(|\xi|^2 + k_1^2\right) \leq \rho \leq 4\left(\xi^2 + k_1^2\right), \\[2mm] \rho - \eta \geq k_1^2, \end{cases} \tag{5.6.160}$$

and thus,

$$\frac{1}{4\sqrt{2}}\frac{1}{\sqrt{|\xi|^2 + k_1^2}} \leq \Re\widehat{G} \leq \sqrt{10}\frac{1}{\sqrt{|\xi|^2 + k_1^2}}, \tag{5.6.161}$$

$$\frac{1}{\sqrt{2}}\frac{k_1^2}{\left(|\xi|^2 + k_1^2\right)^{3/2}} \leq -\Im\widehat{G} \leq 8\frac{k_1^2}{\left(|\xi|^2 + k_1^2\right)^{3/2}}. \tag{5.6.162}$$

This proves the lemma.

Lemma 5.6.2 *For k_1 large enough and η bounded,*

$$
\left\{
\begin{aligned}
&\Re \int_\Gamma \int_\Gamma \frac{e^{(-k_1+ik_2)\,|x-y|}}{4\pi\,|x-y|}\left(\overrightarrow{\mathrm{curl}}_\Gamma \varphi(x)\cdot \overrightarrow{\mathrm{curl}}_\Gamma \overline{\varphi}(y)\right)d\gamma d\gamma \\
&\quad \geq \alpha \left\|\overrightarrow{\mathrm{curl}}_\Gamma \varphi\right\|^2_{TH^{-1/2}_{k_1}(\Gamma)}, \qquad \alpha > 0 \text{ independent of } k_1.
\end{aligned}
\right.
\tag{5.6.163}
$$

$$
\left\{
\begin{aligned}
&-\Im \int_\Gamma \int_\Gamma \frac{e^{(-k_1+ik_2)\,|x-y|}}{4\pi\,|x-y|}\left(\overrightarrow{\mathrm{curl}}_\Gamma \varphi(x)\cdot \overrightarrow{\mathrm{curl}}_\Gamma \overline{\varphi}(y)\right)d\gamma d\gamma \\
&\quad \geq \alpha k_1^2 \left\|\overrightarrow{\mathrm{curl}}_\Gamma \varphi\right\|^2_{TH^{-3/2}_{k_1}(\Gamma)}, \qquad \alpha > 0 \text{ independent of } k_1.
\end{aligned}
\right.
\tag{5.6.164}
$$

$$
\left\{
\begin{aligned}
&\Re \int_\Gamma \int_\Gamma \frac{e^{(-k_1+ik_2)\,|x-y|}}{4\pi\,|x-y|}\left(\nabla_\Gamma \varphi(x)\cdot \nabla_\Gamma \overline{\varphi}(y)\right)d\gamma d\gamma \\
&\quad \geq \alpha \left\|\nabla_\Gamma \varphi\right\|^2_{TH^{-1/2}_{k_1}(\Gamma)}, \qquad \alpha > 0 \text{ independent of } k_1.
\end{aligned}
\right.
\tag{5.6.165}
$$

$$
\left\{
\begin{aligned}
&-\Im \int_\Gamma \int_\Gamma \frac{e^{(-k_1+ik_2)\,|x-y|}}{4\pi\,|x-y|}\left(\nabla_\Gamma \varphi(x)\cdot \nabla_\Gamma \overline{\varphi}(y)\right)d\gamma d\gamma \\
&\quad \geq \alpha k_1^2 \left\|\nabla_\Gamma \varphi\right\|_{TH^{-3/2}_{k_1}(\Gamma)}, \qquad \alpha > 0 \text{ independent of } k_1.
\end{aligned}
\right.
\tag{5.6.166}
$$

Proof

Using the same technique as for Lemma 5.4.7, these integral operators can be localized. Then, the integral is split in the form of a sum of those coming from the charts of diameter bounded by $\mathrm{Log}\,k_1^m/k_1$.

The scalar products are expressed in a local system of axes in the tangent plane of this chart. The terms coming from the difference between the surface and the tangent plane are small compared to the principal coercive terms defined in the tangent plane. Furthermore, by localization, we can prove that the operators

$$
\left\{
\begin{aligned}
&\int_\Gamma \frac{(x-y\cdot n_x)}{|x-y|^3}e^{(-k_1+ik_2)\,|x-y|}\varphi(x)d\gamma(x), \\
&\int_\Gamma \frac{(x-y\cdot n_y)}{|x-y|^3}e^{(-k_1+ik_2)\,|x-y|}\varphi(x)d\gamma(x), \\
&\int_\Gamma \frac{(x-y)\wedge(n_y-n_x)}{|x-y|^3}e^{(-k_1+ik_2)\,|x-y|}\varphi(x)d\gamma(x)
\end{aligned}
\right.
\tag{5.6.167}
$$

are continuous from $H^{-1/2}_{k_1}(\Gamma)$ into $H^{1/2}_{k_1}(\Gamma)$, while the operator

$$
\int_\Gamma \frac{x-y}{|x-y|}e^{(-k_1+ik_2)\,|x-y|}d\gamma(x)
\tag{5.6.168}
$$

is continuous from $H^s_{k_1}(\Gamma)$ into $H^{s+2}_{k_1}(\Gamma)$.

Let us consider the **dielectric problem** with a value of k_e **given and fixed** and a value of k_i **of the form**

$$
\begin{cases}
k_i^2 = \omega^2 \mu_i \tilde{\varepsilon} = \omega^2 \mu_i \varepsilon_i + i\omega\mu_i\sigma = (k_2 + ik_1)^2, \\[2mm]
k_1^2 - k_2^2 = -\omega^2 \mu_i \varepsilon_i, \quad 2k_1 k_2 = \omega\mu_i\sigma, \\[2mm]
k_1 = \sqrt{\frac{\omega\mu_i}{2}\left(\sqrt{2\sigma^2 + \omega^2\varepsilon_i^2} - \omega\varepsilon_i\right)}, \\[2mm]
k_2 = \sqrt{\frac{\omega\mu_i}{2}\left(\sqrt{2\sigma^2 + \omega^2\varepsilon_i^2} + \omega\varepsilon_i\right)}.
\end{cases}
\tag{5.6.169}
$$

We examine the integral equation formulation of this dielectric problem. It is convenient to use the currents j_i and m_i as unknowns, and we denote them by j and m.

We suppose that the incident fields E^{inc} and H^{inc} are the traces on the surface Γ of a field which satisfies the Maxwell equations for a right-hand side in the domain Ω_i, and the parameters (ε_e, μ_e).

This field satisfies the representation Theorem 5.5.1, which applied in the domain Ω_e yields

$$
\begin{cases}
\frac{1}{2}\int_\Gamma \left(E^{\text{inc}} \cdot j^t\right) d\gamma \\[3mm]
\quad = -\frac{i}{\omega}\frac{1}{\tilde{\varepsilon}_e}\int_\Gamma\int_\Gamma G_e(x-y)\,\text{div}_\Gamma\left(H^{\text{inc}} \wedge n(x)\right)\text{div}_\Gamma\, j^t(y)d\gamma d\gamma \\[3mm]
\quad + i\omega\mu_e\int_\Gamma\int_\Gamma G_e(x-y)\left(\left(H^{\text{inc}} \wedge n(x)\right)\cdot j^t(y)\right)d\gamma d\gamma \\[3mm]
\quad + \int_\Gamma\int_\Gamma \left(\nabla_y G_e(x-y)\cdot\left(\left(E^{\text{inc}} \wedge n(x)\right) \wedge j^t(y)\right)\right)d\gamma d\gamma,
\end{cases}
\tag{5.6.170}
$$

$$
\begin{cases}
\frac{1}{2}\int_\Gamma \left(H^{\text{inc}} \cdot m^t\right) d\gamma \\[3mm]
\quad = \frac{i}{\omega}\frac{1}{\mu_e}\int_\Gamma\int_\Gamma G_e(x-y)\,\text{div}_\Gamma\left(E^{\text{inc}} \wedge n(x)\right)\text{div}_\Gamma\, m^t(y)d\gamma d\gamma \\[3mm]
\quad - i\omega\varepsilon_e\int_\Gamma\int_\Gamma G_e(x-y)\left(\left(E^{\text{inc}} \wedge n(x)\right)\cdot m^t(y)\right)d\gamma d\gamma \\[3mm]
\quad + \int_\Gamma\int_\Gamma \left(\nabla_y G_e(x-y)\cdot\left(\left(H^{\text{inc}} \wedge n(x)\right) \wedge m^t(y)\right)\right)d\gamma d\gamma.
\end{cases}
\tag{5.6.171}
$$

It follows from these identities that the integral equations (5.6.60) and (5.6.61) take the form

$$
\begin{cases}
-\dfrac{i}{\omega}\left[\dfrac{1}{\varepsilon}\displaystyle\int_\Gamma\int_\Gamma G_i(x-y)\operatorname{div}_\Gamma j(x)\operatorname{div}_\Gamma j^t(y)d\gamma(x)d\gamma(y)\right.\\[2mm]
\qquad\left.+\dfrac{1}{\varepsilon_e}\displaystyle\int_\Gamma\int_\Gamma G_e(x-y)\operatorname{div}_\Gamma j(x)\operatorname{div}_\Gamma j^t(y)d\gamma(x)d\gamma(y)\right]\\[2mm]
+i\omega\displaystyle\int_\Gamma\int_\Gamma(\mu_iG_i+\mu_eG_e)\left(j(x)\cdot j^t(y)\right)d\gamma d\gamma \qquad\qquad (5.6.172)\\[2mm]
+\displaystyle\int_\Gamma\int_\Gamma\left((\nabla_yG_i+\nabla_yG_e)\cdot\left(m(x)\wedge j^t(y)\right)\right)d\gamma d\gamma\\[2mm]
\qquad=\displaystyle\int_\Gamma\left(E^{\text{inc}}\cdot j^t\right)d\gamma,\quad\forall j^t\in H_{\text{div}}^{-1/2}(\Gamma).
\end{cases}
$$

$$
\begin{cases}
\dfrac{i}{\omega}\displaystyle\int_\Gamma\int_\Gamma\left(\dfrac{1}{\mu_i}G_i+\dfrac{1}{\mu_e}G_e\right)\operatorname{div}_\Gamma m(x)\operatorname{div}_\Gamma m^t(y)d\gamma d\gamma\\[2mm]
-i\omega\left[\displaystyle\int_\Gamma\int_\Gamma\widetilde{\varepsilon}G_i(x-y)\left(m(x)\cdot m^t(y)\right)d\gamma(x)d\gamma(y)\right.\\[2mm]
\qquad\left.+\displaystyle\int_\Gamma\int_\Gamma\varepsilon_eG_e(x-y)\left(m(x)\cdot m^t(y)\right)d\gamma(x)d\gamma(y)\right]\qquad (5.6.173)\\[2mm]
+\displaystyle\int_\Gamma\int_\Gamma\left((\nabla_yG_i+\nabla_yG_e)\cdot\left(j(x)\wedge m^t(y)\right)\right)d\gamma d\gamma\\[2mm]
\qquad=\displaystyle\int_\Gamma\left(H^{\text{inc}}\cdot m^t\right)d\gamma,\quad\forall m^t\in H_{\text{div}}^{-1/2}(\Gamma).
\end{cases}
$$

Theorem 5.6.6 *Consider the solution of the dielectric problem in the form of the system of integral equations (5.6.147) (5.6.148), with an interior electric permittivity of the form*

$$\widetilde{\varepsilon}=\varepsilon_i+i\frac{\sigma}{\omega}\qquad\qquad(5.6.174)$$

(corresponding to a conducting medium with a real conductivity σ). Suppose that E_T^{inc} and H_T^{inc} belongs to $H_{\text{curl}}^{-1/2}(\Gamma)$, and that k_e^2 is not an eigenvalue of the interior Maxwell problem associated to the boundary condition $E\wedge n=0$. When σ tends to infinity, the solution j of equations (5.6.147) (5.6.148) tends weakly in the Hilbert space $H_{\text{div}}^{-1/2}(\Gamma)$ toward the solution j_0 of the exterior perfect conductor problem

$$
\begin{cases}
-\dfrac{i}{\omega\varepsilon_e}\displaystyle\int_\Gamma\int_\Gamma Ge(x-y)\operatorname{div}_\Gamma j_0(x)\operatorname{div}_\Gamma j^t(y)d\gamma(x)d\gamma(y)\\[2mm]
\qquad+i\omega\mu_e\displaystyle\int_\Gamma\int_\Gamma Ge(x-y)\left(j_0(x)\cdot j^t(y)\right)d\gamma(x)d\gamma(y)\qquad (5.6.175)\\[2mm]
\qquad=\displaystyle\int_\Gamma\left(E^{\text{inc}}\cdot j^t\right)d\gamma,\quad\forall j^t\in H_{\text{div}}^{-1/2}(\Gamma).
\end{cases}
$$

Proof

Using the Helmholtz decomposition, we express the currents j and m in the form

$$\begin{cases} j = \nabla_\Gamma q + \overrightarrow{\mathrm{curl}}_\Gamma p, \\ m = \nabla_\Gamma w + \overrightarrow{\mathrm{curl}}_\Gamma v. \end{cases} \tag{5.6.176}$$

The equations for q, p, w and v are deduced from (5.6.172) and (5.6.173). Besides, we have

$$\begin{cases} \Re \tilde{\varepsilon} = \varepsilon_i, \ \Im \tilde{\varepsilon} = \dfrac{\sigma}{\omega}, \ \Re \dfrac{1}{\tilde{\varepsilon}} = \dfrac{\varepsilon_i \omega^2}{\sigma^2 + \varepsilon_i^2 \omega^2} \leq \dfrac{\varepsilon_i \omega^2}{\sigma^2}, \\ -\Im \dfrac{1}{\tilde{\varepsilon}} = \dfrac{\omega \sigma}{\sigma^2 + \omega^2 \varepsilon_i^2} \leq \dfrac{\omega}{\sigma}. \end{cases} \tag{5.6.177}$$

Choosing $q^t = \bar{q}$ in the modified equation (5.6.73), we obtain

$$\begin{cases} -\dfrac{i}{\omega} \left[\dfrac{1}{\tilde{\varepsilon}} \displaystyle\int_\Gamma \int_\Gamma G_i(x-y) \Delta_\Gamma q(x) \Delta \bar{q}(y) d\gamma(x) d\gamma(y) \right. \\ \qquad\qquad \left. + \dfrac{1}{\tilde{\varepsilon}_e} \displaystyle\int_\Gamma \int_\Gamma G_e(x-y) \Delta_\Gamma q(x) \Delta_\Gamma \bar{q}(y) d\gamma(x) d\gamma(y) \right] \\ + i\omega \displaystyle\int_\Gamma \int_\Gamma (\mu_i G_i + \mu_e G_e) \left(\nabla_\Gamma q(x) \cdot \nabla_\Gamma \bar{q}(y) \right) d\gamma(x) d\gamma(y) \\ + i\omega \displaystyle\int_\Gamma \int_\Gamma (\mu_i G_i + \mu_e G_e) \left(\overrightarrow{\mathrm{curl}}_\Gamma p(x) \cdot \nabla_\Gamma \bar{q}(y) \right) d\gamma(x) d\gamma(y) \\ + \displaystyle\int_\Gamma \int_\Gamma \Big((\nabla_y G_i(x-y) + \nabla_y G_e(x-y)) \\ \qquad \cdot \big((\overrightarrow{\mathrm{curl}}_\Gamma v(x) + \nabla_\Gamma w(x)) \wedge \nabla_\Gamma \bar{q}(y) \big) \Big) d\gamma(x) d\gamma(y) \\ \qquad\qquad = \displaystyle\int_\Gamma \left(E^{\mathrm{inc}} \cdot \nabla_\Gamma \bar{q} \right) d\gamma. \end{cases} \tag{5.6.178}$$

Choosing $p^t = \bar{p}$ in the equation in p, we obtain

$$\begin{cases} i\omega \displaystyle\int_\Gamma \int_\Gamma (\mu_i G_i + \mu_e G_e) \left(\overrightarrow{\mathrm{curl}}_\Gamma p(x) \cdot \overrightarrow{\mathrm{curl}}_\Gamma \bar{p}(y) \right) d\gamma d\gamma \\ + i\omega \displaystyle\int_\Gamma \int_\Gamma (\mu_i G_i + \mu_e G_e) \left(\nabla_\Gamma q(x) \cdot \overrightarrow{\mathrm{curl}}_\Gamma \bar{p}(y) \right) d\gamma d\gamma \\ + \displaystyle\int_\Gamma \int_\Gamma \Big((\nabla_y G_i(x-y) + \nabla_y G_e(x-y)) \\ \qquad \cdot \big((\mathrm{curl}_\Gamma\, v(x) + \nabla_\Gamma w(x)) \wedge \overrightarrow{\mathrm{curl}}_\Gamma \bar{p}(y) \big) \Big) d\gamma(x) d\gamma(y) \\ \qquad\qquad = \displaystyle\int_\Gamma \left(E^{\mathrm{inc}} \cdot \overrightarrow{\mathrm{curl}}_\Gamma \bar{p}(y) \right) d\gamma. \end{cases} \tag{5.6.179}$$

Choosing $w^t = \overline{w}$ in the equation in w, we obtain

$$
\begin{cases}
-\dfrac{i}{\omega}\displaystyle\int_\Gamma\int_\Gamma\left(\dfrac{1}{\mu_i}G_i + \dfrac{1}{\mu_e}G_e\right)\Delta_\Gamma w(x)\Delta\overline{w}(y)d\gamma d\gamma \\[3mm]
\quad - i\omega\displaystyle\int_\Gamma\int_\Gamma(\tilde{\varepsilon}G_i + \varepsilon_e G_e)\left(\nabla_\Gamma w(x)\cdot\nabla_\Gamma\overline{w}(y)\right)d\gamma d\gamma \\[3mm]
\quad - i\omega\displaystyle\int_\Gamma\int_\Gamma(\tilde{\varepsilon}G_i + \varepsilon_e G_e)\left(\overrightarrow{\mathrm{curl}}_\Gamma v(x)\cdot\nabla_\Gamma\overline{w}(y)\right)d\gamma d\gamma \\[3mm]
\quad + \displaystyle\int_\Gamma\int_\Gamma\Big((\nabla_y G_i(x-y) + \nabla_y G_e(x-y)) \\[3mm]
\qquad \cdot\left((\overrightarrow{\mathrm{curl}}_\Gamma p(x) + \nabla_\Gamma q(x))\wedge\nabla_\Gamma\overline{w}(y))\right)\Big)d\gamma(x)d\gamma(y) \\[3mm]
\qquad = \displaystyle\int_\Gamma\left(H^{\mathrm{inc}}\cdot\nabla_\Gamma\overline{w}\right)d\gamma.
\end{cases}
\tag{5.6.180}
$$

Choosing $v^t = \overline{v}$ in the equation in w, we obtain

$$
\begin{cases}
-i\omega\displaystyle\int_\Gamma\int_\Gamma(\tilde{\varepsilon}G_i + \varepsilon_e G_e)\left(\overrightarrow{\mathrm{curl}}_\Gamma v(x)\cdot\overrightarrow{\mathrm{curl}}_\Gamma\overline{v}(y)\right)d\gamma d\gamma \\[3mm]
\quad - i\omega\displaystyle\int_\Gamma\int_\Gamma(\tilde{\varepsilon}G_i + \varepsilon_e G_e)\left(\nabla_\Gamma w(x)\cdot\overrightarrow{\mathrm{curl}}_\Gamma\overline{v}(y)\right)d\gamma d\gamma \\[3mm]
\quad + \displaystyle\int_\Gamma\int_\Gamma\Big((\nabla_y G_i(x-y) + \nabla_y G_e(x-y)) \\[3mm]
\qquad \cdot\left((\overrightarrow{\mathrm{curl}}_\Gamma p(x) + \nabla_\Gamma q(x))\wedge\overrightarrow{\mathrm{curl}}_\Gamma\overline{v}(y))\right)\Big)d\gamma(x)d\gamma(y) \\[3mm]
\qquad = \displaystyle\int_\Gamma\left(H^{\mathrm{inc}}\cdot\overrightarrow{\mathrm{curl}}_\Gamma\overline{v}(y)\right)d\gamma.
\end{cases}
\tag{5.6.181}
$$

From (5.6.123) and the above estimates combined with the imaginary part of (5.6.178), we can deduce the estimate

$$
\begin{cases}
\alpha\|\Delta_\Gamma q\|^2_{H^{-1/2}(\Gamma)} \le \dfrac{c}{\sigma}\|\Delta_\Gamma q\|^2_{H^{-1/2}(\Gamma)} + c\|\Delta_\Gamma q\|^2_{H^{-3/2}(\Gamma)} \\[3mm]
\quad + c\left\|\nabla_\Gamma q + \overrightarrow{\mathrm{curl}}_\Gamma p\right\|_{TH^{-1/2}(\Gamma)}\|\nabla_\Gamma q\|_{TH^{-1/2}(\Gamma)} \\[3mm]
\quad + c\left\|\nabla_\Gamma w + \overrightarrow{\mathrm{curl}}_\Gamma v\right\|_{TH^{-1/2}(\Gamma)}\|\nabla_\Gamma q\|_{TH^{-1/2}(\Gamma)} \\[3mm]
\quad + c\sqrt{\sigma}\|\nabla_\Gamma q\|_{TH^{-1/2}_{k_1}(\Gamma)}\left\|\overrightarrow{\mathrm{curl}}_\Gamma v + \nabla_\Gamma q\right\|_{TH^{-1/2}_{k_1}(\Gamma)} \\[3mm]
\quad + c\|E^{\mathrm{inc}}_T\|_{TH^{-1/2}(\Gamma)}\|\nabla_\Gamma q\|_{TH^{-1/2}(\Gamma)}.
\end{cases}
\tag{5.6.182}
$$

Using relation (5.6.97) to bound the term of mixed type where j and m appear, in the part which contains the kernel G_i, the imaginary part of (5.6.179) yields:

$$
\left\{
\begin{aligned}
&\alpha \left(\left\| \overrightarrow{\mathrm{curl}}_\Gamma p \right\|^2_{TH^{-1/2}_{k_1}(\Gamma)} + \left\| \overrightarrow{\mathrm{curl}}_\Gamma p \right\|^2_{TH^{-1/2}(\Gamma)} \right) \\
&\quad \le c \left\| \nabla_\Gamma q \right\|_{TH^{-1/2}(\Gamma)} \left\| \overrightarrow{\mathrm{curl}}_\Gamma p \right\|_{TH^{-1/2}(\Gamma)} \\
&\quad + c \left\| \overrightarrow{\mathrm{curl}}_\Gamma p \right\|^2_{TH^{-3/2}(\Gamma)} + c \left\| \overrightarrow{\mathrm{curl}}_\Gamma p \right\|_{TH^{-1/2}(\Gamma)} \\
&\qquad \times \left(\sqrt{\sigma} \left\| m \right\|_{TH^{-1/2}_{k_1}(\Gamma)} + \left\| \nabla_\Gamma w \right\|_{TH^{-1/2}(\Gamma)} \right) \\
&\quad + c \left\| \mathrm{curl}_\Gamma E^{\mathrm{inc}}_T \right\|_{H^{-1/2}(\Gamma)} \left\| p \right\|_{H^{1/2}(\Gamma)} \\
&\quad + c \left\| \overrightarrow{\mathrm{curl}}_\Gamma v \right\|_{TH^{-1/2}_{k_1}(\Gamma)} \left(\sigma \left\| p \right\|_{TH^{-3/2}_{k_1}(\Gamma)} + \left\| p \right\|_{H^{-3/2}(\Gamma)} \right).
\end{aligned}
\right.
\tag{5.6.183}
$$

Using (5.6.97) and Lemma 5.6.2, the real part of (5.6.181) leads to the estimate

$$
\left\{
\begin{aligned}
&\alpha\sigma \left\| \overrightarrow{\mathrm{curl}}_\Gamma v \right\|^2_{TH^{-1/2}_{k_1}(\Gamma)} \le c\sigma \left\| \overrightarrow{\mathrm{curl}}_\Gamma v \right\|^2_{TH^{-3/2}_{k_1}(\Gamma)} \\
&\quad + c \left\| \overrightarrow{\mathrm{curl}}_\Gamma v \right\|^2_{TH^{-3/2}(\Gamma)} + c\sqrt{\sigma} \left\| \overrightarrow{\mathrm{curl}}_\Gamma v \right\|_{TH^{-1/2}_{k_1}(\Gamma)} \\
&\qquad \times \left(c\sqrt{\sigma} \left\| \nabla_\Gamma w \right\|_{TH^{-1/2}_{k_1}(\Gamma)} + \left\| \nabla_\Gamma q \right\|_{TH^{-1/2}_{k_1}(\Gamma)} \right) \\
&\quad + c \left\| \overrightarrow{\mathrm{curl}}_\Gamma v \right\|_{TH^{-1/2}(\Gamma)} \\
&\qquad \times \left(\left\| \nabla_\Gamma w \right\|_{TH^{-3/2}(\Gamma)} + \left\| \nabla_\Gamma q \right\|_{TH^{-1/2}(\Gamma)} + \left\| p \right\|_{H^{-3/2}(\Gamma)} \right) \\
&\quad + c \left\| v \right\|_{H^{1/2}(\Gamma)} \left\| \mathrm{curl}_\Gamma H^{\mathrm{inc}}_T \right\|_{H^{-1/2}(\Gamma)} \\
&\quad + c\sigma \left\| \overrightarrow{\mathrm{curl}}_\Gamma v \right\|_{TH^{-1/2}_{k_1}(\Gamma)} \left\| p \right\|_{TH^{-3/2}_{k_1}(\Gamma)}.
\end{aligned}
\right.
\tag{5.6.184}
$$

Besides, we have

$$
\left\| p \right\|_{H^{-3/2}_{k_1}(\Gamma)} \le \frac{c}{k_1} \left\| p \right\|_{H^{-1/2}_{k_1}(\Gamma)},
\tag{5.6.185}
$$

from which it follows that

$$
\sqrt{\sigma} \left\| p \right\|_{H^{-3/2}_{k_1}(\Gamma)} \le c \left\| p \right\|_{H^{-1/2}(\Gamma)}.
\tag{5.6.186}
$$

The imaginary part of (5.6.181) and Lemma 5.6.2 yield

$$
\left\{
\begin{aligned}
&\alpha \left[\left\| \overrightarrow{\mathrm{curl}}_\Gamma v \right\|_{TH_{k_1}^{-1/2}(\Gamma)}^2 + \sigma^2 \left\| \overrightarrow{\mathrm{curl}}_\Gamma v \right\|_{TH_{k_1}^{-3/2}(\Gamma)}^2 \right. \\
&\qquad\qquad \left. + \left\| \overrightarrow{\mathrm{curl}}_\Gamma v \right\|_{TH^{-1/2}(\Gamma)}^2 \right] \\
&\quad \leq c \left\| \overrightarrow{\mathrm{curl}}_\Gamma v \right\|_{TH^{-3/2}(\Gamma)}^2 + c\sigma \left\| \overrightarrow{\mathrm{curl}}_\Gamma v \right\|_{TH_{k_1}^{-3/2}(\Gamma)}^2 \\
&\qquad + c\sigma \left\| \overrightarrow{\mathrm{curl}}_\Gamma v \right\|_{TH_{k_1}^{-1/2}(\Gamma)} \left\| \nabla_\Gamma w \right\|_{TH_{k_1}^{-1/2}(\Gamma)} \\
&\qquad + c \left\| \overrightarrow{\mathrm{curl}}_\Gamma v \right\|_{TH^{-1/2}(\Gamma)} \left\| \nabla_\Gamma w \right\|_{TH^{-3/2}(\Gamma)} \\
&\qquad + c\sqrt{\sigma} \left\| \overrightarrow{\mathrm{curl}}_\Gamma v \right\|_{TH_{k_1}^{-1/2}(\Gamma)} \left\| \nabla_\Gamma q \right\|_{TH_{k_1}^{-1/2}(\Gamma)} \\
&\qquad + c \left\| \overrightarrow{\mathrm{curl}}_\Gamma v \right\|_{TH^{-1/2}(\Gamma)} \left\| \nabla_\Gamma q \right\|_{TH^{-1/2}(\Gamma)} \\
&\qquad + c \left\| v \right\|_{H^{1/2}(\Gamma)} \left\| \mathrm{curl}_\Gamma H_T^{\mathrm{inc}} \right\|_{H^{-1/2}(\Gamma)} \\
&\qquad + c\sigma \left\| \overrightarrow{\mathrm{curl}}_\Gamma v \right\|_{TH_{k_1}^{-1/2}(\Gamma)} \left\| p \right\|_{TH_{k_1}^{-3/2}(\Gamma)} \\
&\qquad + c \left\| \overrightarrow{\mathrm{curl}}_\Gamma v \right\|_{TH^{-1/2}(\Gamma)} \left\| p \right\|_{H^{-3/2}(\Gamma)}.
\end{aligned}
\right. \tag{5.6.187}
$$

From (5.6.184) and (5.6.187), we obtain

$$
\left\{
\begin{aligned}
&\alpha \left[\sigma \left\| \overrightarrow{\mathrm{curl}}_\Gamma v \right\|_{TH_{k_1}^{-1/2}(\Gamma)}^2 + \sigma^2 \left\| \overrightarrow{\mathrm{curl}}_\Gamma v \right\|_{TH_{k_1}^{-3/2}(\Gamma)}^2 \right. \\
&\qquad\qquad \left. + \left\| \overrightarrow{\mathrm{curl}}_\Gamma v \right\|_{TH^{-1/2}(\Gamma)}^2 \right] \\
&\quad \leq c \left[\left\| \overrightarrow{\mathrm{curl}}_\Gamma v \right\|_{TH^{-3/2}(\Gamma)}^2 + \frac{1}{\sigma^2} \left\| \nabla_\Gamma w \right\|_{TH^{-3/2}(\Gamma)}^2 \right. \\
&\qquad + \sigma \left\| \nabla_\Gamma w \right\|_{TH_{k_1}^{-1/2}(\Gamma)} \left\| \overrightarrow{\mathrm{curl}}_\Gamma v \right\|_{TH_{k_1}^{-1/2}(\Gamma)} \\
&\qquad + \left\| \nabla_\Gamma q \right\|_{TH^{-1/2}(\Gamma)}^2 + \left\| p \right\|_{H^{-1/2}(\Gamma)}^2 \\
&\qquad \left. + \left\| \mathrm{curl}_\Gamma H_T^{\mathrm{inc}} \right\|_{H^{-1/2}(\Gamma)} \left\| v \right\|_{H^{1/2}(\Gamma)} \right].
\end{aligned}
\right. \tag{5.6.188}
$$

Furthermore, choosing $m^t = \overline{m}$ in equation (5.6.173), it follows that

$$
\left\{
\begin{aligned}
&\frac{i}{\omega} \int_\Gamma \int_\Gamma \left(\frac{1}{\mu_i} G_i + \frac{1}{\mu_e} G_e \right) \Delta_\Gamma w(x) \Delta_\Gamma \overline{w}(y) d\gamma d\gamma \\
&\quad - i\omega \int_\Gamma \int_\Gamma \left(\tilde{\varepsilon} G_i + \varepsilon_e G_e \right) (m(x) \cdot \overline{m}(x)) d\gamma d\gamma \\
&\quad + \int_\Gamma \int_\Gamma \left((\nabla_y G_i(x-y) + \nabla_y G_e(x-y)) \right. \\
&\qquad \left. \cdot \left(\left(\overrightarrow{\mathrm{curl}}_\Gamma p(x) + \nabla_\Gamma q(x) \right) \wedge \overline{m}(y) \right) \right) d\gamma(x) d\gamma(y) \\
&\quad = \int_\Gamma \left(H^{\mathrm{inc}} \cdot m \right) d\gamma.
\end{aligned}
\right.
\tag{5.6.189}
$$

The real part of this equation takes the form

$$
\left\{
\begin{aligned}
&\sigma \int_\Gamma \int_\Gamma \Re \left(G_i(x-y)(m(x) \cdot \overline{m}(y)) \, d\gamma(x) d\gamma(y) \right. \\
&\quad + \omega \int_\Gamma \int_\Gamma \Im \left((\varepsilon_i G_i + \varepsilon_e G_e) (m(x) \cdot \overline{m}(y)) \, d\gamma d\gamma \right. \\
&\quad - \frac{1}{\omega} \int_\Gamma \int_\Gamma \Im \left(\left(\frac{1}{\mu_i} G_i + \frac{1}{\mu_e} G_e \right) \Delta_\Gamma w(x) \Delta_\Gamma \overline{w}(y) \right) d\gamma d\gamma \\
&\quad + \int_\Gamma \int_\Gamma \Re \left((\nabla_y G_i + \nabla_y G_e) \cdot (j(x) \wedge \overline{m}(y)) \right) d\gamma d\gamma \\
&\quad = \Re \int_\Gamma \left(H^{\mathrm{inc}} \cdot \overline{m} \right) d\gamma.
\end{aligned}
\right.
\tag{5.6.190}
$$

The imaginary part of this equation takes the form

$$
\left\{
\begin{aligned}
&\frac{1}{\omega} \int_\Gamma \int_\Gamma \Re \left(\left(\frac{1}{\mu_i} G_i + \frac{1}{\mu_e} G_e \right) \Delta_\Gamma w(x) \Delta_\Gamma \overline{w}(y) \right) d\gamma d\gamma \\
&\quad + \sigma \int_\Gamma \int_\Gamma \Im \left(G_i(x-y)(m(x) \cdot \overline{m}(y)) \, d\gamma(x) d\gamma(y) \right. \\
&\quad - \omega \int_\Gamma \int_\Gamma \Re \left((\varepsilon_i G_i + \varepsilon_e G_e) (m(x) \cdot \overline{m}(y)) \, d\gamma d\gamma \right. \\
&\quad + \int_\Gamma \int_\Gamma \Im \left((\nabla_y G_i + \nabla_y G_e) \cdot (j(x) \wedge \overline{m}(y)) \right) d\gamma d\gamma \\
&\quad = \int_\Gamma \left(H^{\mathrm{inc}} \cdot \overline{m} \right) d\gamma.
\end{aligned}
\right.
\tag{5.6.191}
$$

We treat the terms which contain G_i through the use of the coercivity inequalities of Lemma 5.6.2. Using (5.6.97) to estimate the mixed terms where j and m occur, the real part yields the estimate

$$
\begin{cases}
\alpha \left[\sigma \|\Delta_\Gamma w\|^2_{TH_{k_1}^{-3/2}(\Gamma)} + \sigma \|m\|^2_{TH_{k_1}^{-1/2}(\Gamma)} \right] \\[2mm]
\quad \leq c \Big[\sigma \|m\|^2_{TH_{k_1}^{-3/2}(\Gamma)} + \|m\|^2_{TH^{-3/2}(\Gamma)} \\[2mm]
\qquad + \left(\sqrt{\sigma} \|m\|_{TH_k^{-1/2}(\Gamma)} + \|m\|_{TH^{-1/2}(\Gamma)} \right) \\[2mm]
\qquad \times \left(\|\nabla_\Gamma q\|_{TH^{-1/2}(\Gamma)} + \|p\|_{H^{-1/2}(\Gamma)} \right) \\[2mm]
\qquad + \|m\|_{H_{\mathrm{div}}^{-1/2}(\Gamma)} \|H_T^{\mathrm{inc}}\|_{H_{\mathrm{curl}}^{-1/2}(\Gamma)} \Big].
\end{cases}
\tag{5.6.192}
$$

The imaginary part gives

$$
\begin{cases}
\alpha \left[\|\Delta_\Gamma w\|^2_{H_{k_1}^{-1/2}(\Gamma)} + \|\Delta_\Gamma w\|^2_{H^{-1/2}(\Gamma)} \right] \\[2mm]
\quad \leq c \Big[\sigma^2 \|m\|^2_{TH_{k_1}^{-3/2}(\Gamma)} + \|m\|^2_{TH_{k_1}^{-1/2}(\Gamma)} + \|\Delta_\Gamma w\|^2_{H^{-5/2}(\Gamma)} \\[2mm]
\qquad + \left(\sqrt{\sigma} \|m\|_{TH_{k_1}^{-1/2}(\Gamma)} + \|m\|_{TH^{-1/2}(\Gamma)} \right) \\[2mm]
\qquad \times \left(\|\nabla_\Gamma q\|_{TH^{-1/2}(\Gamma)} + \|p\|_{H^{-1/2}(\Gamma)} \right) \\[2mm]
\qquad + \|m\|_{H_{\mathrm{div}}^{-1/2}(\Gamma)} \|H_T^{\mathrm{inc}}\|_{H_{\mathrm{curl}}^{-1/2}(\Gamma)} \Big].
\end{cases}
\tag{5.6.193}
$$

The terms on the right-hand side where m appears are decomposed in terms where only w and v occur.

We want an estimate where the right-hand side contains only compact terms with q and p and constants, excluding terms with v or w. Thus, the disturbing terms on the right-hand side are the quadratric ones with v or w. The terms with q and p in the right-hand side with a k_1 indexed norm, can be estimated by the same terms with the usual norm. The quadratric term with m, w or v with a k_1 indexed norm, are bounded using the estimate

$$
\sigma \|m\|^2_{TH_{k_1}^{-3/2}(\Gamma)} \leq c \|m\|^2_{H_{k_1}^{-1/2}(\Gamma)}.
\tag{5.6.194}
$$

Furthermore, we use

$$
\|m\|_{TH^{-1/2}(\Gamma)} \leq c\sqrt{\sigma} \|m\|_{TH_{k_1}^{-1/2}(\Gamma)}
\tag{5.6.195}
$$

and

$$
\|m\|_{TH^{-3/2}(\Gamma)} \leq c\sqrt{\sigma} \|m\|_{TH_{k_1}^{-3/2}(\Gamma)} \leq c \|m\|_{H_{k_1}^{-1/2}(\Gamma)}.
\tag{5.6.196}
$$

By addition of (5.6.192) and (5.6.193) with "ad hoc" weights, for k_1 large enough, we obtain

$$\begin{cases} \|\Delta_\Gamma w\|^2_{H^{-1/2}_{k_1}(\Gamma)} + \|\Delta_\Gamma w\|^2_{H^{-1/2}(\Gamma)} \\ \qquad\qquad + \sigma \|\Delta_\Gamma w\|^2_{H^{-3/2}_{k_1}(\Gamma)} + \sigma \|m\|^2_{TH^{-1/2}_{k_1}(\Gamma)} \qquad\qquad (5.6.197) \\ \qquad \le c \left[\|\nabla_\Gamma q\|^2_{TH^{-1/2}(\Gamma)} + \|p\|^2_{H^{-1/2}(\Gamma)} + \|H^{inc}_\Gamma\|^2_{H^{-1/2}_{curl}(\Gamma)} \right]. \end{cases}$$

The term $\|\Delta_\Gamma w\|^2_{H^{-5/2}(\Gamma)}$, which is associated to the smooth kernel $\Im G_e$, is dominated by the term $\sigma \|\Delta_\Gamma w\|^2_{H^{-3/2}_{k_1}(\Gamma)}$.

Adding a combination with "ad hoc" weights of equations (5.6.182) (5.6.183) (5.6.188) and (5.6.197), for k_1 large enough, we obtain

$$\begin{cases} \|\Delta_\Gamma q\|^2_{H^{-1/2}(\Gamma)} + \left\|\overrightarrow{curl}_\Gamma p\right\|^2_{TH^{-1/2}_{k_1}(\Gamma)} + \left\|\overrightarrow{curl}_\Gamma p\right\|^2_{TH^{-1/2}(\Gamma)} \\ \quad + \|\Delta_\Gamma w\|^2_{H^{-1/2}_{k_1}(\Gamma)} + \|\Delta_\Gamma w\|^2_{H^{-1/2}(\Gamma)} + \sigma \|\Delta_\Gamma w\|^2_{H^{-3/2}_{k_1}(\Gamma)} \\ \quad + \sigma \left\|\overrightarrow{curl}_\Gamma v\right\|^2_{TH^{-1/2}_{k_1}(\Gamma)} + \left\|\overrightarrow{curl}_\Gamma v\right\|^2_{TH^{-1/2}(\Gamma)} \\ \quad + \sigma^2 \left\|\overrightarrow{curl}_\Gamma v\right\|^2_{TH^{-3/2}_{k_1}(\Gamma)} + \sigma \|m\|^2_{TH^{-1/2}_{k_1}(\Gamma)} \qquad\qquad (5.6.198) \\ \quad \le c \Big[\|\nabla_\Gamma q\|^2_{TH^{-1/2}(\Gamma)} + \|p\|^2_{H^{-1/2}(\Gamma)} \\ \qquad\qquad + \|E^{inc}_T\|^2_{H^{-1/2}_{curl}(\Gamma)} + \|H^{inc}_T\|^2_{H^{-1/2}_{curl}(\Gamma)} \Big]. \end{cases}$$

We now proceed by contradiction. **Either** the right hand of (5.6.198) is bounded, **or** some subsequence tends to infinity. In the **first case**, the currents j and m are bounded in $H^{-1/2}_{div}(\Gamma)$. They converge weakly in these spaces:

$$j \to j_0, \quad \text{in } H^{-1/2}_{div}(\Gamma),$$

$$m \to m_0, \quad \text{in } H^{-1/2}_{div}(\Gamma).$$

Moreover, the above inequality (5.6.198) proves that $\sqrt{\sigma} m$ is bounded in $H^{-1/2}_{k_1}(\Gamma)$.

We take the limit in (5.6.147) for any test current j^t in $H^{-1/2}_{div}(\Gamma) \cap TL^2(\Gamma)$. The non-usual terms are those where G_i occurs. The first one is

$$\begin{cases} \left| \frac{1}{\bar{\varepsilon}} \int_\Gamma \int_\Gamma G_i(x - y) \, \text{div}_\Gamma \, j(x) \, \text{div}_\Gamma \, j^t(y) d\gamma(x) d\gamma(y) \right| \\ \qquad\qquad \le \frac{c}{\sigma} \|j^t\|_{TH^{-1/2}(\Gamma)}. \end{cases} \qquad (5.6.199)$$

The second one is

$$
\left\{
\begin{aligned}
&\left| \int_\Gamma \int_\Gamma G_i(x-y)(j(x) \cdot j^t(y)) d\gamma(x) d\gamma(y) \right| \\
&\qquad\qquad \leq c \, \|j\|_{TH_{k_1}^{-1/2}(\Gamma)} \, \|j^t\|_{TH_{k_1}^{-1/2}(\Gamma)} \\
&\qquad\qquad \leq \frac{c}{\sqrt{\sigma}} \, \|j\|_{TH^{-1/2}(\Gamma)} \, \|j^t\|_{TL^2(\Gamma)} .
\end{aligned}
\right.
\tag{5.6.200}
$$

The third one is

$$
\left\{
\begin{aligned}
&\left| \int_\Gamma \int_\Gamma (\nabla_y G_i(x-y) \cdot (m(x) \wedge j^t(y))) \, d\gamma(x) d\gamma(y) \right| \\
&\qquad\qquad \leq \frac{c}{\sqrt{\sigma}} \sqrt{\sigma} \, \|m\|_{TH_{k_1}^{-1/2}(\Gamma)} \, \|j^t\|_{TL^2(\Gamma)} .
\end{aligned}
\right.
\tag{5.6.201}
$$

Thus, it follows that j_0 is a solution of the perfect conducting problem

$$
\left\{
\begin{aligned}
&-\frac{i}{\omega}\frac{1}{\bar{\varepsilon}_e} \int_\Gamma \int_\Gamma G_e(x-y) \, \mathrm{div}_\Gamma \, j_0(x) \, \mathrm{div}_\Gamma \, j^t(y) d\gamma(x) d\gamma(y) \\
&\quad + i\omega\mu_e \int_\Gamma \int_\Gamma G_e(x-y) \, (j_0(x) \cdot j^t(y)) \, d\gamma(x) d\gamma(y) \\
&\quad = \int_\Gamma (E^{\mathrm{inc}} \cdot j^t) \, d\gamma, \quad \forall j^t \in H_{\mathrm{div}}^{-1/2}(\Gamma).
\end{aligned}
\right.
\tag{5.6.202}
$$

In the **second case**, the right-hand side of (5.6.198) is not bounded and tends to infinity. Let λ_σ be its value. Then, $\tilde{j} = j/\lambda_\sigma$ and $\tilde{m} = m/\lambda_\sigma$ are bounded in $H_{\mathrm{div}}^{-1/2}(\sigma)$. As in the first case, we take the limit and show that $\tilde{j}$ converges weakly towards a j_0 solution of the perfect conducting problem (5.6.199) with a zero right-hand side. If k_e^2 is not an eigenvalue of the interior Maxwell problem, this integral formulation is equivalent to the associated exterior problem and thus $j_0 = 0$. From (5.6.198) and the compact imbedding of $H^{3/2}(\Gamma)$ into $H^{1/2}(\Gamma)$ (for q) and of $H^{1/2}(\Gamma)$ into $H^{-1/2}(\Gamma)$ (for p), there exists a subsequence which converges strongly. This is contradictory since the λ_σ has a unit norm and tends to zero.

We have proved the weak convergence of j toward j_0. $\blacksquare$

Comment

This result of convergence gives a first justification to the perfect conducting model. It would be nice to know if the strong convergence is true in this setting. It is verified at least in a weaker space, and we can use this result to give a precise description of the skin effect and an estimation of the interior boundary layer. It would be also useful to obtain some error estimates in terms of the large conductivity parameter.

5.7   The Far Field

5.7.1   Far field and scattering amplitude

We present in this section some properties of the far field. We suppose that, outside a given ball, the constants ε and μ are those of the vacuum. We denote by $G(r)$ **the outgoing fundamental** of the Helmholtz equation

$$G(r) = \frac{1}{4\pi r} e^{ikr}, \quad k = \omega\sqrt{\varepsilon\mu}. \tag{5.7.1}$$

Let j and m be the electric and magnetic currents tangent to a surface Γ which contains the above ball. These currents are the following traces on Γ of the fields E and H,

$$j = -H \wedge n, \tag{5.7.2}$$

$$m = -E \wedge n. \tag{5.7.3}$$

Then, for any point exterior to Γ, the fields E and H admit the integral representation

$$\begin{cases} E(y) = i\omega\mu \int_\Gamma G(x-y)j(x)d\gamma(x) \\[2mm] \qquad + \dfrac{i}{\omega\varepsilon}\nabla\int_\Gamma G(x-y)\operatorname{div}_\Gamma j(x)d\gamma(x) \\[2mm] \qquad + \operatorname{curl}\int_\Gamma G(x-y)m(x)d\gamma(x), \quad y \notin \Gamma, \end{cases} \tag{5.7.4}$$

$$\begin{cases} H(y) = -i\omega\varepsilon \int_\Gamma G(x-y)m(x)d\gamma(x) \\[2mm] \qquad - \dfrac{i}{\omega\mu}\nabla\int_\Gamma G(x-y)\operatorname{div}_\Gamma m(x)d\gamma(x) \\[2mm] \qquad + \operatorname{curl}\int_\Gamma G(x-y)j(x)d\gamma(x), \quad y \notin \Gamma. \end{cases} \tag{5.7.5}$$

In order to find the expression of the far field, we expand the function G for large arguments. We have

$$\frac{1}{|x-y|} = \frac{1}{|y|}\left(1 + \frac{(x\cdot y)}{|y|^2} + \mathcal{O}\left(\frac{1}{|y|^2}\right)\right), \tag{5.7.6}$$

$$e^{ik|x-y|} = e^{ik|y|}e^{-ik\left(x\cdot\frac{y}{|y|}\right)}\left(1 + \mathcal{O}\left(\frac{1}{|y|}\right)\right), \tag{5.7.7}$$

$$\frac{e^{ik|x-y|}}{|x-y|} = \frac{e^{ik|y|}}{|y|}e^{-ik\left(x\cdot\frac{y}{|y|}\right)}\left(1 + \mathcal{O}\left(\frac{1}{|y|}\right)\right). \tag{5.7.8}$$

We also need an asymptotic expansion for the gradient and the second derivatives of the function G. They follow from (5.7.8) using the identities

$$\frac{d}{dr}G = \left(ik - \frac{1}{|y|}\right)G, \tag{5.7.9}$$

$$\nabla_y G = \left(ik - \frac{1}{|y-x|}\right)G\left(|y-x|\right)\frac{y-x}{|y-x|}, \tag{5.7.10}$$

$$\frac{d^2}{dr^2}G = -k^2G - \frac{2}{|y|}\frac{d}{dr}G, \tag{5.7.11}$$

$$\begin{cases}
\dfrac{\partial^2}{\partial y_i \partial y_j}G = \dfrac{d^2}{dr^2}G\left(|y-x|\right)\dfrac{(y_i - x_i)(y_j - x_j)}{|y-x|^2} \\[2mm]
\quad + \dfrac{1}{|y-x|}\dfrac{d}{dr}G\left(|y-x|\right)\left[\delta_i^j - \dfrac{(y_i - x_i)(y_j - x_j)}{|y-x|^2}\right] \\[3mm]
\quad = -k^2 G\left(|y-x|\right)\dfrac{(y_i - x_i)(y_j - x_j)}{|y-x|^2} \\[2mm]
\quad + \dfrac{1}{|y-x|}\dfrac{d}{dr}G\left(|y-x|\right)\left[\delta_i^j - 3\dfrac{(y_i - x_i)(y_j - x_j)}{|y-x|^2}\right].
\end{cases} \tag{5.7.12}$$

The main terms of ∇G and $D^2 G$ are

$$\begin{cases}
\dfrac{\partial}{\partial y_i}G \sim \dfrac{ik}{4\pi}\dfrac{e^{ik|y|}}{|y|}e^{-ik(x\cdot\frac{y}{|y|})}\dfrac{y_i}{|y|}, \\[4mm]
\dfrac{\partial^2}{\partial y_i \partial y_j}G \sim -\dfrac{k^2}{4\pi}\dfrac{e^{ik|y|}}{|y|}e^{-ik(x\cdot\frac{y}{|y|})}\dfrac{y_i\, y_j}{|y|^2}.
\end{cases} \tag{5.7.13}$$

From (5.7.4) et (5.7.5), it follows that the main terms of E and H are

$$\begin{cases}
E(y) \sim \dfrac{1}{4\pi}\dfrac{e^{ik|y|}}{|y|} \\[3mm]
\quad \times \left[i\omega\mu\displaystyle\int_\Gamma e^{-ik(x\cdot y)/|y|}\left[j(x) + \dfrac{i}{k}\dfrac{y}{|y|}\,\mathrm{div}_\Gamma\, j(x)\right]d\gamma(x)\right] \\[3mm]
\quad + \dfrac{1}{4\pi}\dfrac{e^{ik|y|}}{|y|}ik\displaystyle\int_\Gamma e^{-ik(x\cdot y)/|y|}\left(\dfrac{y}{|y|}\wedge m(x)\right)d\gamma(x) \\[3mm]
\quad = \dfrac{e^{ik|y|}}{4\pi\,|y|}\left[i\omega\mu\displaystyle\int_\Gamma e^{-ik(x\cdot y)/|y|}\left[j(x) - \left(j(x)\cdot\dfrac{y}{|y|}\right)\right]d\gamma(x)\right. \\[3mm]
\quad + ik\displaystyle\int_\Gamma e^{-ik(x\cdot y)/|y|}\left(\dfrac{y}{|y|}\wedge m(x)\right)d\gamma(x)\Big],
\end{cases} \tag{5.7.14}$$

$$\begin{cases}
H(y) \sim \dfrac{1}{4\pi} \dfrac{e^{ik|y|}}{|y|} \\[2mm]
\qquad \times \left[-i\omega\varepsilon \displaystyle\int_\Gamma e^{-ik(x \cdot y)/|y|} \left[m(x) + \dfrac{i}{k}\dfrac{y}{|y|} \operatorname{div}_\Gamma m(x) \right] d\gamma(x) \right] \\[2mm]
\qquad + \dfrac{1}{4\pi} \dfrac{e^{ik|y|}}{|y|} ik \displaystyle\int_\Gamma e^{-ik(x \cdot y)/|y|} \left(\dfrac{y}{|y|} \wedge j(x) \right) d\gamma(x) \\[2mm]
= \dfrac{1}{4\pi} \dfrac{e^{ik|y|}}{|y|} \left[i\omega\varepsilon \displaystyle\int_\Gamma e^{-ik(x \cdot y)/|y|} \right. \\[2mm]
\qquad \times \left[\left(m(x) \cdot \dfrac{y}{|y|} \right) - m(x) \right] d\gamma(x) \\[2mm]
\qquad \left. + ik \displaystyle\int_\Gamma e^{-ik(x \cdot y)/|y|} \left(\dfrac{y}{|y|} \wedge j(x) \right) d\gamma(x) \right].
\end{cases} \tag{5.7.15}$$

These two expressions describe the far field. Let θ, φ denote the two angles defined by the vector $y/|y|$ in spherical coordinates. We define the **scattering amplitude** A by

$$E(y) \sim \frac{e^{ik|y|}}{|y|} A(\theta, \varphi), \tag{5.7.16}$$

and thus, A is given by

$$\begin{cases}
A(\theta, \varphi) = \left[\dfrac{i\omega\mu}{4\pi} \displaystyle\int_\Gamma e^{-ik(x \cdot y)/|y|} \left[j(x) - \left(j(x) \cdot \dfrac{y}{|y|} \right) \right] d\gamma(x) \right. \\[2mm]
\qquad \left. + \dfrac{ik}{4\pi} \displaystyle\int_\Gamma e^{-ik(x \cdot y)/|y|} \left(\dfrac{y}{|y|} \wedge m(x) \right) d\gamma(x) \right].
\end{cases} \tag{5.7.17}$$

The field H has the expansion

$$H(y) \sim -\frac{e^{ik|y|}}{|y|} \sqrt{\frac{\varepsilon}{\mu}} \left(A(\theta, \varphi) \wedge \frac{y}{|y|} \right), \tag{5.7.18}$$

The **complex Poynting vector** is defined as

$$S = \frac{c}{8\pi} (E \wedge \overline{H}). \tag{5.7.19}$$

From the expression of the scattering amplitude A, we can deduce the expression of the **scattered energy** which is the limit of the integral on the sphere S_R of the real part of the flux of the Poynting vector as R tends to infinity. The following quantity is proportional to this energy:

$$\begin{cases}
I = \lim_{R \to \infty} - \displaystyle\int_{S_R} \Re \left(\left(E \wedge \dfrac{y}{|y|} \right) \cdot \overline{H} \right) d\sigma \\[2mm]
= 4\pi \sqrt{\dfrac{\varepsilon}{\mu}} \displaystyle\int_S \left| A(\theta, \varphi) \wedge \dfrac{y}{|y|} \right|^2 d\sigma.
\end{cases} \tag{5.7.20}$$

Lemma 5.7.1 *Consider an exterior domain Ω_e whose boundary is Γ, and let the real coefficients ε and μ be given. Consider fields E and H which solve the Maxwell system in Ω_e and satisfy the outgoing Silver-Müller radiation condition. We have*

$$4\pi\sqrt{\frac{\varepsilon}{\mu}}\int_S \left| A(\theta,\varphi) \wedge \frac{y}{|y|} \right|^2 d\sigma = \int_\Gamma \Re\left((E \wedge n) \cdot \overline{H}\right) d\gamma. \qquad (5.7.21)$$

Proof

Let R be large enough and S_R be the sphere of radius R. Let E, H and F, L be two solutions of the Maxwell system. From the Green formula (5.4.33) (the minus sign is related to the orientation of the normal n to Γ), it follows that

$$\begin{cases} \int_{S_R} ((E \wedge n) \cdot L) - ((F \wedge n) \cdot H)\, d\sigma \\ \qquad = -\int_\Gamma ((E \wedge n) \cdot L) - ((F \wedge n) \cdot H)\, d\gamma. \end{cases} \qquad (5.7.22)$$

From the Silver-Müller radiation condition, choosing $F = \overline{E}$ and $L = \overline{H}$, the right-hand side has a limit when R tends to infinity which is the quantity 2 I, given by (5.7.20). ∎

We will now rewrite these properties in two special cases, where the incident field which creates the scattered field is a solution of the Maxwell system in the domain Ω_e but does not satisfy the outgoing radiation condition.

Let us denote by $(E_{\text{inc}}, H_{\text{inc}})$ the incident field. The scattered field which satisfies the outgoing Silver-Müller radiation condition is denoted by (E_d, H_d). The total field is

$$E = E_{\text{inc}} + E_d, \qquad (5.7.23)$$

$$H = H_{\text{inc}} + H_d \qquad (5.7.24)$$

and is subject to a boundary condition on Γ. We consider the case where this boundary condition is that of the perfect conductor

$$E \wedge n|_\Gamma = 0. \qquad (5.7.25)$$

From the Green formula (5.7.22), it follows that

$$\Re \int_{S_R} \left((E \wedge n) \cdot \overline{H}\right) d\sigma = 0. \qquad (5.7.26)$$

Notice that all the following properties are still valid for more general boundary conditions, as long as we have (5.7.26). This is the case for non-dissipative boundary conditions.

The first special case that we study is the one where the incident field $(E_{\text{inc}}, H_{\text{inc}})$ is an **incoming spherical wave**, given as a sum of **transverse**

electric multipoles and **transverse magnetic multipoles**. When the origin is chosen in the domain Ω_i, it takes the form

$$
\begin{cases}
E_{\text{inc}}(x) = \sum_{\ell=1}^{\infty} \sum_{m=-\ell}^{\ell} \left[\alpha_\ell^m \frac{h_\ell^{(2)}(kr)}{h_\ell^{(2)}(k)} T_\ell^m(\theta, \varphi) + \frac{ik}{\omega\varepsilon} \beta_\ell^m \right. \\
\left. \times \left[\frac{\ell+1}{2\ell+1} \frac{h_{\ell-1}^{(2)}(kr)}{h_\ell^{(2)}(k)} I_{\ell-1}^m(\theta, \varphi) + \frac{\ell}{2\ell+1} \frac{h_{\ell+1}^{(2)}(kr)}{h_\ell^{(2)}(k)} N_{\ell+1}^m(\theta, \varphi) \right] \right],
\end{cases}
\tag{5.7.27}
$$

$$
\begin{cases}
H_{\text{inc}}(x) = \sum_{\ell=1}^{\infty} \sum_{m=-\ell}^{\ell} \left[\beta_\ell^m \frac{h_\ell^{(2)}(kr)}{h_\ell^{(2)}(k)} T_\ell^m(\theta, \varphi) - \frac{ik}{\omega\mu} \alpha_\ell^m \right. \\
\left. \times \left[\frac{\ell+1}{2\ell+1} \frac{h_{\ell-1}^{(2)}(kr)}{h_\ell^{(2)}(k)} I_{\ell-1}^m(\theta, \varphi) + \frac{\ell}{2\ell+1} \frac{h_{\ell+1}^{(2)}(kr)}{h_\ell^{(2)}(k)} N_{\ell+1}^m(\theta, \varphi) \right] \right],
\end{cases}
\tag{5.7.28}
$$

and thus $E_{\text{inc}}, H_{\text{inc}}$ satisfy the Maxwell system in Ω_e.

The scattered field (E_d, H_d) has the form

$$
\begin{cases}
E_d(x) = \sum_{\ell=1}^{\infty} \sum_{m=-\ell}^{\ell} \left[\gamma_\ell^m \frac{h_\ell^{(1)}(kr)}{h_\ell^{(1)}(k)} T_\ell^m(\theta, \varphi) + \frac{ik}{\omega\varepsilon} \delta_\ell^m \right. \\
\left. \times \left[\frac{\ell+1}{2\ell+1} \frac{h_{\ell-1}^{(1)}(kr)}{h_\ell^{(1)}(k)} I_{\ell-1}^m(\theta, \varphi) + \frac{\ell}{2\ell+1} \frac{h_{\ell+1}^{(1)}(kr)}{h_\ell^{(1)}(k)} N_{\ell+1}^m(\theta, \varphi) \right] \right],
\end{cases}
\tag{5.7.29}
$$

$$
\begin{cases}
H_d(x) = \sum_{\ell=1}^{\infty} \sum_{m=-\ell}^{\ell} \left[\delta_\ell^m \frac{h_\ell^{(1)}(kr)}{h_\ell^{(1)}(k)} T_\ell^m(\theta, \varphi) - \frac{ik}{\omega\mu} \gamma_\ell^m \right. \\
\left. \times \left[\frac{\ell+1}{2\ell+1} \frac{h_{\ell-1}^{(1)}(kr)}{h_\ell^{(1)}(k)} I_{\ell-1}^m(\theta, \varphi) + \frac{\ell}{2\ell+1} \frac{h_{\ell+1}^{(1)}(kr)}{h_\ell^{(1)}(k)} N_{\ell+1}^m(\theta, \varphi) \right] \right].
\end{cases}
\tag{5.7.30}
$$

Lemma 5.7.2 *When the incident field admits, for some real positive s, a trace on the unit sphere in the space $TH^s(S)$, it holds that*

$$
\begin{cases}
\lim_{R \to \infty} \int_{S_R} |E_{\text{inc}} \wedge n|^2 d\sigma = \lim_{R \to \infty} \int_{S_R} |E_d \wedge n|^2 d\sigma \\
\qquad\qquad = \int_S \left| A(\theta, \varphi) \wedge \frac{y}{|y|} \right|^2 d\sigma,
\end{cases}
\tag{5.7.31}
$$

$$
\begin{cases}
\lim_{R \to \infty} \int_{S_R} |H_{\text{inc}} \wedge n|^2 d\sigma = \lim_{R \to \infty} \int_{S_R} |H_d \wedge n|^2 d\sigma \\
\qquad\qquad = \frac{\varepsilon}{\mu} \int_S \left| A(\theta, \varphi) \wedge \frac{y}{|y|} \right|^2 d\sigma.
\end{cases}
\tag{5.7.32}
$$

This can be expressed in the following way: **The operator which asso-ciates to the scattering amplitude of an incoming spherical wave, the scattering amplitude of the corresponding outgoing spherical wave is unitary.**

Proof

From (5.7.26), it follows that

$$\Re \int_{S_R} \left(((E_{\text{inc}} + E_d) \wedge n) \cdot (\overline{H}_{\text{inc}} + \overline{H}_d) \right) d\sigma = 0. \tag{5.7.33}$$

The theorem comes from (5.7.21) and the outgoing Silver-Müller radiation condition, taking the limit in the equality (5.7.33). ∎

In the second special case we shall consider, the **incident field** is a **plane wave**. It is defined by a unit propagation vector y and a unit polarization vector β orthogonal to the vector y. Thus, the incident field is

$$\begin{cases} E_{\text{inc}} = \beta\, e^{ik(y \cdot x)}, \quad |y| = 1, \quad (y \cdot \beta) = 0, \\ H_{\text{inc}} = \sqrt{\varepsilon/\mu}\,(y \wedge \beta)\, e^{ik(y \cdot x)}. \end{cases} \tag{5.7.34}$$

We **denote by** $A(x; y, \beta)$ **the scattering amplitude** of the scattered field in the direction of the unit vector x. Its component along the direction of a unit vector α orthogonal to x (called the **polarization in the direction of the vector** α) is $(A(x; y, \beta) \cdot \alpha)$ which is **denoted by** $A(x, \alpha; y, \beta)$. Let $(F_{\text{inc}}, L_{\text{inc}})$ be another incident plane wave

$$\begin{cases} F_{\text{inc}} = \gamma\, e^{ik(z \cdot x)}, \quad |z| = 1, \quad (z \cdot \gamma) = 0, \\ L_{\text{inc}} = \sqrt{\varepsilon/\mu}\,(z \wedge \gamma)\, e^{ik(z \cdot x)}. \end{cases} \tag{5.7.35}$$

The associated scattered field is denoted by (F_d, L_d). The incident field whose vector of propagation is $-z$, and which has the same polarization vector, is $(\overline{F}_{\text{inc}}, -\overline{L}_{\text{inc}})$.

Lemma 5.7.3 *We have*

$$\begin{cases} \lim_{R \to \infty} \int_{S_R} \left[\left((\overline{F}_{\text{inc}} \wedge \frac{x}{|x|}) \cdot H_d \right) + \left((E_d \wedge \frac{x}{|x|}) \cdot \overline{L}_{\text{inc}} \right) \right] d\sigma \\ \qquad = 4\pi \sqrt{\varepsilon/\mu} A(z, \gamma; y, \beta). \end{cases} \tag{5.7.36}$$

Proof

We expand (5.7.36) in the form

$$\begin{cases} \int_{S_R} \left[\left((E_d \wedge \frac{x}{|x|}) \cdot \overline{L}_{\text{inc}} \right) + \left((\overline{F}_{\text{inc}} \wedge \frac{x}{|x|}) \cdot H_d \right) \right] d\sigma \\ = \int_{S_R} \left(\left(\left(E_d - \sqrt{\mu/\varepsilon} \left(H_d \wedge \frac{x}{|x|} \right) \right) \wedge \frac{x}{|x|} \right) \cdot \overline{L}_{\text{inc}} \right) d\sigma \\ + \int_{S_R} \left(\left(\left(\overline{F}_{\text{inc}} + \sqrt{\mu/\varepsilon} \left(\overline{L}_{\text{inc}} \wedge \frac{x}{|x|} \right) \right) \wedge \frac{x}{|x|} \right) \cdot H_d \right) d\sigma. \end{cases} \tag{5.7.37}$$

From (5.7.8) and (5.7.12), it follows that

$$\left| \left(E_d - \sqrt{\frac{\mu}{\varepsilon}} \left(H_d \wedge \frac{x}{|x|} \right) \right) \wedge \frac{x}{|x|} \right| \leq \frac{c}{r^3}, \quad \text{for large } r. \qquad (5.7.38)$$

Thus, the first term in the right-hand side of (5.7.37) has a zero limit.

We study the limit of the second term in the right-hand side, in the frame of S_R associated to the two angles θ, φ, with the following choice: the propagation vector z is $\vec{e_3}$, while the polarization vector γ is $\vec{e_1}$. From (5.7.18), it follows that

$$\begin{cases} \displaystyle\int_{S_R} \left(\left(\left(\overline{F}_{\text{inc}} + \sqrt{\frac{\mu}{\varepsilon}} \left(\overline{L}_{\text{inc}} \wedge \frac{x}{|x|} \right) \right) \wedge \frac{x}{|x|} \right) \cdot H_d \right) d\sigma \\[2mm] \quad = -\sqrt{\frac{\varepsilon}{\mu}} \frac{e^{ikR}}{R} R^2 \displaystyle\int_S e^{-ikR\cos\theta} \\[2mm] \quad \times \left(\left(\gamma - (\gamma \wedge z) \wedge \frac{x}{|x|} \right) \cdot A(\theta, \varphi; y, \beta) \right) \sin\theta d\theta d\varphi, \end{cases} \qquad (5.7.39)$$

$$\begin{cases} \left(\left(\gamma - (\gamma \wedge z) \wedge \frac{x}{|x|} \right) \cdot A(\theta, \varphi; y, \beta) \right) \\[2mm] \quad = (\gamma \cdot A(\theta, \varphi; y, \beta)) \\[2mm] \qquad + (\gamma \cdot A(\theta, \varphi; y, \beta)) \left(z \cdot \frac{x}{|x|} \right) - (z \cdot A(\theta, \varphi; y, \beta)) \left(\gamma \cdot \frac{x}{|x|} \right) \\[2mm] \quad = (1 + \cos\theta)(\gamma \cdot A(\theta, \varphi; y, \beta)) + \cos\theta\cos\varphi \, (z \cdot A(\theta, \varphi; y, \beta)), \end{cases} \qquad (5.7.40)$$

$$\begin{cases} \displaystyle\int_{S_R} \left(\left(\left(\overline{F}_{\text{inc}} + \sqrt{\frac{\mu}{\varepsilon}} \left(\overline{L}_{\text{inc}} \wedge \frac{x}{|x|} \right) \right) \wedge \frac{x}{|x|} \right) \cdot H_d \right) d\sigma \\[2mm] \quad = -\sqrt{\frac{\varepsilon}{\mu}} \frac{e^{ikR}}{R} R^2 \displaystyle\int_S e^{-ikR\cos\theta} \\[2mm] \quad \times (1 + \cos\theta)(\gamma \cdot A(\theta, \varphi; y, \beta)) \sin\theta d\theta d\varphi. \end{cases} \qquad (5.7.41)$$

Using an integration by parts with respect to the variable θ, we obtain

$$\begin{cases} \displaystyle\int_0^\pi e^{-ikR\cos\theta} \varphi(\theta) \sin\theta d\theta = -\frac{1}{ikR} \\[2mm] \quad \times \left[\displaystyle\int_0^\pi e^{-ikR\cos\theta} \frac{d\varphi}{d\theta}(\theta) d\theta + \left[e^{-ikR}\varphi(0) - e^{ikR}\varphi(\pi) \right] \right]. \end{cases} \qquad (5.7.42)$$

(5.7.41) and (5.7.42) yield

$$
\left\{
\begin{aligned}
& Re^{ikR} \int_S e^{-ikR\cos\theta} \\
& \qquad \times (ik(1+\cos\theta))\,(\gamma\cdot A(\theta,\varphi;y,\beta))\sin\theta\,d\theta\,d\varphi \\
& = \frac{e^{ikR}}{ik}\int_0^\pi \int_0^{2\pi} e^{-ikR\cos\theta}\Big[ik\sin\theta\,(\gamma\cdot A(\theta,\varphi;y,\beta)) \\
& \qquad - ik(1+\cos\theta)\frac{d}{d\theta}\,(\gamma\cdot A(\theta,\varphi;y,\beta))\Big]d\theta\,d\varphi \\
& \qquad - 2\int_0^{2\pi}(\gamma\cdot A(0,\varphi;y,\beta))\,d\varphi.
\end{aligned}
\right.
\tag{5.7.43}
$$

The **stationary phase** technique proves that the first term of the right-hand side tends to zero as $1/\sqrt{R}$. Using (5.7.39) and (5.7.43), and taking the limit, we obtain

$$
\left\{
\begin{aligned}
& \lim_{R\to\infty}\int_{S_R}\left(\left(\left(\overline{F}_{\text{inc}}+\sqrt{\frac{\mu}{\varepsilon}}\left(\overline{L}_{\text{inc}}\wedge\frac{x}{|x|}\right)\right)\wedge\frac{x}{|x|}\right)\cdot H_d\right)d\sigma \\
& = 4\pi\sqrt{\frac{\varepsilon}{\mu}}\,(\gamma\cdot A(z;y,\beta)),
\end{aligned}
\right.
\tag{5.7.44}
$$

which proves the lemma.   ∎

Theorem 5.7.1 *The scattering amplitude satisfies the* **reciprocity principle**, *which is expressed by the identity*

$$
A(y,\beta;z,\gamma)=A(-z,\gamma;-y,\beta),
\tag{5.7.45}
$$

Equivalently: **The polarization along the vector γ of the scattering amplitude created by an incident incoming plane wave with propagation vector y and polarization vector β, in the outgoing direction of the vector z, is equal to the polarization along the vector β of the scattering amplitude created by an incident incoming plane wave with propagation vector z and polarization vector γ, in the outgoing direction of the vector y.**

Proof
The previous lemma shows that

$$
\left\{
\begin{aligned}
& 4\pi\sqrt{\frac{\varepsilon}{\mu}}A(y,\beta;z,\gamma) \\
& = \lim_{R\to\infty}\int_{S_R}\left[\left(\overline{E}_{\text{inc}}\wedge\frac{x}{|x|}\right)\cdot L_d\right)+\left(\left(F_d\wedge\frac{x}{|x|}\right)\cdot\overline{H}_{\text{inc}}\right)\right]d\sigma.
\end{aligned}
\right.
\tag{5.7.46}
$$

The field $(\overline{E}_{\text{inc}},-\overline{H}_{\text{inc}})$ is a plane wave with propagation vector $-y$ and polarization vector β, and thus it follows from (5.7.22) that

$$\left\{ \begin{array}{l} \displaystyle\int_{S_R} \left[\left((\overline{E}_{\text{inc}} \wedge \frac{x}{|x|}) \cdot L_d \right) + \left((F_d \wedge \frac{x}{|x|}) \cdot \overline{H}_{\text{inc}} \right) \right] d\sigma \\ \displaystyle = -\int_\Gamma \left[((\overline{E}_{\text{inc}} \wedge n) \cdot L_d) + ((F_d \wedge n) \cdot \overline{H}_{\text{inc}}) \right] d\gamma. \end{array} \right. \tag{5.7.47}$$

Let (E_d, H_d) be the solution of the Maxwell system associated to the incident field $(\overline{E}_{\text{inc}}, -\overline{H}_{\text{inc}})$. From the boundary conditions, (5.7.47) is also

$$\left\{ \begin{array}{l} \displaystyle -\int_\Gamma \left[((\overline{E}_{\text{inc}} \wedge n) \cdot L_d) + ((F_d \wedge n) \cdot \overline{H}_{\text{inc}}) \right] d\gamma \\ \displaystyle = \int_\Gamma \left[((E_d \wedge n) \cdot L_d) + ((F_{\text{inc}} \wedge n) \cdot \overline{H}_{\text{inc}}) \right] d\gamma. \end{array} \right. \tag{5.7.48}$$

Besides, using (5.7.22), we have

$$\int_\Gamma \left[((\overline{E}_{\text{inc}} \wedge n) \cdot L_{\text{inc}}) + ((F_{\text{inc}} \wedge n) \cdot \overline{H}_{\text{inc}}) \right] d\gamma = 0, \tag{5.7.49}$$

from which follows

$$\left\{ \begin{array}{l} \displaystyle\int_{S_R} \left[\left((\overline{E}_{\text{inc}} \wedge \frac{x}{|x|}) \cdot L_d \right) + \left((F_d \wedge \frac{x}{|x|}) \cdot \overline{H}_{\text{inc}} \right) \right] d\sigma \\ \displaystyle = \int_\Gamma \left[((E_d \wedge n) \cdot L_d) - ((\overline{E}_{\text{inc}} \wedge n) \cdot L_{\text{inc}}) \right] d\gamma, \end{array} \right. \tag{5.7.50}$$

which can be written

$$\left\{ \begin{array}{l} \displaystyle\int_\Gamma \left[((E_d \wedge n) \cdot L_d) - ((\overline{E}_{\text{inc}} \wedge n) \cdot L_{\text{inc}}) \right] d\gamma \\ \displaystyle = \int_\Gamma \left[((E_d \wedge n) \cdot L_d) - ((F_d \wedge n) \cdot H_d) \right] d\gamma \\ \displaystyle \quad -\int_\Gamma \left[((F_{\text{inc}} \wedge n) \cdot H_d) - ((E_d \wedge n) \cdot L_{\text{inc}}) \right] d\gamma. \end{array} \right. \tag{5.7.51}$$

Moreover, we have

$$\left\{ \begin{array}{l} \displaystyle -\int_\Gamma \left[((E_d \wedge n) \cdot L_d) - ((F_d \wedge n) \cdot H_d) \right] d\gamma \\ \displaystyle = \int_{S_R} \left[\left((E_d \wedge \frac{x}{|x|}) \cdot L_d \right) - \left((F_d \wedge \frac{x}{|x|}) \cdot H_d \right) \right] d\sigma, \end{array} \right. \tag{5.7.52}$$

which shows that the limit of this last term when R tends to infinity is zero. Using again (5.7.22), we obtain

$$\left\{ \begin{array}{l} \displaystyle \lim_{R\to\infty} \int_{S_R} \left[\left((\overline{E}_{\text{inc}} \wedge \frac{x}{|x|}) \cdot L_d \right) + \left((F_d \wedge \frac{x}{|x|}) \cdot \overline{H}_{\text{inc}} \right) \right] d\sigma \\ \displaystyle = \lim_{R\to\infty} \int_{S_R} \left[\left((F_{\text{inc}} \wedge \frac{x}{|x|}) \cdot H_d \right) - \left((E_d \wedge \frac{x}{|x|}) \cdot L_{\text{inc}} \right) \right] d\sigma, \end{array} \right. \tag{5.7.53}$$

which is the reciprocity principle.   ∎

Lemma 5.7.4 *The scattering amplitude satisfies*

$$
\begin{cases}
2\pi \left[A(z, \gamma; y, \beta) + \overline{A}(y, \beta; z, \gamma) \right] \\
\\
\quad = \displaystyle\int_S \left(A(x; y, \beta) \cdot \overline{A}(x; z, \gamma) \right) d\sigma(x).
\end{cases}
\tag{5.7.54}
$$

Proof

Let (E_d, H_d) be the solution of the Maxwell system associated to the incident field $(E_{\text{inc}}, H_{\text{inc}})$ and (F_d, L_d) be the solution of the Maxwell system associated to the incident field $(F_{\text{inc}}, L_{\text{inc}})$. From the Green formula (5.7.22) and the use of the boundary condition follows

$$
\begin{cases}
\displaystyle\int_{S_R} \left(((E_{\text{inc}} + E_d) \wedge n) \cdot (\overline{L}_{\text{inc}} + \overline{L}_d) \right) d\sigma \\
\\
\quad + \displaystyle\int_{S_R} \left(((\overline{F}_{\text{inc}} + \overline{F}_d) \wedge n) \cdot (H_{\text{inc}} + H_d) \right) d\sigma = 0.
\end{cases}
\tag{5.7.55}
$$

Besides, we have

$$
\int_{S_R} \left[((\overline{F}_{\text{inc}} \wedge n) \cdot H_{\text{inc}}) + ((E_{\text{inc}} \wedge n) \cdot \overline{L}_{\text{inc}}) \right] d\sigma = 0,
\tag{5.7.56}
$$

and expanding (5.7.55), we obtain

$$
\begin{cases}
\displaystyle\int_{S_R} \left[((E_{\text{inc}} \wedge n) \cdot \overline{L}_d) + ((\overline{F}_d \wedge n) \cdot H_{\text{inc}}) \right] d\sigma \\
\\
\quad + \displaystyle\int_{S_R} \left[((E_d \wedge n) \cdot \overline{L}_{\text{inc}}) + ((\overline{F}_{\text{inc}} \wedge n) \cdot H_d) \right] d\sigma \\
\\
\quad + \displaystyle\int_{S_R} \left[((E_d \wedge n) \cdot \overline{L}_d) + ((\overline{F}_d \wedge n) \cdot H_d) \right] d\sigma = 0.
\end{cases}
\tag{5.7.57}
$$

Lemma 5.7.3 provides the limit of the two first terms. Using the expressions (5.7.16) and (5.7.18) of the far field, we obtain the limit of the last term. This shows (5.7.54).   ■

Let a density $\varphi(x, \alpha)$ be defined for any point x on the sphere S and any unit vector α orthogonal to x, i.e., for any point of the unit circle of the tangent plane at x to the sphere S. We denote by TS, the set constituted of the points (x, α) where x is a point on the sphere S and α is a point of the unit circle of the tangent plane at x to the sphere S. We associate to the scattering amplitude the integral operator A whose kernel is $A(y, \beta; x, \alpha)$:

$$
A\varphi(y, \beta) = \frac{1}{2\pi^2} \int_S \int_{C(x)} A(y, \beta; x, \alpha) \varphi(x, \alpha) d\sigma(x) d\theta(\alpha).
\tag{5.7.58}
$$

Theorem 5.7.2 *The operator $I - A$, where A is given by (5.7.58), is an isometry in the space $L^2(TS)$, i.e.,*

$$|(I - A)\varphi|^2_{L^2(TS)} = |\varphi|^2_{L^2(TS)}. \tag{5.7.59}$$

Proof

Using the above definition of the polarization vector, we have

$$\begin{cases} \frac{1}{\pi} \int_{C(x)} \left(A(x,\alpha;y,\beta) \cdot \overline{A}(x,\alpha;z,\gamma) \right) d\theta \\ \qquad\qquad = \left(A(x;y,\beta) \cdot \overline{A}(x;z,\gamma) \right). \end{cases} \tag{5.7.60}$$

We then easily check that the identity (5.7.59) is equivalent to (5.7.54). ∎

5.7.2   Integral equations and far field

We show how the far field can used as an auxiliary unknown when solving the integral equations associated to the Maxwell system, following an original idea from B. Desprès [70]. We consider the perfect conductor problem in a homogeneous domain. The integral representation is expressed in terms of the tangent electric current j. This electric current j is determined by the perfect conductor boundary condition on the surface Γ, which is stated in (5.5.12). This gives the integral equation (5.6.7), which we prefer to write in the form (P_Γ stands for the projection operator on the tangent plane to Γ)

$$\begin{cases} -P_\Gamma E_{\text{inc}}(y) = i\omega\mu_0 P_\Gamma \int_\Gamma G(x - y)j(x)d\gamma(x) \\ \qquad\qquad + \frac{i}{\omega\varepsilon_0} \nabla_\Gamma \int_\Gamma G(x - y) \operatorname{div}_\Gamma j(x)d\gamma(x). \end{cases} \tag{5.7.61}$$

We decompose the function G in its real and imaginary parts

$$G(x - y) = \frac{1}{4\pi} \left[\frac{\cos(k|x - y|)}{|x - y|} + i\frac{\sin(k|x - y|)}{|x - y|} \right]. \tag{5.7.62}$$

From (5.4.29), the imaginary part of the fundamental solution G satisfies

$$\frac{\sin(k|x - y|)}{4\pi k|x - y|} = \frac{1}{(4\pi)^2} \int_S e^{ik(y - x \cdot z)} d\sigma(z). \tag{5.7.63}$$

Differentiating, we obtain

$$\nabla_y \frac{\sin(k|x - y|)}{4\pi k|x - y|} = \frac{ik}{(4\pi)^2} \int_S e^{ik(y - x \cdot z)} z \, d\sigma(z). \tag{5.7.64}$$

The far field is given by

$$E(y) \sim \frac{e^{ik|y|}}{|y|} A\left(\frac{y}{|y|}\right), \tag{5.7.65}$$

$$H(y) \sim -\frac{e^{ik|y|}}{|y|}\sqrt{\frac{\varepsilon_0}{\mu_0}}\left(A(\frac{y}{|y|})\wedge\frac{y}{|y|}\right), \qquad (5.7.66)$$

where the scattering amplitude A is given by

$$A(\frac{y}{|y|}) = \frac{i\omega\mu_0}{4\pi}\int_\Gamma e^{-ik\left(x\cdot\frac{y}{|y|}\right)}\left[j(x)-\left(j(x)\cdot\frac{y}{|y|}\right)\right]d\gamma(x). \qquad (5.7.67)$$

Lemma 5.7.5 *The integral equation (5.7.61) is equivalent to the following system, with unknowns j and A:*

$$\left\{\begin{array}{l}
i\omega\mu_0 P_\Gamma \displaystyle\int_\Gamma \frac{\cos(k\,|x-y|)}{4\pi\,|x-y|}j(x)d\gamma(x) \\[2em]
\quad + \displaystyle\frac{i}{\omega\varepsilon_0}\nabla_\Gamma \int_\Gamma \frac{\cos(k\,|x-y|)}{4\pi\,|x-y|}\,\mathrm{div}_\Gamma\,j(x)d\gamma(x) \\[2em]
\quad + \displaystyle\frac{ik}{4\pi}P_\Gamma \int_S e^{ik(y\cdot z)}A(z)\,d\sigma(z) = -P_\Gamma\,E_{\mathrm{inc}}(y), \\[2em]
-\displaystyle\frac{ik}{4\pi}\int_\Gamma e^{-ik\,(x\cdot z)}\,[j(x)-(j(x)\cdot z)\,z]\,d\gamma(x)+\omega\varepsilon_0\,A(z) = 0.
\end{array}\right. \qquad (5.7.68)$$

This system admits a variational formulation with $j \in H_{\mathrm{div}}^{1/2}(\Gamma)$ and $A \in TL^2(S)$ which is

$$\left\{\begin{array}{l}
i\omega\mu_0 \displaystyle\int_\Gamma\int_\Gamma \frac{\cos(k\,|x-y|)}{4\pi\,|x-y|}\,(j(x)\cdot j^t(y))\,d\gamma(x)d\gamma(y) \\[2em]
\quad - \displaystyle\frac{i}{\omega\varepsilon_0}\int_\Gamma\int_\Gamma \frac{\cos(k\,|x-y|)}{4\pi\,|x-y|}\,\mathrm{div}_\Gamma\,j(x)\,\mathrm{div}_\Gamma\,j^t(y)d\gamma(x)d\gamma(y) \\[2em]
\quad + \displaystyle\frac{ik}{4\pi}\int_\Gamma\int_S e^{ik(y\cdot z)}\,(A(z)\cdot j^t(y))\,d\sigma(z)d\gamma(y) \\[2em]
\qquad\qquad = -\displaystyle\int_\Gamma (E_{\mathrm{inc}}\cdot j^t(y))\,d\gamma(y) \\[2em]
\quad - \displaystyle\frac{ik}{4\pi}\int_\Gamma\int_S e^{-ik\,(y\cdot z)}\,(j(y)\cdot A^t(z))\,d\sigma(z)d\gamma(y) \\[2em]
\qquad\qquad + \omega\varepsilon_0 \displaystyle\int_S (A(z)\cdot A^t(z))\,d\sigma(z) = 0,
\end{array}\right. \qquad (5.7.69)$$

for any $j^t \in H_{\mathrm{div}}^{1/2}(\Gamma)$ and any $A^t \in TL^2(S)$.

Proof

From (5.7.63) and (5.7.64), we write

$$
\begin{cases}
i\omega\mu_0 P_\Gamma \displaystyle\int_\Gamma \frac{sin(k\,|x-y|)}{|x-y|} j(x)d\gamma(x) \\[4mm]
\qquad +\dfrac{i}{\omega\varepsilon_0}\nabla_\Gamma \displaystyle\int_\Gamma \frac{sin(k\,|x-y|)}{|x-y|}\,\mathrm{div}_\Gamma\, j(x)d\gamma(x) \\[4mm]
= \dfrac{ik\omega\mu_0}{4\pi} P_\Gamma \displaystyle\int_\Gamma\!\!\int_S e^{ik(y-x\,\cdot\,z)}\left[j(x)-\dfrac{1}{ik}\,\mathrm{div}_\Gamma\, j(x)\, z\right]d\sigma d\gamma, \\[4mm]
= \dfrac{ik\omega\mu_0}{4\pi} P_\Gamma \displaystyle\int_\Gamma\!\!\int_S e^{ik(y-x\,\cdot\,z)}\left[j(x)-(j(x)\cdot z)\,z\right]d\sigma d\gamma.
\end{cases}
\tag{5.7.70}
$$

From the expression of the scattering amplitude A, we deduce (5.7.68). ∎

References

[1] **Abboud, T.**
Etude mathématique et numérique de quelques problèmes de diffraction d'ondes électromagnétiques
Thèse. Ecole Polytechnique (1991)

[2] **Abboud, T. and Nédélec, J.C.**
Electromagnetic waves in inhomogeneous medium
J. Math. Anal. Appl. **164** (1992), 40–58

[3] **Abboud, T., Nédélec, J.C. and Ribbe, J.**
Simulation of Piezoelectric Surface Acoustic Wave devices using a coupling of Integral Equations, Finite Elements and Fourier Modes
Proceedings of the 3rd European Conference on Numerical Mathematics and Advanced Applications. Jyvaskyla, Finland (1999)
P. Neittaanmaki, T. Tiihonen and P. Tarvainen, World Scientific, Singapore, (2000) 188–197

[4] **Abboud, T., Mathis, V. and Nédélec, J.C.**
Diffaction of an electromagnetic travelling wave by a periodic structure
Mathematical and numerical aspects of wave propagation, Mandelieu-La Naplouse (1995)
SIAM, Philadelphia, (1995), 412–421

[5] **Abboud, T., Nédélec, J.C. and Zhou, B.**
Méthode des équations intégrales pour les hautes fréquences
C. R. Acad. Sci. Paris, Série I, **318** (1994), 165–170

[6] **Abboud, T., Nédélec, J.C. and Zhou, B.**
Improvement of the integral equation method for high frequency problems
Mathematical and numerical aspects of wave propagation, Mandelieu-La Naplouse (1995), SIAM, Philadelphia, (1995), 178–187

[7] **Abboud, T., Nédélec, J.C. and Zhou, B.**
Recents developments on a phase separated integral equation method for high frequency scattering problems
Numerical methods in mechanics, Concepcion (1995)
Pitman Res. Notes Math. Ser., **371** (1997), 1–12

[8] **Adams, R.A.**
Sobolev Spaces
Academic Press; New-York (1975)

[9] **Ahner, J.F.**
The exterior Dirichlet problem for the Helmholtz equation
J. Math. Anal. Appl. **52** (1975), 415–429

[10] **Ahner, J.F.**
The exterior Robin problem
J. Math. Anal. Appl. **66** (1978), 37–54

[11] **Ahner, J.F.**
The exterior Dirichlet problem for the Laplace equation
Math. Meth. Appl. Sci. **7** (1985), 461–469

[12] **Ahner, J.F. and Kleinman, R.E.**
The exterior Neumann problem for the Helmholtz equation
Arch. Rat. Mech. Anal. **52** (1973), 26–43

[13] **Ammari, H.**
Diffraction d'ondes par des structures périodiques
Thèse. Ecole Polytechnique (1995)

[14] **Ammari, H., Béreux, N. and Nédélec, J.C.**
Resonance for Maxwell's equation in a periodic structure
C. R. Acad. Sci. Paris, Série I, **325** (1997) 211–215

[15] **Ammari, H., Béreux, N. and Nédélec, J.C.**
Resonant frequencies for a narrow strip grating
Math. Meth. Appl. Sci., **22** (1999) 1121–1152

[16] **Ammari, H., Béreux, N. and Nédélec, J.C.**
Resonances for Maxwell's equations in a periodic structure
Japan J. Indust. Appl. Math., **22** (1999) 149–198

[17] **Ammari, H., Buffa, A. and Nédélec, J.C.**
A justification of eddy currents model for the Maxwell equations
SIAM J. Appl. Math., **60** (2000) 1805–1823

[18] **Ammari, H., Hamdache, K. and Nédélec, J.C.**
Chirality in the Maxwell equations by the dipole approximation
SIAM J. of Appl. Math., **59** (1999) 2045–2059

[19] **Ammari, H., Laouadi, M. and Nédélec, J.C.**
Low frequency behavior of solutions to electromagnetic scattering problems in chiral media
SIAM J. of Appl. Math., **58** (1998) 1022–1042

[20] **Ammari, H., Latiri-Grouz, C. and Nédélec, J.C.**
Approximation de Leontovitch pour les équations de Maxwell à coefficients variables: un problème de perturbation singulière
C. R. Acad. Sci. Paris, Série I, **325** (1997) 677–681

[21] **Ammari, H., Latiri-Grouz, C. and Nédélec, J.C.**
The Leontovitch boundary value problem for the time-harmonic Maxwell equations
Asymp. Anal., **18** (1998) 33–47

[22] **Ammari, H., Latiri-Grouz, C. and Nédélec, J.C.**
The Leontovitch boundary value problem for the time-harmonic Maxwell equations in an inhomogeneous medium: a singular pertubation problem
SIAM J. of Appl. Math., **59** (1999) 1322–1334

[23] **Ammari, H. and Nédélec, J.C.**
Sur les conditions d'impédance généralisées
C. R. Acad. Sci. Paris, Série I, **322** (1996) 995–1000

[24] **Ammari, H. and Nédélec, J.C.**
Equation intégrale et diffraction dans un milieu chiral
C. R. Acad. Sci. Paris, Série I, **322** (1996) 1087–1091

[25] **Ammari, H. and Nédélec, J.C.**
Time-harmonic electromagnetic fields in chiral media
Modern Mathematical Methods in Diffraction Theory and its Applications in Engineering, Peter Lang, (1997) 174–202

[26] **Ammari, H. and Nédélec, J.C.**
Time-harmonic electromagnetic fields in thin chiral curved layers
SIAM J. of Math. Anal., **29** (1998) 395–423

[27] **Ammari, H. and Nédélec, J.C.**
Small chirality behavior of solutions to electromagnetic fields in chiral media
Math. Meth in the Appl. Sci., **21** (1998) 327–359

[28] **Ammari, H. and Nédélec, J.C.**
Propagation d'ondes électromagnétiques à basses fréquences
C. R. Acad. Sci. Paris, Série I, **325** (1997) 797–802

[29] **Ammari, H. and Nédélec, J.C.**
Propagation d'ondes électromagnétiques à basses fréquences
J. Math. Pures Appl., **77** (1998) 839–849

[30] **Ammari, H. and Nédélec, J.C.**
Generalized impedance boundary conditions for the Maxwell equations as singular pertubation problems
Comm. Part. Diff. Equat., **24** (1999) 821–849

[31] **Ammari, H. and Nédélec, J.C.**
low frequency electromagnetic scattering
SIAM J. Math. Anal., **31** (2000) 836–861

[32] **Ammari, H. and Nédélec, J.C.**
Full low-frequency asymptotics for the reduced wave equation
Appl. Math. Lett. **12** (1999), 127–131

[33] **Ammari, H. and Nédélec, J.C.**
Coupling of finite and boundary element methods for the time-harmonic Maxwell equations. A symmetric formulation
The Mazya anniversary collection, Vol 2, Rostock (1998)
Oper. Theory Adv. Appl. **110** (1999), 23–32

304   References

[34] **Ammari, H. and Nédélec, J.C.**
Couplage éléments finis-équations intégrales pour la résolution des équations de Maxwell dans un milieu hétérogène
Equations aux Dérivées Partielles et Applications (dédié à J.L. Lions)
Elsevier, (1998)

[35] **Angel, T.S., Kleinman, R.E. and Hettlich, F.**
The resistive and conductive problem for the exterior Helmholtz equation
SIAM J. Appl. Math. **50** (1990), 1607–1622

[36] **Angel, T.S. and Kleinman, R.E.**
Boundary integral equations for the Helmholtz equation:
The third boundary value problem
Math. Meth. Appl. Sci. **4** (1982), 164–193

[37] **Angel, T.S. and Kress, R.**
L^2*-Boundary integral equations for the Robin problem*
Math. Meth. Appl. Sci. **6** (1984), 345–352

[38] **Angel, T.S. and Kirsch, A.**
The conductive boundary condition for Maxwell's equations
SIAM J. Appl. Math. **52** (1992), 1597–1610

[39] **Arnold, D.N. and Wendland, W.L.**
On the asymptotic convergence of collocation methods
Math. Comp. **41** (1983), 349–381

[40] **Arnold, D.N. and Wendland, W.L.**
The convergence of spline collocation for strongly elliptic equations on curves
Numer. Math. **47** (1985), 313–341

[41] **Bachelot, A. and Pujols, A.**
Equations intégrales espace-temps pour le système de Maxwell
C.R. Acad. Sci. Paris Sér. I **314** (1992), 639–644

[42] **Bamberger, A. and Ha Duong, T.**
Formulation variationnelle pour le calcul de la diffraction d'une onde acoustique par une surface rigide
Math. Methods Appl. Sci. **8** (1986), 598–608

[43] **Bécache, E.**
Résolution par une méthode d'équations intégrales d'un problème de diffraction d'ondes élastiques transitoires par une fissure
Thèse. Ecole Polytechnique (1991)

[44] **Bécache, E., Nédélec, J.C. and Nishimura, N.**
Regularization in 3D for anisotropic elastodynamic crack and obstacle problems
J. Elasticity **31** (1993), 25–46

[45] **Bendali, A.**
Approximation par éléments finis de surface de problèmes de diffraction des ondes électromagnétiques
Thèse. Université de Paris VI (1984)

[46] **Bendali, A.**
Numerical analysis of the exterior boundary value problem for the time-harmonic Maxwell equations by a boundary finite element method.

Part 1 : The continuous problem
Math. Comp. **43** (1984), 29–46

[47] **Bendali, A.**
Numerical analysis of the exterior boundary value problem for the time-harmonic Maxwell equations by a boundary finite element method.
Part 2 : The discrete problem
Math. Comp. **43** (1984), 47–68

[48] **Bendali, A. and Lemrabet, K.**
The effect of a thin coating on the scattering of a time-harmonic wave for the Helmholtz equation
SIAM J. App. Math. **56** (1996), 1664–1693

[49] **Benjelloun Touimi El-Dabaghi, Z.**
Diffraction par un réseau 1-périodique de R3
Thèse. Université de Paris-Nord (1988)

[50] **Benjelloun, Z., Nédélec, J.C. and Starling, F.**
Diffraction d'ondes électromagnétiques par des réseaux de conducteurs et méthodes d'équations intégrales
C. R. Acad. Sci. Paris, Série I, **306** (1988) 547–550

[51] **Brakhage, H. and Werner, P.**
Über das Dirichletsche Ausssenraum Problem für die Helmholtzsche Schwingungsgleichung
Arch. Math. **16** (1965), 325–329

[52] **Brezzi, F., Johnson, C. and Nédélec, J.C.**
On the coupling of boundary integral and finite element methods
Proceeding of the fourth Symposium on Basic Problem in Numerical Mathematics, Charles Univ. Prague (1978), 103–114

[53] **Calderon, A.**
The multipole expansion of radiation fields
Arch. Rat. Mech. Anal. **3** (1954), 523–537

[54] **Christiansen, S. and Nédélec, J.C.**
Des préconditionneurs pour la résolution numérique des équations intégrales de frontière de l'acoustique
C. R. Acad. Sci. Paris, Série I, **330** (2000) 617–622

[55] **Christiansen, S. and Nédélec, J.C.**
Des préconditionneurs pour la résolution numérique des équations intégrales de frontière de l'électromagnétisme
C. R. Acad. Sci. Paris, Série I, **331** (2000) 733–738

[56] **Colton, P. and Kress, R.**
Integral equation methods in scattering theory
John Wiley, New York (1983)

[57] **Colton, P. and Kress, R.**
Inverse acoustic and electromagnetic scattering theory
Springer-Verlag, Berlin (1992)

[58] **Colton, P. and Kress, R.**
Time harmonic electromagnetic waves in an inhomogeneous medium
Proc. Royal Soc. Edinburgh A **116** (1990), 279–293

[59] **Colton, P. and Kress, R.**
The impedance boundary value problem for the time harmonic Maxwell equations
Math. Meth. Appl. Sci. **3** (1981), 475–487

[60] **Costabel, M.**
Principles of boundary element methods
Comput. Phys. Rep. **6** (1987), 243–274

[61] **Costabel, M.**
A Remark on the regularity of solutions of Maxwell's equations on Lipschitz domains
Math. Meth. in Appl. Sci. **4** (1990), 365–368

[62] **Costabel, M.**
A coercive bilinear form for Maxwell's equations
J. Math. Anal. Appl. **2** (1991), 527–541

[63] **Costabel, M. and Stephan, E.P.**
Integral equations for transmission problems in linear elasticity
J. Int. Equa. Appl. **27** (1990), 211–223

[64] **Costabel, M. and Stephan, E.P.**
A direct boundary integral equations method for transmission problems
J. Math. Anal. Appl. **106** (1985), 367–413

[65] **Costabel, M. and Stephan, E.P.**
Coupling of finite and boundary element methods for an elastoplastic interface problem
SIAM J. Numer. Anal. **27** (1990), 1212–1226

[66] **de la Bourdonnaye, A.**
Accélération du traitement numérique de l'équation de Helmholtz par équations intégrales et parallèlisation
Thèse. Ecole Polytechnique (1991)

[67] **De La Bourdonnaye, A.**
Some formulations coupling finite element and integral equation methods for Helmholtz equation and electromagnetism
Numer. Math. **69** (1995), 257–268

[68] **De La Bourdonnaye, A.**
Décomposition de $TH_{div}^{-1/2}(\Gamma)$ et nature de l'opérateur de Steklov-Poincaré du problème extérieur de l'électromagnétisme
C. R. Acad. Sci. Paris, Série I t.**316** (1993), 369–372

[69] **A. Da Costa Sequeira, A.**
Couplage entre la méthode des éléments finis et la méthode des équations intégrales - Application au problème extérieur de Stokes stationnaire dans le plan
Thèse. Université de Paris VI (1981)

[70] **Desprès, B.**
Fonctionnelle quadratique et équations intégrales pour les problèmes d'onde harmonique en domaine extérieur
RAIRO M2AN, **31**, 6 (1997), 679–732

[71] **Ding, Y.**
Méthodes numériques sur l'équation intégrale aux bords pour le problème des ondes acoustiques diffractées par une surface rigide en 3 D
Thèse. Université de Paris 11 (1989)

[72] **Dolph, C.L.**
The integral equation method in scattering theory
Problems in Analysis (R.C. Gunning Ed.) Princeton University Press, Princeton, N.J. (1970), 201–277

[73] **Engquist, B. and Nédélec, J.C.**
Effective boundary conditions for acoustic and electromagnetic scattering in thin layers
Rapport interne du C.M.A.P., Ecole Polytechnique, **278** (1993)

[74] **Ervin, V.J. and Stephan, E.P.**
A boundary element Galerkin method for a hypersingular integral equation on open surfaces
Math. Meth. Appl. Sci. **13** (1990), 281–289

[75] **Èskin, G.I.**
Boundary value problems for elliptic pseudodifferential equations
Translations of Mathematical Monographs, Vol. 52,
Amer. Math. Soc., Providence, R.I. (1981)

[76] **Filipe, M.**
Etude mathématique et numérique d'un problème d'interaction fluide - structure dépendant du temps par la méthode de couplage - Eléments finis - Equations intégrales
Thèse. Ecole Polytechnique (1994)

[77] **Fredholm, I.**
Sur une classe d'équations fonctionnelles
Acta Math. **27** (1903), 365–390

[78] **Giraud, G.**
Equations à intégrales principales
Ann. Sci. Ec. Norm. Sup. **51**, 3 and 4 (1934), 251–272

[79] **Giroire, J.**
Etude de quelques problèmes aux limites extérieurs et résolution par équations intégrales
Thèse. Université de Paris VI (1987)

[80] **Giroire, J. and Nédélec, J.C.**
Numerical solution of an exterior Neumann problem using a double-layer potential
Math. Comp. **32** (1978), 973–990

[81] **Giroire, J. and Nédélec, J.C.**
A new system of boundary integral equations for plates with free edges
Math. Meth. Appl. Sci. **18** (1995), 755–772

[82] **Gray, G.A. and Kleinman, R.E.**
The integral equation method in electromagnetic scattering
J. Math. Anal. Appl. **107** (1985), 455–477

[83] **Ha Duong, T.**
Equations intégrales pour la résolution numérique de problèmes de diffraction d'ondes acoustiques dans R3
Thèse. Université de Paris VI (1987)

[84] **Hähner, P.**
An exterior boundary value problem for the Maxwell equations with boundary data in Sobolev space
Proc. Royal Soc. Edinburgh A **109** (1988), 213–224

[85] **Hamdi, M.A.**
Une formulation variationnelle par équations intégrales pour la résolution de l'équation de Helmholtz avec des conditions aux limites mixtes
C.R. Acad. Sci. Paris Sér. II **292** (1981), 17–20

[86] **Hariharan, S.I. and MacCamy, R.C.**
Low frequency acoustic and electromagnetic scattering
Appl. Numer. Math. **2** (1986), 23–35

[87] **Hsiao, G. and Wendland, W.L.**
A finite element method for some integral equations of the first kind
J. Math. Anal. Appl. **58** (1977), 449–481

[88] **Johnson, C. and Nédélec, J.C.**
On the coupling of boundary integral and finite element methods
Math. Comp. **35** (1980), 1063–1079

[89] **Kellogg, O.D.**
Foundations of potential theory
Springer-Verlag, Berlin (1929)

[90] **Kleinman, R. and Vainberg, R.E.**
Full low-frequency asymptotic expansion for the second-order elliptic equations in two dimensions
Math. Meth. Appl. Sci. **17** (1994), 989–1004

[91] **Kleinman, R. and Wendland, R.E.**
On Neumann's method for the exterior Neumann problem for the Helmholtz equation
J. Math. Anal. Appl. **57** (1977), 170–202

[92] **Kress, R.**
On the limiting behaviour of solutions to boundary integral equations associated with time harmonic wave equations for small frequencies
Math. Meth. in Appl. Sci. **1** (1979), 89–100

[93] **Kress, R.**
A singular perturbation problem for linear operators with application to the limiting behaviour of stationary electromagnetic wave fields for small frequencies
Meth. Verf. Math. Phys. **21** (1981), 5–30

[94] **Kress, R.**
Linear Integral Equations
Springer-Verlag, Berlin (1989)

[95] **Kress, R.**
On the low wave number asymptotics for the two-dimensional exterior Dirichlet problem for the reduced wave equation
Math. Meth. Appl. Sci. **9** (1987), 335–341

[96] **Knauff, W. and Kress, R.**
On the exterior boundary value problem for the time harmonic equations
J. Math. Anal. Appl. **72** (1979), 215–235

[97] **Leontovitch, M.A.**
Investigations of propagation of radio waves
Soviet Radio, Moscow, (1948)

[98] **Levillain, V.**
Couplage éléments finis-équations intégrales pour la résolution des équations de Maxwell en milieu hétérogène
Thèse. Ecole Polytechnique (1991)

[99] **Le Roux, M.N.**
Approximation par éléments finis d'une équation intégrale singulière
Thèse. Université de Rennes (1974)

[100] **Le Roux, M.N.**
Méthode d'éléments finis par la résolution numérique de problèmes extérieurs en dimension 2
RAIRO Anal. Numér. **11** (1977), 27–60

[101] **Mathis, V.**
Etude de la diffraction d'ondes électromagnétiques par des réseaux dans le domaine temporel
Thèse. Ecole Polytechnique (1996)

[102] **Mazari, A.**
Détermination par une méthode d'équations intégrales du champ électromagnétique rayonné par une structure filiforme
Thèse. Ecole Polytechnique (1991)

[103] **Mikhlin, S.G. and Prössdorf, S.**
Singular integral operators
Springer-Verlag, Berlin (1986)

[104] **Morelot, A.**
Etude d'une méthode numérique de simulation de la diffraction d'une onde électromagnétique par un réseau bi-périodique
Thèse. Ecole Polytechnique (1992)

[105] **Müller, C.**
Foundations of the mathematical theory of electromagnetic waves
Springer-Verlag, New York (1969)

[106] **Müller, C. and Niemeyer, H.**
Greensche tensoren und asymptotische gesetze der elektromagnetischen hohlraumschwingungen
Arch. Rat. Mech. Anal. **7** (1961), 305–348

[107] **Muskhelishvili, N.I.**
Singular integral equations
Noordhoff, Groningen (1953)

[108] **Nédélec, J.C.**
Curved finite element methods for the solution of singular integral equations on surfaces in R^3
Comp. Methods Appl. Mech. Engrg. **8** (1976), 61–80

[109] **Nédélec, J.C.**
Calcul par éléments finis des courants de Foucault sur une surface conductrice de R3
Compte rendu deuxième Colloque CRM-IRIA (Univ. Montréal), (1976)
Ann. Sci. Math. Québec, **1** (1977), 297–328

[110] **Nédélec, J.C.**
Computation of eddy currents on a surface in R3 by finite element methods
SIAM J. Numer. Anal. **15** (1978), 580–595

[111] **Nédélec, J.C.**
Approximation par potentiel de double couche du problème de Neumann extérieur
C. R. Acad. Sci. Paris, Série A, **286** (1978), 103–106

[112] **Nédélec, J.C.**
Numerical approximations for singular integral equations
The use of finite element and finite difference method in geophysics, Proc. Summer School, Liblice, **1** (1977), 207–227
Cesk. Akad. V. ed, Prague (1978)

[113] **Nédélec, J.C.**
La méthode des éléments finis appliquée aux équations intégrales de la physique
Actes du 1er Colloque AFCET-SMF, Ecole Polytechnique, **1** (1978), 181–190

[114] **Nédélec, J.C.**
Formulations variationnelles de quelques équations intégrales faisant intervenir des parties finies
Proc. Symp. on Innovative Numerical Analysis in Applied Engineering Science, University Press of Virginia (1981)

[115] **Nédélec, J.C.**
Integral equations with non integrable kernels
Integral Equations Operator Theory, **5** (1982), 562–572

[116] **Nédélec, J.C.**
Approximation of integral equations with nonintegrable kernels
The mathematics of finite elements and applications, VI, Uxbridge (1987), 343–352

[117] **Nédélec, J.C.**
Le potentiel de double couche pour les ondes élastiques
Rapport interne du C.M.A.P., Ecole Polytechnique, **99** (1983)

[118] **Nédélec, J.C.**
The double layer potential for periodic elastic waves in R3
Boundary elements, Beijing, (1986), 439–448

[119] **Nédélec, J.C.**
Finite Elements for exterior problems using integral equations
Internat. J. Numer. Methods Fluids, **7** (1987), 1229–1234

[120] **Nédélec, J.C.**
Quelques propriétés des dérivées logarithmiques des fonctions de Hankel
C. R. Acad. Sci. Paris, Série I, **314** (1992), 507–510

[121] **Nédélec, J.C.**
New trends in the use and analysis of integral equations
Proc. Sympos. Appl. Math., A.M.S., Math. Comp., **48** (1994), 151–176

[122] **Nédélec, J.C. and Planchard, J.**
Une méthode variationnelle d'éléments finis pour la résolution numérique d'un problème extérieur dans R^3
RAIRO R3 **7** (1973), 105–129

[123] **Nédélec, J.C. and Starling, F.**
Integral equation methods in a quasi-periodic diffraction problem for the time-harmonic Maxwell's equations
SIAM J. Math. Anal., **22** (1991), 1679–1702

[124] **Nédélec, J.C. and Vérité, J.C.**
Computation of eddy currents on a surface in R3 by finite element methods
Computing methods in applied scienece and engineering, Proc. Third Internat. Sympos., Versailles (1977), 111–126
Lecture Notes in Math., Springer Verlag, **704** (1979)

[125] **Nédélec, J.C. and Wolf, S.**
Homogenéisation des équations de Maxwell dans un transformateur
C. R. Acad. Sci. Paris, Série II, **286** (1987), 103–106.

[126] **Nédélec, J.C. and Wolf, S.**
Homogenization of the problem of eddy-current in a transformer core
SIAM J. Numer. Anal., **26** (1989), 1407–1424

[127] **Nishimura, N. and Kobayashi, S.**
A regularized boundary integral equation method for elastodynamic crack problems
Comput. Mech. **4** (1989), 319–328

[128] **Panick, O.I.**
On the question of the solvability of the exterior boundary value problems for the wave equation and Maxwell's equations
Russian Math. Surveys **20** (1965) 221–226

[129] **Paquet, L.**
Problèmes mixtes pour le système de Maxwell
Ann. Fac. Sci. Toulouse **4** (1982)

[130] **Picard, R.**
On the low frequency asymptotics in electromagnetic theory
J. Reine Angew. Math. **354** (1984), 50–73

[131] **Picard, R.**
The frequency limit for the time harmonic acoustics waves
Math. Meth. Appl. Sci. **8** (1986), 436–450

[132] **Picard, R.**
On boundary value problems for electro-and magnetostatics
Proc. Roy. Soc. Edinburgh A **92** (1982), 165–174

[133] **Prössdorf, S. and Rathsfeld, A.**
A finite element collocation method for singular integral equations
Math. Nachr. **100** (1981), 33–60

[134] **Rao, S.M. and Wilton, D.R.**
Transient scattering by conducting surfaces of arbitrary shape
IEEE Trans. Antennas and Propagation **39** (1991), 56–61

[135] **Rogier, F.**
Problèmes mathématiques et numériques liés à l'approximation de la géométrie d'un corps diffraction dans les équations de l'électromagnétisme
Thèse. Université de Paris VI (1989)

[136] **Rumsey, V.H.**
Reaction concept in electromagnetic theory
Physical Review, **94** (1954), 1438–1491

[137] **Saranen, J.**
On electric and magnetic problems for vector fields in anistropic non homogeneous media
J. Math. Anal. Appl. **91** (1983), 254–275

[138] **Seeley, R.T.**
Singular integrals on compact manifolds
Amer. J. Math. **81** (1959), 658–690

[139] **Seeley, R.T.**
Topics in pseudo-differential operators,(in: Pseudo-Differential Operators (ed. L. Nirenberg))
Edizioni Cremonese, Roma (1969), 169–305

[140] **Sommerfeld, W.**
Die Greensche Funktion der schwingungsgleichung
Math. Verein. **21** (1912), 309–353

[141] **Starling, F.**
Etude mathématique de quelques problèmes de diffraction en électromagnétisme
Thèse. Ecole Polytechnique (1991)

[142] **Stein, E.M.**
Singular integrals and differentiability properties of functions
Princeton University Press,Princeton N.J. (1970)

[143] **Stephan, E.P.**
Boundary integral equations for screen problems in R^3
Integral Equations and Operator Theory **10** (1987), 236–257

[144] **Terrasse, I.**
Résolution mathématique et numérique des équations de Maxwell instationnaires par une méthode de potentiels retardés
Thèse. Ecole Polytechnique (1993)

[145] **Wilcox, C.H.**
Scattering theory for diffraction gratings
App. Math. Sc. Vol. 46, Springer-Verlag, New-York, (1984)

Index

A, 290

D, 116, 142

D^*, 116, 142

$H(\text{div})$, 209

H, 102

$H(\text{curl})$, 209

$H^1(S)$, 41

$H^m(\Omega_i)$, 48

$H_0^m(\Omega_i)$, 54

$H^s(S)$, 41

$H^s(\mathbb{R}^2)$, 49

$H^s(\Gamma)$, 206

$H^s_{\text{curl}}(S)$, 203

$H^{-1/2}_{\text{curl}}(S)$, 191

$H^{-1/2}_{\text{curl}}(\Gamma)$, 207

$H^s_{\text{div}}(S)$, 203

$H^{-1/2}_{\text{div}}(\Gamma)$, 207

$H^{-1/2}_{\text{div}}(S)$, 191

N, 116, 137, 143

S, 116, 141

T, 97

$TH^s(\Gamma)$, 207

$TH^s(S)$, 191

$TL^2(S)$, 191

$TL^2(\Gamma)$, 207

T_R, 104

$W^1(B_e)$, 44

$W^k(\Omega_e)$, 59

$W^k(B_e)$, 42

$W^{k,\alpha}(\Omega_e)$, 61

$W_0^{k,\alpha}(\Omega_e)$, 61

$W^{m,p}(\mathbb{R})$, 157

X, 193, 205, 213

$\nabla_S Y_\ell^m$, 191

$\mathcal{N}$, 206

$\mathcal{P}$, 68

$\mathcal{T}$, 200

$\mathcal{T}_R$, 215

$\mathcal{D}(S)$, 41

$\mathcal{D}'(S)$, 41

$\mathcal{S}(\Gamma)$, 50

$\mathcal{S}(\mathbb{R}^2)$, 49

$\mathcal{S}'(\mathbb{R}^2)$, 49

$\mathcal{S}'(\Gamma)$, 50

$\overrightarrow{\text{curl}}_S Y_\ell^m$, 191

acoustic equation, 3

addition theorem, 27

Ampère–Maxwell law, 178

area element, 69
associated Legendre functions, 22, 24
atlas, 47

Bessel equation, 12

Calderon projectors, 116, 240
Calderon relations, 116
capacity operator, 97, 104, 200, 202, 215
charts, 47
circular plane wave, 5
compact, 55
conducting media, 178
conductivity, 178
contravariant vector, 72
coupled variational formulation, 226, 229, 233
covariant vector, 71
covering, 47
curvature operator, 69

dielectric case, 253
Dirac mass, 12, 100, 179
Dirichlet problem, 10, 55
Dirichlet-to-Neumann operator, 46
double layer potential, 113, 270
duality operator, 193, 208

eigenfunctions Y_i, 205
elastic wave, 6
electric current, 235, 254, 288
electric field, 4, 177
electric induction, 178
electric permittivity, 4, 177
electromagnetic waves, 177
Euler identity, 17
even kernel, 161
existence theorem, 105
exterior Dirichlet problem, 42, 64
exterior domain, 9
exterior harmonic Maxwell problem, 184
exterior Neumann problem, 46, 65

exterior normal derivative, 136
exterior problem, 11
exterior problem for the Helmholtz equation, 102

far field, 122, 288, 298
Faraday law, 178
field of normals, 68
finite part, 158
Fourier transform, 49
Fredholm alternative, 11, 142, 146, 218, 245
frequency, 3
fundamental solution, 12, 179
fundamental solution of Helmholtz, 12
fundamental solution of Maxwell, 179

Gauss curvature, 69

Hankel function, 12
Hardy's inequality, 59
harmonic Maxwell equations, 5, 178
harmonic polynomial, 15, 28
harmonic solution, 3, 10
Helmholtz decomposition, 249, 260
Helmholtz equation, 3, 9, 84, 187
high-frequency asymptotic, 124
Hilbert transform, 150
Hodge operator, 73
homogeneous kernel, 158
homogeneous kernel of class $-m$, 168
homogeneous polynomials, 15
Hölder condition, 150

impedance, 230, 233
impedance condition, 185, 226
incident field, 125, 291
infinite conductivity limit, 266
integral equation, 110, 117, 243, 244, 298

integral representation, 110, 234, 288
interior Dirichlet problem, 42
interior domain, 9
interior harmonic Maxwell problem, 185
interior Neumann problem, 46
interior normal derivative, 136
interior problem, 10
inverse capacity operator, 203
isotropic dielectric medium, 177

Jacobi relation, 24

kinetic moments, 23

Lamé coefficients, 6
Laplace equation, 40
Laplace operator, 14
Laplace-Beltrami operator, 14, 16, 23, 73, 205
Lax-Milgram theorem, 56
Legendre operator, 20
Legendre polynomials, 17
Leontovitch condition, 185
lifting, 50, 209, 268
Lorentz gauge, 180, 238

magnetic current, 235, 254, 288
magnetic field, 4, 177
magnetic induction, 178
magnetic permeability, 4, 177
Marcinkiewicz, 152
Maxwell equations, 4, 177, 234
Maxwell exterior problem, 226
mean curvature, 69
metric tensor, 71
multipole solution, 185

Neumann problem, 10, 57

odd kernel, 158
outgoing fundamental solution, 235
outgoing wave condition, 11

parallel surfaces, 68
partition of unity, 47
perfect conductor, 214, 243, 266
perfectly conducting medium, 178
phase plane, 5
physical optics approximation, 130
plane wave, 3, 4, 7, 97, 179
Poincaré's inequality, 57
polarization vector, 293
Poynting vector, 290
pressure waves, 7
principal curvatures, 69
principal value, 150
progressive wave, 11
property of the wronskian, 100
pseudo-homogeneous kernel, 173
pulsation, 3, 178

reciprocity principle, 127, 295
relations of commutation, 23
relative permeability, 177
relative permittivity, 177
representation theorem, 110, 234
resonance values, 108
Riesz transforms, 160
Rodrigues formula, 18
Rumsey principle, 245, 266

saddle-point formulation, 245, 247
scalar potential, 180, 238
scalar surfacic rotational, 39
scattered energy, 290
scattered field, 97, 125, 291
scattering amplitude, 123, 126, 127, 288, 290, 293, 297
shearing waves, 8
Silver-Müller condition, 182, 186, 204, 226, 234, 291
simply connected, 206
single layer potential, 113, 269
singular integral operator, 150
singular operator, 157
Sobolev spaces, 47
Sommerfeld condition, 11, 85, 181
spectral decomposition, 10

speed of light, 4, 177
speed of wave, 177
spherical Bessel equation, 86
spherical Bessel functions, 85, 89
spherical coordinates, 13
spherical Hankel functions, 86
spherical harmonics, 14
spherical wave, 3
stationary phase, 128, 295
stationary wave solution, 10
Steklov eigenvalues, 108
Stokes formula, 210
Stokes identities, 73
strain tensor, 6
stress tensor, 6
surfaces, 47
surfacic divergence, 39, 73
surfacic divergence of a
        contravariant vector, 75
surfacic gradient, 14, 39
surfacic rotational, 73, 145
surfacic rotational of a covariant
        vector, 75

tangential gradient, 68
tangential rotational, 69
theorem of interpolation, 155
total field, 125, 291
trace, 209, 268
trace theorem, 50
transmission conditions, 178

transverse electric, 6
transverse electric multipole, 186,
        292
transverse magnetic, 6
transverse magnetic multipole,
        186, 292

uniqueness theorem, 102
unit normal, 10

variational formulation, 56, 58, 64,
        65, 137, 141, 211, 214, 224,
        257
vector basis in the cotangent
        plane, 71
vector basis in the tangent plane,
        71
vector potential, 180, 238
vectorial harmonic polynomials, 35
vectorial Laplace-Beltrami
        operator, 39, 191, 205
vectorial spherical harmonics, 35,
        185
vectorial surfacic rotational, 39

wave number, 3, 10, 180
weak type (p, q), 152
weighted Sobolev spaces, 61

zero frequency limit, 250

Applied Mathematical Sciences

(continued from page ii)

61. *Sattinger/Weaver:* Lie Groups and Algebras with Applications to Physics, Geometry, and Mechanics.
62. *LaSalle:* The Stability and Control of Discrete Processes.
63. *Grasman:* Asymptotic Methods of Relaxation Oscillations and Applications.
64. *Hsu:* Cell-to-Cell Mapping: A Method of Global Analysis for Nonlinear Systems.
65. *Rand/Armbruster:* Perturbation Methods, Bifurcation Theory and Computer Algebra.
66. *Hlaváček/Haslinger/Necasl/Lovisek:* Solution of Variational Inequalities in Mechanics.
67. *Cercignani:* The Boltzmann Equation and Its Applications.
68. *Temam:* Infinite-Dimensional Dynamical Systems in Mechanics and Physics, 2nd ed.
69. *Golubitsky/Stewart/Schaeffer:* Singularities and Groups in Bifurcation Theory, Vol. II.
70. *Constantin/Foias/Nicolaenko/Temam:* Integral Manifolds and Inertial Manifolds for Dissipative Partial Differential Equations.
71. *Catlin:* Estimation, Control, and the Discrete Kalman Filter.
72. *Lochak/Meunier:* Multiphase Averaging for Classical Systems.
73. *Wiggins:* Global Bifurcations and Chaos.
74. *Mawhin/Willem:* Critical Point Theory and Hamiltonian Systems.
75. *Abraham/Marsden/Ratiu:* Manifolds, Tensor Analysis, and Applications, 2nd ed.
76. *Lagerstrom:* Matched Asymptotic Expansions: Ideas and Techniques.
77. *Aldous:* Probability Approximations via the Poisson Clumping Heuristic.
78. *Dacorogna:* Direct Methods in the Calculus of Variations.
79. *Hernández-Lerma:* Adaptive Markov Processes.
80. *Lawden:* Elliptic Functions and Applications.
81. *Bluman/Kumei:* Symmetries and Differential Equations.
82. *Kress:* Linear Integral Equations, 2nd ed.
83. *Bebernes/Eberly:* Mathematical Problems from Combustion Theory.
84. *Joseph:* Fluid Dynamics of Viscoelastic Fluids.
85. *Yang:* Wave Packets and Their Bifurcations in Geophysical Fluid Dynamics.
86. *Dendrinos/Sonis:* Chaos and Socio-Spatial Dynamics.
87. *Weder:* Spectral and Scattering Theory for Wave Propagation in Perturbed Stratified Media.
88. *Bogaevski/Povzner:* Algebraic Methods in Nonlinear Perturbation Theory.
89. *O'Malley:* Singular Perturbation Methods for Ordinary Differential Equations.
90. *Meyer/Hall:* Introduction to Hamiltonian Dynamical Systems and the N-body Problem.
91. *Straughan:* The Energy Method, Stability, and Nonlinear Convection.
92. *Naber:* The Geometry of Minkowski Spacetime.
93. *Colton/Kress:* Inverse Acoustic and Electromagnetic Scattering Theory, 2nd ed.
94. *Hoppensteadt:* Analysis and Simulation of Chaotic Systems, 2nd ed.
95. *Hackbusch:* Iterative Solution of Large Sparse Systems of Equations.
96. *Marchioro/Pulvirenti:* Mathematical Theory of Incompressible Nonviscous Fluids.
97. *Lasota/Mackey:* Chaos, Fractals, and Noise: Stochastic Aspects of Dynamics, 2nd ed.
98. *de Boor/Höllig/Riemenschneider:* Box Splines.
99. *Hale/Lunel:* Introduction to Functional Differential Equations.
100. *Sirovich (ed):* Trends and Perspectives in Applied Mathematics.
101. *Nusse/Yorke:* Dynamics: Numerical Explorations, 2nd ed.
102. *Chossat/Iooss:* The Couette-Taylor Problem.
103. *Chorin:* Vorticity and Turbulence.
104. *Farkas:* Periodic Motions.
105. *Wiggins:* Normally Hyperbolic Invariant Manifolds in Dynamical Systems.
106. *Cercignani/Illner/Pulvirenti:* The Mathematical Theory of Dilute Gases.
107. *Antman:* Nonlinear Problems of Elasticity.
108. *Zeidler:* Applied Functional Analysis: Applications to Mathematical Physics.
109. *Zeidler:* Applied Functional Analysis: Main Principles and Their Applications.
110. *Diekmann/van Gils/Verduyn Lunel/Walther:* Delay Equations: Functional-, Complex-, and Nonlinear Analysis.
111. *Visintin:* Differential Models of Hysteresis.
112. *Kuznetsov:* Elements of Applied Bifurcation Theory, 2nd ed.
113. *Hislop/Sigal:* Introduction to Spectral Theory: With Applications to Schrödinger Operators.
114. *Kevorkian/Cole:* Multiple Scale and Singular Perturbation Methods.
115. *Taylor:* Partial Differential Equations I, Basic Theory.
116. *Taylor:* Partial Differential Equations II, Qualitative Studies of Linear Equations.
117. *Taylor:* Partial Differential Equations III, Nonlinear Equations.

(continued on next page)

Applied Mathematical Sciences

(continued from previous page)

118. *Godlewski/Raviart:* Numerical Approximation of Hyperbolic Systems of Conservation Laws.
119. *Wu:* Theory and Applications of Partial Functional Differential Equations.
120. *Kirsch:* An Introduction to the Mathematical Theory of Inverse Problems.
121. *Brokate/Sprekels:* Hysteresis and Phase Transitions.
122. *Gliklikh:* Global Analysis in Mathematical Physics: Geometric and Stochastic Methods.
123. *Le/Schmitt:* Global Bifurcation in Variational Inequalities: Applications to Obstacle and Unilateral Problems.
124. *Polak:* Optimization: Algorithms and Consistent Approximations.
125. *Arnold/Khesin:* Topological Methods in Hydrodynamics.
126. *Hoppensteadt/Izhikevich:* Weakly Connected Neural Networks.
127. *Isakov:* Inverse Problems for Partial Differential Equations.
128. *Li/Wiggins:* Invariant Manifolds and Fibrations for Perturbed Nonlinear Schrödinger Equations.
129. *Müller:* Analysis of Spherical Symmetries in Euclidean Spaces.
130. *Feintuch:* Robust Control Theory in Hilbert Space.
131. *Ericksen:* Introduction to the Thermodynamics of Solids, Revised ed.
132. *Ihlenburg:* Finite Element Analysis of Acoustic Scattering.
133. *Vorovich:* Nonlinear Theory of Shallow Shells.
134. *Vein/Dale:* Determinants and Their Applications in Mathematical Physics.
135. *Drew/Passman:* Theory of Multicomponent Fluids.
136. *Cioranescu/Saint Jean Paulin:* Homogenization of Reticulated Structures.
137. *Gurtin:* Configurational Forces as Basic Concepts of Continuum Physics.
138. *Haller:* Chaos Near Resonance.
139. *Sulem/Sulem:* The Nonlinear Schrödinger Equation: Self-Focusing and Wave Collapse.
140. *Cherkaev:* Variational Methods for Structural Optimization.
141. *Naber:* Topology, Geometry, and Gauge Fields: Interactions.
142. *Schmid/Henningson:* Stability and Transition in Shear Flows.
143. *Sell/You:* Dynamics of Evolutionary Equations.
144. *Nédélec:* Acoustic and Electromagnetic Equations: Integral Representations for Harmonic Problems.